Localized Energy Transition in the 4th Industrial Revolution

This book presents a holistic view on localized energy transition while addressing current challenges associated with the production of biofuels, introducing new materials to produce solar photovoltaic (PV) panels, and digital systems for sustainable energy monitoring on a small scale, carbon capture, and sequestration. Also, each chapter of the book addresses specific aspects of the renewable and sustainable energy space while focusing more on energy improvement and storage technologies that are practical focused.

Features:

- Offers useful information on new forms of renewable energy generation with reference to Industry 4.0.
- Illustrates practical approaches to energy transition.
- Provides guidance on renewable energy sources and energy storage systems.
- Discusses the application of the Fourth Industrial Revolution (4IR)-related approaches to emerging energy storage technologies.
- Includes studies that reveal approaches to realizing productivity, profitability, and increased return on investment (ROI).

This book is aimed at graduate students and researchers in mechanical, chemical, and mechatronics engineering, and renewable energy systems.

Energy Transition in the 21st Century

Series Editors: Opeyeolu Timothy Laseinde and Andrew C. Eloka-Eboka

Localized Energy Transition in the 4th Industrial Revolution
Edited by Opeyeolu Timothy Laseinde and Andrew C. Eloka-Eboka

For more information about this series, please visit: www.routledge.com/Energy-Transition-in-the-21st-Century/book-series/ET21C

Localized Energy Transition in the 4th Industrial Revolution

Edited by
Opeyeolu Timothy Laseinde and Andrew C. Eloka-Eboka

CRC Press
Taylor & Francis Group
Boca Raton London New York

CRC Press is an imprint of the
Taylor & Francis Group, an **informa** business

First edition published 2025
by CRC Press
2385 NW Executive Center Drive, Suite 320, Boca Raton FL 33431

and by CRC Press
4 Park Square, Milton Park, Abingdon, Oxon, OX14 4RN

CRC Press is an imprint of Taylor & Francis Group, LLC

ISBN: 9781032538792 (hbk)
ISBN: 9781032651972 (pbk)
ISBN: 9781032651958 (ebk)

DOI: 10.1201/9781032651958

Typeset in Times
by Newgen Publishing UK

Contents

Chapter 1   The Effect of Specifications of Solar Photovoltaic Modules on Array Peak Power, Quantity, and Weight in Residential Buildings

Sogo Mayokun Abolarin, Manasseh Babale Shitta, Emmanuel Aghogho Metuaghan, Blessing Precious Nwosu, Michael Chuks Aninyem, Louis Lagrange, and Daniel Raphael Ejike Ewim

Chapter 2   A Systematic Approach to Exploring Recent Improvements in the Sustainability of Biodiesel Production

Ochuko Felix Orikpete and Daniel Raphael Ejike Ewim

Chapter 3   In-Tube Condensation Heat Transfer Correlations for Horizontal Smooth Macro Tubes

S. C. Blose, D. R. E. Ewim, A. C. Eloka-Eboka, and A. O. Adelaja

Chapter 4   A Review of Numerical Tools for Solar Cells

George Chukwuebuka Enebe, Kingsley Ukoba, and Tien-Chien Jen

Chapter 5   Performance Evaluation of a Batch: Reactor for Catalytic Depolymerization of Polymeric Waste

O. L. Rominiyi, M. A. Akintunde, E. I Bello, L. Lajide, and O. M. Ikumapayi

Chapter 6   Analysis of a Hybrid Solar Power System as a Potential Substitute in Selected Southern Nigerian Economic Activity Areas

Zubairu Ismaila, Olugbenga A. Falode, Chukwuemeka J. Diji, Rasaq A. Kazeem, Omolayo M. Ikumapayi, Tien-Chien Jen, Opeyeolu Timothy Laseinde, and Esther T. Akinlabi

Chapter 7   From Domestic Sewage Sludge to Clean Energy, Useful Resources, Zero Waste and Circular Economy Approach

J. K. Bwapwa

Preface

The book aims to create a bigger picture of the localized energy transition's impact on meeting the global energy deficit in line with the Sustainable Development Goal 7 (SDG 7). In addition, the book shall address current challenges in large-scale energy production.

The book focuses on producing biofuels, introducing new materials for manufacturing solar PV panels, digital systems for sustainable energy monitoring on a small scale, emerging carbon capture technologies, and sequestration. Also covered in this book are all forms of energy improvements, innovations, generation, conversion, and storage technologies.

Acknowledgments

The editors wish to appreciate all the authors and organizations that sponsored the research carried out by the chapter authors. Special appreciation goes to the Royal Academy of Engineering (RAEng) and Lloyd's Foundation, United Kingdom, whose SEEL Champion grant award supported the research. The Universities that endorsed the book publication are duly acknowledged (the University of Johannesburg and the North-West University, South Africa).

About the Editors

Opeyeolu Timothy Laseinde is a teaching and research Associate Professor of Mechanical and Industrial Engineering at the University of Johannesburg. He is a National Research Foundation (NRF)-rated established researcher, has authored/co-authored over 90 peer-reviewed articles, and has served as an external reviewer for notable publishers. He has vast experience interfacing with stakeholders in the industry, manufacturing sector, development agencies, and academia. As a proponent of the Sustainable Development Goals (SDGs), he has received notable awards and grants in line with innovative research involving multidisciplinary works.

Timothy has partnered with the British Council as a Programme Manager in a University Staff Doctorate Programme (USDP) award, which has yielded significant results. He also partners with other international institutions, including the Manchester Metropolitan University (MMU), United Kingdom. He is currently a "Safer End of Engineered Life Champion (SEEL)", an ENGINEERING X initiative by the Royal Academy of Engineering (RAEng), United Kingdom, and Lloyd's Registered Foundation aimed at sustainable net-zero emissions. He has excelled as an academic researcher in previous internationally funded projects requiring new product design and manufacturing.

He is the energy steering committee representative (2022–2024) at the Engineering for Change (E4C), supported by the American Society of Mechanical Engineers (ASME). With 17 cumulative years of work experience in extraordinarily skilled endeavours, he provides advisory services to small- and medium-sized industries towards upskilling and improving their processes in a bid to achieve net-zero emissions and green production systems that will guarantee a safer, sustainable, and carbon-free future. He is currently one of the Series Editors for the series, *Energy Transition in the 21st Century*.

Andrew C. Eloka-Eboka is an emerging dynamic researcher, chemical, and mechanical engineer with research incursion in the niche area of Food-Energy-Water-Climate Nexus, Bioenergy, Biofuels, Process Engineering, and the Environment. He has a PhD in Mechanical Engineering from the University of KwaZulu-Natal, South Africa, a Master's Degree in Mechanical Engineering specializing in Energy Studies, and a Bachelor's Degree (Hons) in Chemical Engineering/Technology. He is the Programme Manager and Associate Professor in the School of Chemical and Mineral Engineering at the North-West University, Potchefstroom, South Africa, and is leading a Research Group.

He has delivered well-accepted and seasoned research papers at different local and international conferences and has won best paper awards in 2013 and 2018. He had several research fellowships in The World Academy of Science (TWAS) Central and South Asia Regional Partner (TWAS ROCASA), Centre for European Policy Studies (CEPS), and EU/AU policy at the European Joint Research Centre, Ispra, Italy, The World Academy of Sciences for the Advancement of Science in Developing Countries, the Deutsche Forschungsgemeinschaft (TWAS DFG), and EU/AU Intra Africa Mobility at the University of Tlemcen, Algeria, among others. He was among the selected Africans who attended the European Science Forum in 2016. He has several grants, awards, technical reports, research collaborations, and more than 50 peer-reviewed journal publications, over 40 conference proceedings, five book chapters, and one book. He is an NRF-Rated researcher and an Affiliate of the African Academy of Sciences. He is currently one of the Series Editors for the series, *Energy Transition in the 21st Century*.

Contributors

Sogo Mayokun Abolarin
Department of Engineering Sciences
University of the Free State
Bloemfontein, South Africa

Adekunle Omolade Adelaja
Mechanical Engineering Department
University of Lagos
Lagos, Nigeria

Damola S. Adelekan
The Energy and Environment Research Group
  (TEERG)
Mechanical Engineering Department
Covenant University
Ogun State, Nigeria

Esther Titilayo Akinlabi
Department of Mechanical and Construction
  Engineering
Faculty of Engineering and Environment
Northumbria University
Newcastle, United Kingdom

Mutalubi Aremu Akintunde
Department of Mechanical Engineering
Federal University of Technology
  Akure (FUTA)
Ondo State, Nigeria

Jephtha Akporhuarho
Department of Mechanical Engineering
Federal University of Petroleum Resources
Effurun
Delta State, Nigeria

Michael Chuks Aninyem
National Centre for Energy Efficiency and
  Conservation
Energy Commission of Nigeria
Faculty of Engineering, University of Lagos
Lagos, Nigeria

Emmanuel Ibitola Bello
Department of Mechanical Engineering
Federal University of Technology
  Akure (FUTA)
Ondo State, Nigeria

Hakim Benyelles
Institute of Water and Energy
  Sciences – PAUWES
Pan African University
Tlemcen, Algeria

Sibongakonke Cebolenkosi Blose
Department of Mechanical Engineering
Durban University of Technology
Kwazulu-Natal, South Africa

Samir Brairi
Institute of Water and Energy
  Sciences – PAUWES
Pan African University
Tlemcen, Algeria

Joseph Kapuku Bwapwa
Department of Civil Engineering
Mangosuthu University of Technology
Umlazi, Durban,
Kwazulu-Natal, South Africa

Emanuela Colombo
Politecnico di Milano Piazza Leonardo da Vinci
Milano
Northern Italy

Giacomo Crevani
Politecnico di Milano Piazza Leonardo
  da Vinci
Milano
Northern Italy

Erdem Cuce
Laboratory of Low/Zero Carbon Energy
  Technologies
Engineering and Architecture Faculty
Recep Tayyip Erdogan University
Rize, Turkey

Pinar Mert Cuce
Laboratory of Low/Zero Carbon Energy
  Technologies
Engineering and Architecture Faculty,
  Architecture, Zihni Derin Campus
Recep Tayyip Erdogan University
Rize, Turkey

Amazigh Dib
Institute of Water and Energy
    Sciences – PAUWES
Pan African University
Tlemcen, Algeria

Chukwuemeka Jude Diji
Department of Mechanical Engineering
Kampala International University
Uganda

Andrew C. Eloka-Eboka
Centre of Excellence in Carbon-based Fuels
School of Chemical and Mineral
    Engineering
North-West University, South Africa

George Chukwuebuka Enebe
Mechanical Engineering Science Department
University of Johannesburg
Auckland Park campus
Johannesburg, South Africa

Christopher C. Enweremadu
Department of Mechanical Engineering
University of South Africa, Science Campus
Johannesburg, South Africa

Ivrogbo D. Eseoghene
Department of Mechanical Engineering
Federal University of Petroleum Resources
Effurun
Delta State, Nigeria

Daniel Raphael Ejike Ewim
Department of Mechanical Engineering
Durban University of Technology
Durban
Kwazulu-Natal, South Africa

Victor U. Ezekiel
The Energy and Environment Research Group
    (TEERG)
Mechanical Engineering Department
Covenant University
Ogun State, Nigeria

Olugbenga Adebanjo Falode
Department of Mineral, Petroleum, Energy
    Economics and Law
University of Ibadan
Ibadan, Nigeria

Tamer Guclu
Bayburt University, Engineering Faculty,
    Mechanical Engineering. Bayburt
Turkey
and
Laboratory of Low/Zero Carbon Energy
    Technologies
Engineering and Architecture Faculty
Recep Tayyip Erdogan University
Rize, Turkey

Uchechukwu Emerald Henry
The Energy and Environment Research Group
    (TEERG)
Mechanical Engineering Department
Covenant University
Ogun State, Nigeria

Akinrodoye Ibidapo
Department of Mechanical Engineering
Federal University of Petroleum Resources
Effurun
Delta State, Nigeria

Omolayo Michael Ikumapayi
Department of Mechanical and Industrial
    Engineering Technology
University of Johannesburg
Doornfontein Campus
Johannesburg, South Africa
and
Department of Mechanical and Mechatronics
    Engineering
Afe Babalola University
Ado Ekiti
Ekiti State, Nigeria

Zubairu Ismaila
Department of Mineral, Petroleum, Energy
    Economics and Law
University of Ibadan
Ibadan, Nigeria

Tien-Chien Jen
Mechanical Engineering Science Department
University of Johannesburg
Auckland Park campus
Johannesburg, South Africa

Rasaq Adebayo Kazeem
Department of Mechanical Engineering
University of Ibadan, Ibadan
Ibadan, Nigeria
and
Department of Mechanical Engineering Science
University of Johannesburg
Auckland Park
Johannesburg, South Africa

Louis Lagrange
Department of Engineering Sciences
University of the Free State
Free State, South Africa

Labunmi Lajide
Department of Chemistry
Federal University of Technology
    Akure (FUTA)
Ondo State, Nigeria

Opeyeolu Timothy Laseinde
Department of Mechanical and Industrial
    Engineering Technology
University of Johannesburg
Doornfontein Campus
Johannesburg, South Africa

Olaniran J. Matthew
Institute of Ecology and Environmental
    Studies
Obafemi Awolowo University, Ile-Ife
Osun State, Nigeria

Merwan Messaoudi
Institute of Water and Energy
    Sciences – PAUWES
Pan African University
Tlemcen, Algeria

Emmanuel Aghogho Metuaghan
National Centre for Energy Efficiency and
    Conservation
Energy Commission of Nigeria
Faculty of Engineering, University of Lagos
Lagos, Nigeria

Mutali Nepfumbada
Harmattan Renewables
Rosebank
Johannesburg, South Africa

Blessing Precious Nwosu
National Centre for Energy Efficiency and
    Conservation
Energy Commission of Nigeria
Faculty of Engineering, University of Lagos
Lagos, Nigeria

Olayinka S. Ohunakin
The Energy and Environment Research Group
    (TEERG)
Mechanical Engineering Department, Covenant
    University
Ogun State, Nigeria
and
Faculty of Engineering & The Built Environment
University of Johannesburg
Johannesburg, South Africa

Samuel Abiodun Olatubosun
Department of Mechanical and Aerospace
    Engineering
The Ohio State University, Columbus
Ohio, United States

Ochuko Felix Orikpete
Bristow Helicopters Nigeria Limited
ExxonMobil Qua Iboe Terminal, Eket
Akwa Ibom State, Nigeria

Ogaga Ovuedhere
Department of Mechanical Engineering
Federal University of Petroleum Resources
Effurun
Delta State, Nigeria

Oluwasina Lawan Rominiyi
Department of Mechanical and Mechatronics
    Engineering
Afe Babalola University Ado – Ekiti,
    (ABUAD) Nigeria

Olusegun David Samuel
Department of Mechanical Engineering
Federal University of Petroleum Resources
Effurun
Delta State, Nigeria
and
Department of Mechanical Engineering
University of South Africa, Science Campus
Florida, South Africa

Windmanagda Sawadogo
University of Augsburg
Institute of Geography
Augsburg, Germany

Temiloluwa Olatunji Scott
Department of Education
Vaal University of Technology
Vanderbijlpark, South Africa

Chakib Seladji
Institute of Water and Energy
    Sciences – PAUWES
Pan African University
Tlemcen, Algeria

Harun Sen
Dokuz Eylul University
Bergama Vocational School
Heavy Equipment Operator Program.
İzmir, Turkey

Manasseh Babale Shitta
National Centre for Energy Efficiency and
    Conservation, Energy Commission of
    Nigeria, Faculty of Engineering, University
    of Lagos
Lagos, Nigeria

Kingsley Ukoba
Mechanical Engineering Science Department
University of Johannesburg
Auckland Park campus
Johannesburg, South Africa

Abdellatif Zerga
Institute of Water and Energy
    Sciences – PAUWES
Pan African University
Tlemcen, Algeria

Introduction

The book *Localized Energy Transition in the 4th Industrial Revolution* is the first treatise in the series *Energy Transition in the 21ˢᵗ Century*, it is edited by Professor Opeyeolu Timothy Laseinde and Professor Andrew C. Eloka-Eboka, both of whom are South Africa distinguished academics and researchers. The book provides information and interesting insights on the various types of nascent renewable energies, some of which are still evolving at various research and developmental levels. In addition, modern trends and applications in energy monitoring, efficiency improvement, and energy storage are succinctly described. Smart and sustainable energy systems in the Fourth Industrial Revolution (4IR) era were delved into and X-rayed. The overarching aim is to create a bigger picture of the localized energy transition while addressing current challenges associated with the production of biofuels, introducing new materials to produce solar PV panels, and digital systems for sustainable energy monitoring on a small scale, carbon capture and sequestration. Each chapter of the book addresses specific aspects of the renewable and sustainable energy system, strategies for devolving, applications, designs and implementation, industrial outlook, and future direction while focusing more on energy improvement and storage technologies.

Focusing on digitization and the challenges of global climate change, the energy industry is experiencing momentous and enormous bottlenecks, which calls for more intentional and transitional approach to energy operation, management, generation, storage, and distribution. Localized energy transition in the 4IR era is therefore essential and indeed a *sine qua non* for the renewable and sustainable energy nexus. The approach has the potential of providing technical energy management solutions to mitigate the challenges faced within the energy sector. At the same time, the localized energy transition makes an effort towards mitigating environmental challenges through global climate actions. The underlining purpose is to create a regenerative amplification of the solution and the impact thereof of these localized energies in their transitional thrust towards meeting the global energy deficit in line with the Sustainability Development Goal 7 (SDG 7). Furthermore, there are inherent solutions to addressing current challenges in large-scale energy production which have been highlighted in the chapters. Cursory incursion presents focus on producing biofuels, introducing new materials for manufacturing solar PV panels, digital systems for sustainable energy monitoring on a small scale, emerging carbon capture technologies, and carbon sequestration. Further elucidated are energy improvement and storage technologies.

The book consists of 16 Chapters categorized appropriately to address specific aspects within the renewable and sustainable energy themes and scenarios. The book gives in-depth insights into the current thrust on various renewable energy research activities, helping researchers, energy scientists, technocrats, and investors to develop and commercialize the technologies. Trends in energy monitoring, efficiency improvement, and carbon capture technologies are critical tools in these ventures. Investors will have access to a wide range of scalable technologies they can consider investing in, especially research currently in their gestation stage. Start-offs (upstream, midstream, and downstream) stages, small-scale firms, and multi-national in the arena of production, distribution, and monitoring of energy in the 4IR era have a manufacturing mantra in this piece.

1 The Effect of Specifications of Solar Photovoltaic Modules on Array Peak Power, Quantity, and Weight in Residential Buildings

Sogo Mayokun Abolarin, Manasseh Babale Shitta,
Emmanuel Aghogho Metuaghan, Blessing Precious Nwosu,
Michael Chuks Aninyem, Louis Lagrange, and
Daniel Raphael Ejike Ewim

NOMENCLATURE

Symbol	Description	Unit
EC	Energy consumption	kWh
g	Acceleration due to gravity	m/s^2
m_{up}	Unit mass of a selected solar module	kg
N_{sp}	Quantity (number) of solar PV modules	
P_{sp}	Solar PV array peak power	kW
P_{up}	Unit capacity of a selected solar module	W
S_{sh}	Solar sun hour	h
W_s	Weight of the solar PV array	N

Subscript

i	Index for the days in a week	
max	maximum	
min	Minimum	

1.1 INTRODUCTION

Fossil fuels, including coal, oil, and gas, are the most consumed energy sources for many years to meet the needs for energy in various sectors of the economy worldwide. The increasing demand for energy has led to the overconsumption of these non-renewable resources, causing a depletion in their reserves. Moreover, fossil fuels are known to release greenhouse gases, such as CO_2 and methane. The CO_2 accounts for approximately 80% of all greenhouse gas emissions. The release of these gases has largely contributed to global warming and climate change. These emissions trap heat in the Earth's atmosphere, leading to a rise in global temperatures and other adverse impacts, such as rising sea levels as a result of the melting of glaciers and more frequent severe weather conditions. Additionally, the extraction, transportation, and processing of fossil fuels can also result in environmental damage, including oil spills, air and water pollution, and habitat destruction. Therefore,

DOI: 10.1201/9781032651958-1

"""

the continued reliance on fossil fuels for energy production poses significant environmental and economic risks. It has now become imperative for all energy consumers to consider transitioning to non-fossil energy sources. The transition to clean energy has become a critical priority in the face of global warming, climate change, and the Fourth Industrial Revolution (4IR). The implementation of energy efficiency measures, such as improving insulation and upgrading appliances, can reduce energy demand, promote a sustainable and resilient energy future, create new job opportunities, and foster economic growth. Additionally, the transition to clean energy sources, would further reduce the impact of greenhouse gas emissions, making it imperative to prioritize the shift towards clean energy [1–6].

In addition, the reserves where the fossil fuels are explored are in most cases not globally spread, thus either limiting direct access or leading to increased cost [7]. All these challenges call for reliance on viable alternatives to renewable and sustainable sources of energy such as solar, wind, tides, geothermal energy, etc. [8–13]. Among these alternative sources, solar energy has been identified as one of the most abundant and readily deployable to meet energy needs in a sustainable manner across the different economic sectors worldwide [14, 15]. Considering the abundance of this energy resource, many countries around the world have been transitioning from the use of fossil fuels to solar energy. Solar photovoltaic (PV) technology is a solution that can be deployed to meet the increasing energy demands while reducing greenhouse gas emissions. In the year 2017, the world recorded a global solar PV installed capacity of about 400 GW, and this is estimated to increase to 4500 GW by the year 2050 [16], as many more countries implement their sustainable development goals strategies in the form of providing affordable and clean energy as well in climate change actions.

The International Energy Agency (IEA) in 2020 forecast that the addition of solar photovoltaic installed capacity for all the major sectors (commercial, industrial, off-grid, residential, and utility) of the economy globally may reach 107 GW. This additional capacity is expected to increase by 23 GW by the year 2023 and thereafter by 35 GW by the year 2025. This expansion in the installed solar photovoltaic capacity is expected to be driven by improvement in the global economy, the signing of bilateral contracts, and power purchase agreements [17]. According to the IEA Renewable Report of 2020 [17], solar photovoltaics addition in countries in Africa are also expected to increase by different capacities by 2025. For example, 170 MW more solar photovoltaic power is expected to be added to Ethiopia through bidding. Feed-in-Tariff (FiT) is expected to add an extra 120 MW of capacity in Kenya by 2025, while grants are expected to add 30 MW of solar photovoltaic capacity in Tanzania. In the same vein, through a power purchase agreement, Nigeria plans to add 113 MW to its energy supply. This extra capacity is needed to make up for the difference between how much energy is needed and how much is available, as well as the big difference between how much energy comes from fossil fuels and how much comes from clean energy sources. Additional lessons on the transition to clean energy sources, most especially renewable sources, are available in the work of Iychettira [18].

A high solar radiation level is received in Nigeria daily, for instance according to the work of Oghogho, Sulaimon, Egbune, and Abanihi [19] and Nnaji and Unachukwue [20], Nigeria is blessed with 1500 PJ of sunshine annually. On a daily basis, the average solar radiation received in Nigeria varies from 3.5 to 7 kWh/m^2 [21], from the Southern to the Northern parts of the country, and so, efforts are being channelled towards utilizing this abundant solar radiation. Taking advantage of the abundant solar radiation will enable the country's energy end-users harness the green energy to meet their energy needs in a sustainable manner and as such reduce load shedding and the emission of greenhouse gases.

Solar energy is generally harvested with the use of solar PV or solar thermal solutions. These solutions have been identified to contribute little to no adverse effects on the environment, generate no CO_2 emissions at the point of use, and have low maintenance requirements [22]. According to Fuster-Palop, Prades-Gil, Masip, Viana-Fons, and Payá [23], solar PV technologies have the propensity to reduce emissions associated with energy consumption, while Manfren, Nastasi, Tronchin, Groppi, and Garcia [24] opined that for transition and decarbonization in the built environment, smart energy technologies are needed. This is because the installation of solar PV technologies will contribute to the reduction in global warming thus achieving the 1.5 °C temperature goal [25]. While not only reducing adverse environmental impacts [26], renewable energy generation from a solar source also has the inherent benefits of increasing energy access through a reduction in load-shedding particularly in energy-poor communities [27], improving energy security [28], and power supply safety [29].

Solar PV technology has largely become popular being a sustainable source of electrical energy. As a result, governments, private individuals, researchers, and industries have generated tremendous interest in the design as well as installation of solar PV system in buildings [30, 31] in many countries.

However, in the designs and installations, the specifications of the different modules need to be evaluated and compared to make informed decisions on the module that could produce the needed energy yield with the least module quantities and weight possible. This is important because, solar PV modules have been estimated to have a useful life of about 25 years [32], investigating the least quantity of solar PV to meet a particular daily energy requirement could contribute to the reduction of the waste that is expected to be generated at the end-of-life of the panels [16]. This is because according to Xu, Li, Tan, Peters, and Yang [33], an estimated tonnes of solar PV modules that will be reaching their end-of-life is about 9.57 million.

Due to the low efficiencies of today's solar PV modules, many modules are usually needed particularly for meeting medium to high energy needs. In most cases particularly in the residential, commercial, and industrial buildings, these modules are installed on the roofs of buildings [23, 34]. In the work of Zhong, Zhang, Chen, Zhang, Zhou, Zhu, Wang, Lü, and Yan [35], determining the solar energy potential on the rooftop of a building is required for sustainable energy planning. Using a sampling strategy developed from deep learning that involved spatial optimization, the study was able to extract the total rooftop area of buildings in Nanjing to be 330.36 km^2 and thereafter estimated, with a high level of accuracy of about 92%, that the area can accommodate solar PV modules of about 66 GW and in turn generate an annual clean energy of 311,853 GWh.

During the design of existing buildings, the weight these modules would exact might not have been provided for and so, the choice of solar PV modules in this modern-day world needs to be given a priority. Several measures have been proposed to reduce the weight of solar PV modules. One such is using an acrylic that is hard-coated with thermal conductive properties in place of the conventional front glass cover; this is estimated to reduce the weight by 20% [36]. Another is the case of electric vehicles carrying their own solar PV modules. According to Kim, Holz, Park, Yoon, Cho, and Yi [37], using polymeric material composed of tempered glass with low iron will make these electric vehicles lightweight and thus improve their efficiencies. However, these measures are mostly applicable during the manufacturing processes and not at the point of installation. Investigation of the specifications of solar PV modules with respect to the variables such as least quantity and weight that have the propensity of impacting space and load (additional allowable weight) could enable clean energy enthusiasts, developers, and energy end users to make the right decision.

The specifications for solar photovoltaics generally include module efficiency, performance tolerance, short circuit current, working temperature, unit peak power, maximum voltage, open circuit voltage, length, width, weight, etc. [9]. In addition to these variables, there is a need to consider the land, environment, costs, etc. [38] for the soptimal design of solar PV systems. Most of the time, making decisions based on these specifications is a major problem that needs to be simplified.

To help simplify the challenges surrounding decisions on these specifications, this study investigates the effect of unit peak power choice on the array of peak power, quantity, and weight of modules that are needed to satisfy the maximum daily energy consumption of two obvious energy-consuming appliances, which are lighting and air conditioning units in a residential building in Lagos, Nigeria. This work considered and compared 20 modules with unit peak power specifications of 280–400 W. The motivation for this study stems from the need to identify the solar PV array to generate the maximum amount of energy while minimizing the system's weight and cost. It must be noted that selecting the suitable solar PV module for a particular facility remains a complex task, especially in residential complexes with limited space and cost constraints. Therefore, this study contributes a valuable understanding of the design and optimization of solar PV systems in the 4IR context, leading to a more sustainable and energy-efficient future.

1.2 METHODOLOGY

In this study, an energy audit was conducted to determine the daily energy consumption of lighting use as well as that of air conditioning units installed in a five-bedroom residential complex. The data collected during the energy audit procedure were assessed and analysed following the method prescribed in the work of Lagrange [39], as indicated in Figure 1.1.

In general, this method showcases an audit plan informed by a preliminary assessment of energy consumption, metrics and implementation of energy audit levels according to the American Society of Heating Refrigerating and Air Conditioning Engineers (ASHRAE) energy audit levels 1 and 2 [40]. The Level 1 involved the preliminary energy audit, which enables the energy auditors to obtain the building owner's consent prior to data collection, definition of the scope, objective and boundaries of the energy audit and walk-through energy survey. The level 2 involved the energy survey and analysis of the energy audit procedures.

The preliminary and walk-through energy audit levels allowed the collection of a series of data, such as the appliance type, quantity, hour of use, and power ratings of the appliances. These energy uses and the corresponding daily peak energy consumption are summarized in Table 1.1 for lighting with ratings 10–100 W, while the rated input power demands were from 1.16 kW to 2.1 kW.

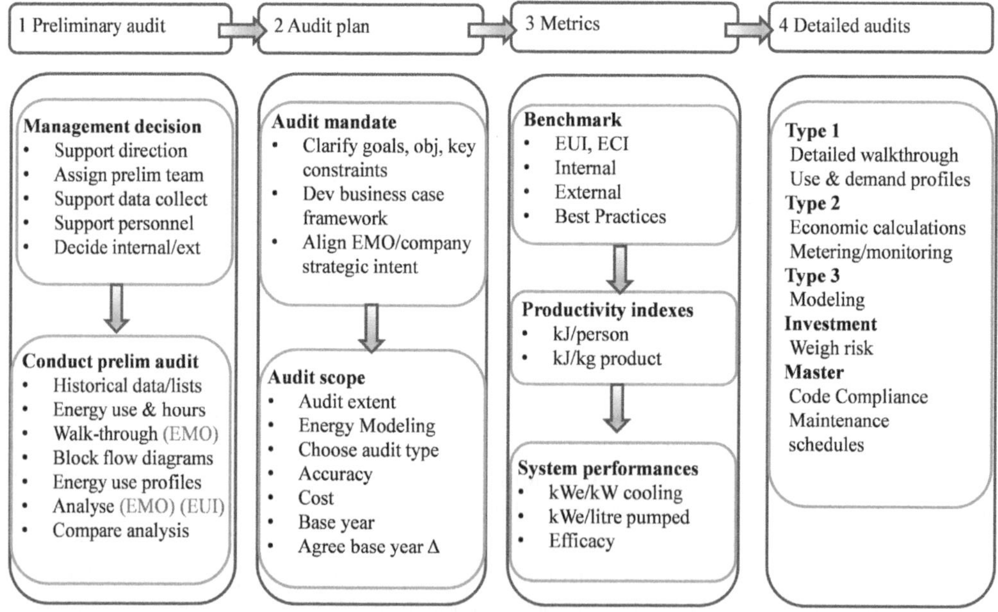

FIGURE 1.1 The energy use assessments in the work of Lagrange [21].

TABLE 1.1
The Residence Appliances and Daily Maximum Energy Consumption

	Appliances	
	Lighting	Air Conditioning Unit
Daily maximum energy consumption [kWh]	48.02	213.73

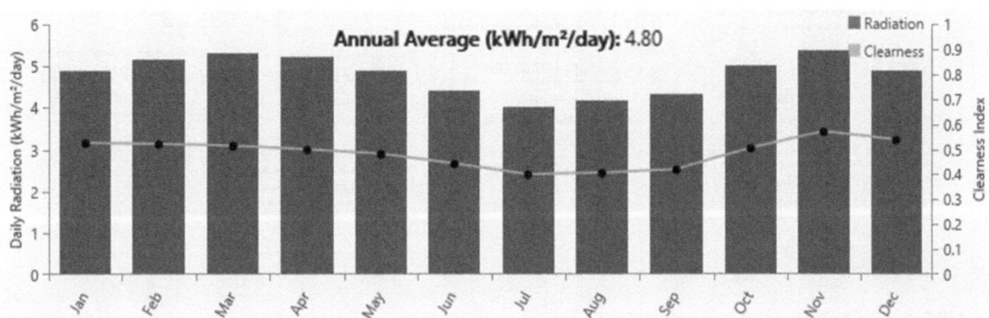

FIGURE 1.2 The global horizontal irradiance at the building's location obtained from HOMER Pro software.

The daily hour of use was collected for 5 days, while other variables did not change during the period of energy audit data collection. The daily peak energy consumption for lighting use was 48.02 kWh, and air conditioning use was 213.73 kWh.

The maximum daily energy consumption of the appliances and the building location's solar sun hours were used to determine the peak power, quantity and weight of the PV array needed to produce clean energy required to meet the demand. The 20 solar PV modules were assigned with pseudonyms ("A"-"T"). Each module specification obtained from the specification sheet of the manufacturer was used in the data reduction.

The average daily horizontal solar radiation of the building's location was obtained from the HOMER Pro software (version 3.14.5) as 4.8 kWh/m²/day, as shown in Figure 1.2.

These data were stored, and the reduction method was programmed into an Excel spreadsheet Visual Basic for Application (VBA) called Electronic Energy Audit and Solar siZing (e-EASZ) developed for energy audit analysis and solar PV sizing. The flowchart of the tool is summarized in Figure 1.3.

1.3 DATA REDUCTION

The study analysed the energy consumption-related data collected during the energy audit. It determined the daily maximum energy consumption, $[EC]_{max}$ of each of the two appliances in the residential building. The total size of solar PV, which is also the same as the solar PV array peak power, P_{sp} for each appliance was determined as a function of the maximum daily energy consumption of each appliance and average irradiation (solar sun hours) of the building's location, S_{sh}

$$P_{sp} = 1.3 \times \left(\frac{\left[EC_i \right]_{max}}{S_{sh}} \right)$$

(1.1)

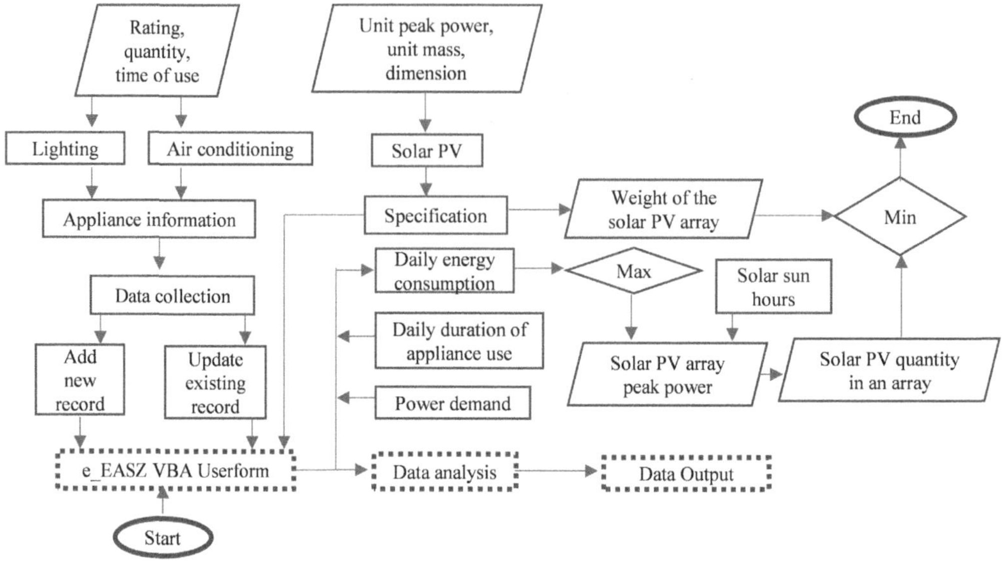

FIGURE 1.3 Flow chart of the methodology.

To meet the daily energy requirement and account for any unintended losses that may occur due to the intermittency of solar radiation, the solar PV peak power was raised by 30% [41]. Thus, accurately predicting the power of solar PV technology is vital towards the development of mini-grid projects [42].

The solar PV modules quantity, N_{sp} that would be acquired and installed was determined using the ratio of the solar PV peak power to the unit capacity of each pseudonym solar module, P_{up}:

$$N_{sp} = \text{Even}\left[\text{Roundup}\left(\frac{P_{sp}}{P_{up}}, 0 \right) \right] \tag{1.2}$$

The actual rated capacity of the solar PV array (P_{spr}) was determined based on the product of the choice commercially available unit module peak power (P_{up}) and the number of the solar modules in the array (N_{sp}) as:

$$P_{spr} = N_{sp} \times P_{up} \tag{1.3}$$

The weight of the solar PV array, W_{sp}, was determined as the product of the number of the modules, N_{sp}, and unit mass of each module, m_{up}, from the modules' manufacturers specification sheets, and the acceleration due to gravity ($g = 9.81$ m/s^2) as:

$$W_{sp} = N_{sp} \times m_{up} \times g \tag{1.4}$$

The unit peak power of these solar PV modules considered in this work ranged from 280 to 400 W. It was programmed into an Excel spreadsheet, a VBA tool called e-EASZ tool. This tool accepts the data from the energy as input, stored, and analysed the data and generate results, which are discussed in this study.

1.4 RESULTS AND DISCUSSION

1.4.1 THE ARRAY PEAK POWER

1.4.1.1 The Lighting Case

The results illustrate that the rated array peak power determined for each choice pseudonym module to meet the lighting needs varies from one array to the other, as shown in Figure 1.4.

Figure 1.4 indicates that for modules "A" and "B", the array peak power is 13.44 kW, and for "C" it is 13.34 kW. Each of the modules "D", "E", and "F", respectively has a rated array peak power of 13.20 kW, 13.02 kW, and 13.44 kW. Modules "G" and "H" have array power of 13.20 kW each, "I" and "J" each have array power of 13.60 kW and 13.30 kW, respectively while each of "K" and "L", "M" and "N", "O" and "P", "Q" and "R", and "S" and "T", respectively, have rated array peak power of 13.68 kW, 13.32 kW, 13.68 kW, 13.26 kW, and 13.60 kW.

Figure 1.5 illustrates the rated solar PV array peak power needed to be installed to meet the daily energy consumption of lighting as a function of each commercially available unit module peak power choice.

Based on the manufacturers' specifications of the modules considered in this study, it was found that the array peak power required reduces as the unit peak power increases within the range of 280–310 W. Furthermore, the result shows that for modules from 300 W, the rated peak power

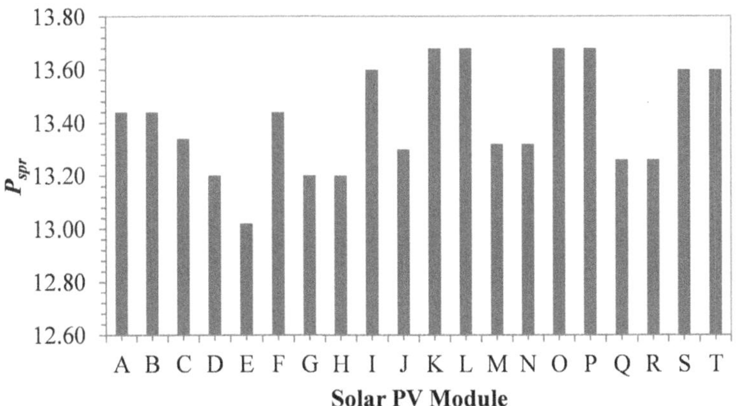

FIGURE 1.4   The lighting case' array peak power (kW) against the pseudonym modules.

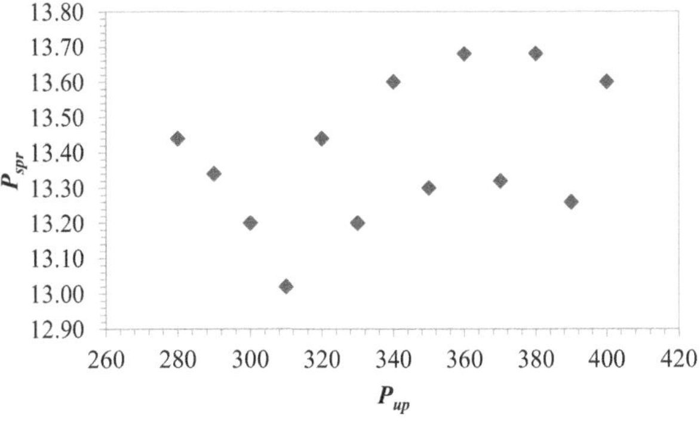

FIGURE 1.5   The rated solar PV peak power against the unit peak power for the lighting.

varied significantly because of the choice unit module capacity selected, and the number of modules required. The rated array peak power for choosing the modules with unit peak powers of 300, 320, 340, 360, 380, and 400 W were higher than those of the modules with unit peak powers of 310, 330, 350, 370, and 390 W. This could be attributed to the fact that as the peak power increases the quantity of modules required to meet the same energy reduces [43].

1.4.1.2 The Air-Conditioning Case

The rated array peak power for the air conditioning case is presented in Figure 1.6.

As expected, the magnitude of the array peak powers for the lighting load is lower than those of the air conditioning load. This is because the power demand of air conditioning units is higher than that of lighting [44]. According to Sutikno, Aldina, Sari, and Handogo [45], the high-power demand for air conditioning units contributes to their high energy consumption. Another factor that contributes to high energy consumption is the climate region in which the building is located. For instance, the building under consideration is in the tropic region. Pokhrel, Walker, and González [14] reported that the warm and humid climates as well as extreme weather (heat) conditions of the tropics contribute to the need for the use of air-conditioning units to achieve cooling and maintain indoor occupancy comfort all year round.

In the year 2010, energy consumption associated to Heating, Cooling, and Air-Conditioning in the residential buildings was about 61% [46], this has since increased to 63% as reported in the work of Pokhrel, Walker, and González [14]. Cooling energy consumption is projected to increase further by 15% in the 21st century due to the growth in the economy and the increased impact of climate change [47] and by 20% when the weather conditions are extreme [48]. This means that as the energy consumption is needed to size and determine suitable modules, high energy consumption will lead to a high array of peak power of the modules.

For the air conditioning case in Figure 1.6, the array of each combination of modules "A" and "B" each has a rated array peak power of 58.24 kW. Each array of modules "C", "D", "E", and "F" is of the peak power 58, 58.20, 58.28, and 58.24 kW, respectively. The modules "G" and "H" have the same rated capacity of 58.08 kW. The rated array peak power of each module from "I" and "J" is 58.48 kW and 58.10 kW, respectively, while that of modules groups "K" and "L", "M" and "N", "O" and "P", "Q" and "R" as well as "S" and "T" are respectively, 58.32, 58.46, 58.52, 58.50, and 58.40 kW.

Figure 1.7 shows that the rated peak array power for the pseudonym modules differs based on the choice of the unit module capacity and the number of solar PV modules required to meet the

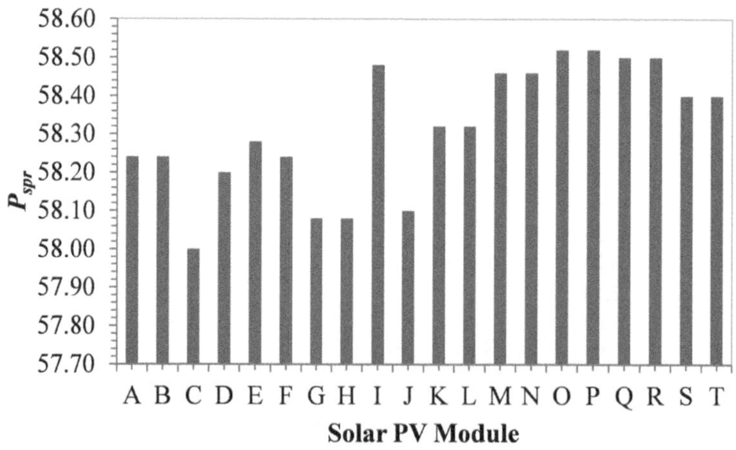

FIGURE 1.6 Rated array peak power against the pseudonym modules for the air conditioning case.

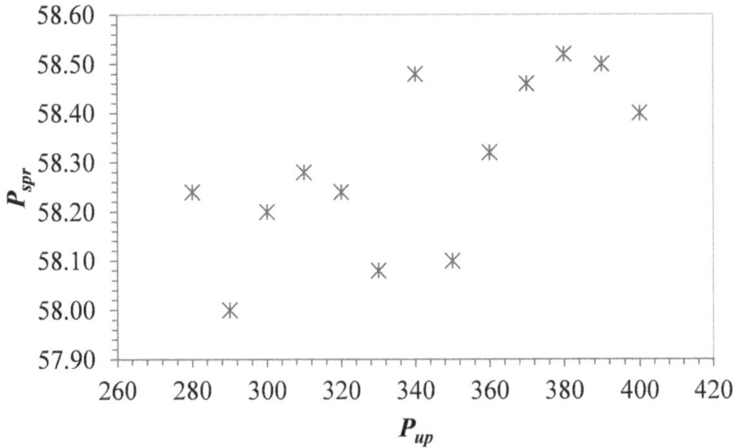

FIGURE 1.7 The rated peak array power as a function of the unit peak power of the PV modules.

air-conditioning needs. For instance, the "A" and "B" each with a unit peak power of 280 W formed an array with a rated power of 58.24 kW. Each of the modules "G" and "H" with a unit capacity of 330 W formed an array with a rated peak power of 58.08 kW, while each of the modules "O" and "P" formed an array power of 58.52 kW from unit peak power of 380 W. On the other hand, the modules "S" and "T" with a unit power of 400 W, formed an array with a peak power of 58.40 kW. These results showed that the choice of different unit peak power of solar modules produces different array peak power needed to meet the same energy needs.

The determination and comparison of the rated solar array peak powers of the different solar PV modules considered in this study are necessary for the determination of the number of solar PV modules, the total weight of each solar PV array, and the selection of a suitable solar PV module needed to meet the energy needs of the two energy uses (lighting and air conditioning units) being investigated.

1.4.2 NUMBER OF SOLAR PV MODULES

To enable the selection of a solar PV module that its array can yield clean energy that meets the daily energy requirements while minimizing space, the specifications of the 20 pseudonym modules were compared. These modules' specifications (unit peak power, unit area, and unit weight) are summarized in Table 1.2.

1.4.2.1 The Lighting Case

Table 1.2 shows that the number of solar PV modules required to meet the same lighting and air conditioning energy needs (Table 1.1) varies with the pseudonym modules (Figure 1.8), unit peak power (Figure 1.9), and dimension of the modules (Table 1.2). An increase in the unit peak power of each solar PV module led to a reduction in the required quantity of modules as illustrated in Figure 1.9.

As summarized in Table 1.1, the building's daily lighting energy requirement is 48.02 kWh, while that of the air-conditioning is 213.73 kWh. The results from the different modules in Table 1.1 reveal that the number of modules required was reduced from 48 to 34 for lighting as the unit peak power increased, as shown in Figure 1.9.

This reduction in the quantity of modules translates to a significant reduction in the surface area of the solar PV array required to meet the same energy needs for lighting purposes.

TABLE 1.2
Specifications of the 20 Solar PV Modules

Pseudonyms	A	B	C	D	E	F	G	H	I	J
P_{up} (W)	280	280	290	300	310	320	330	330	340	350
A_{up} (m²)	1.64	1.64	1.64	1.94	1.94	1.63	1.85	1.94	1.79	1.94
N_{sp}-lighting	48	48	46	44	42	42	40	40	40	38
N_{sp} – AC	208	208	200	194	188	182	176	176	172	166
W_{up}	182	191	191	228	228	182	207	221	195	212
Pseudonyms	K	L	M	N	O	P	Q	R	S	T
P_{up} (W)	360	360	370	370	380	380	390	390	400	400
A_{up} (m²)	1.98	1.98	2.01	1.64	1.97	2.01	1.98	2.01	2.01	2.01
N_{sp}-lighting	38	38	36	36	36	36	34	34	34	34
N_{sp} – AC	162	162	158	158	154	154	150	150	146	146
W_{up}	222	221	216	221	219	216	226	216	221	223

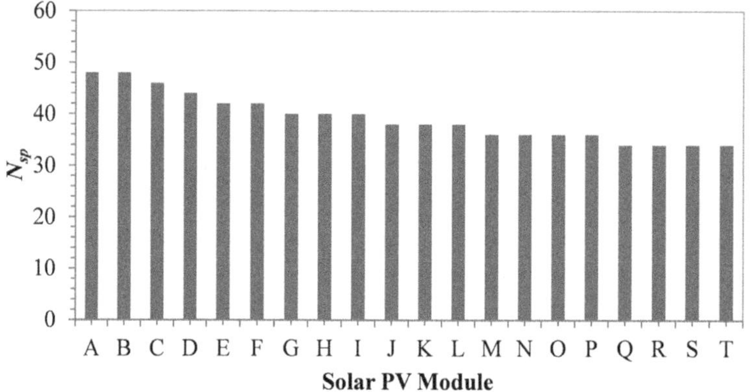

FIGURE 1.8 Variation of quantity with the different solar PV modules for the lighting case.

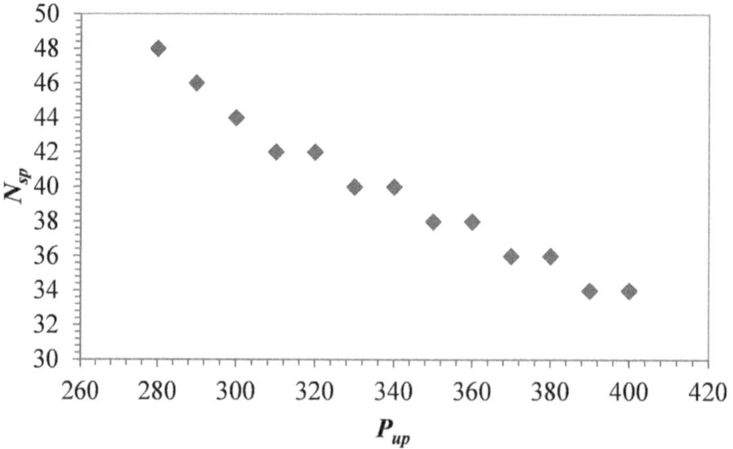

FIGURE 1.9 The comparison of the required solar module quantity with the unit peak power for the lighting case.

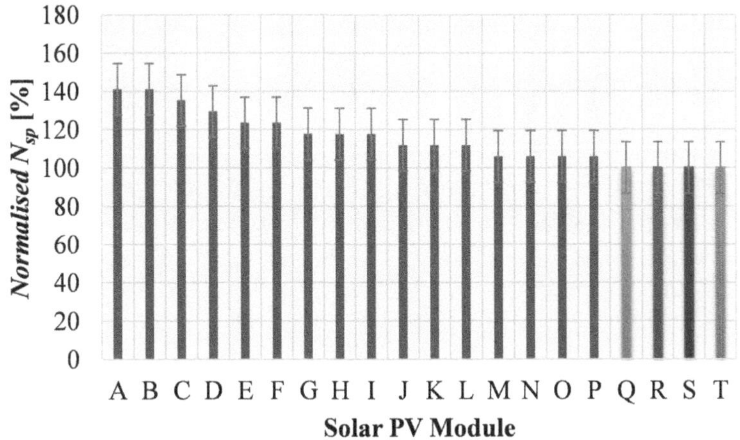

FIGURE 1.10 Normalized quantity with the different solar PV modules for the lighting case.

Figures 1.6 and 1.9 show the pseudonym modules "A"–"T" that have been individually evaluated to determine the respective peak power and the module quantity needed to produce the yield required to meet the maximum daily energy needs as previously indicated in Table 1.1. From the list, the module(s) capable of meeting the daily energy needs at the least module quantity would be selected. Based on this condition, the suitable solar PV modules selected are "Q", "R", "S", and "T" for the lighting energy consumption. This means that, in terms of quantity of modules, for "Q", "R", "S", and "T", only 34 module units are required for the lighting case. Other arrays, like "A" and "B" each require a total of 48, while "C" and "D" require 46 and 44 modules, respectively. The solar PV modules in each of the groups "E" and "F", "G", "H", and "I" need a total of 42 and 40 modules each. The arrays "J", "K", and "L" require 38 modules each, while arrays with the pseudonyms "M", "N", "O", and "P" need 36 quantity of solar PV modules, respectively. These variations in solar PV module quantities reveal the savings accruable when considerations are given to the solar PV specifications when designing solar PV systems for clean energy transition in the residential, commercial, and industrial sectors of the economy.

Figure 1.10 shows savings in the quantity of solar PV required to meet the daily lighting energy demand. The savings are represented in a normalized bar chart (Figure 1.10), which shows the percentage increase in other modules relative to selected reference modules "Q", "R", "S", or "T". As illustrated, the normalized results indicate that selecting either of the modules "Q", "R", "S", or "T" leads to significant savings in module quantity. As illustrated, savings of 41%, 35%, 29%, and 24% are achieved for not selecting the modules "A" and "B", "C" "D", and "E" and "F", respectively. When compared with modules "G" to "I", "J" to "L", and "M" to "P", selecting either of the modules "Q", "R", "S", or "T" for the lighting over the other modules resulted in percentage savings of 18%, 12%, and 6%, respectively.

1.4.2.2 The Air Conditioning Case

The quantity of the solar PV modules required to meet the air conditioning needs reduces, as shown in Figure 1.11 for the pseudonym modules and with an increase in unit peak power as illustrated in Figure 1.12. As illustrated for the air conditioning, the number of modules varied from 208 to 146 as the unit peak capacity of the modules was reduced.

Figure 1.11 illustrates the quantity of solar PV modules required to meet the daily air conditioning energy needs in the residential complex being considered in this study. The selection criterion is to select an array that could generate clean energy that meets the required cooling energy

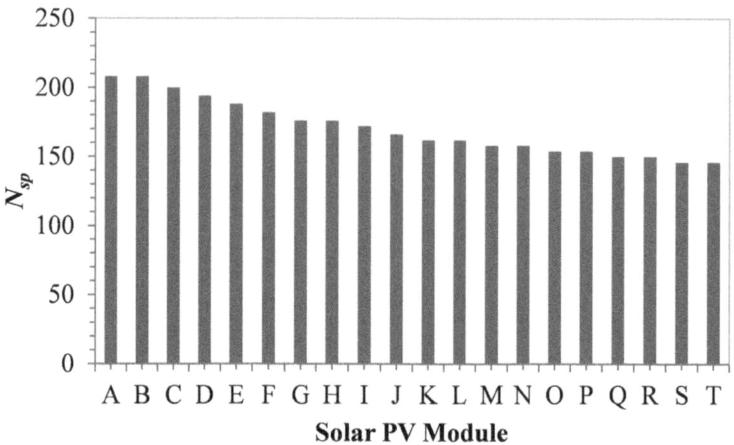

FIGURE 1.11 Variation of quantity with the different solar PV modules for the air conditioning case.

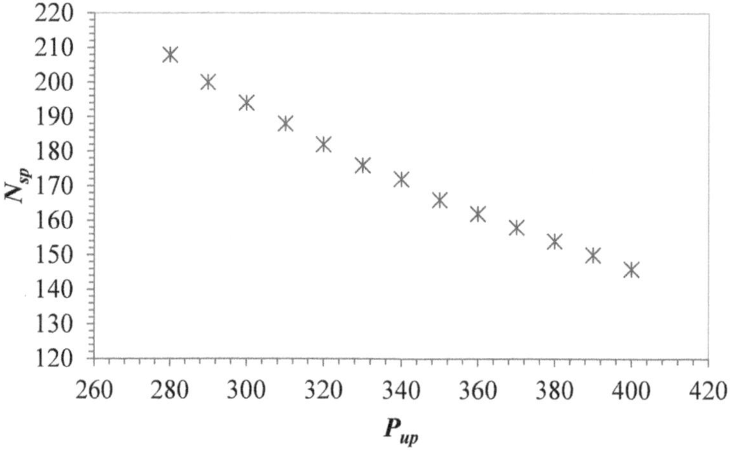

FIGURE 1.12 The comparison of the required solar module quantity with the unit peak power for the air-conditioning case.

with the least quantity out of the 20 pseudonym modules. Figure 1.11 shows that only 146 modules are required to produce the yield that meets the air conditioning need when either of the modules "S", or "T" is selected. This is important because each of these modules has the characteristics of generating the required energy yield at the highest unit peak power and the best module efficiency. Figure 1.11 also shows that other arrays such as "A" and "B", "C", "D", "E", and "F" each require total module quantities of 208, 200, 194, 188, and 182, respectively. Other range of modules such as "G" and "H", "I", "J", "K" and "L", "M" and "N", "O" and "P", "Q" and "R" each require a total number of 176, 172, 166, 162, 158, 154, and 150 solar PV panels, respectively.

Considering the specifications, as in the case of the air conditioning energy use is beneficial, as illustrated in Figure 1.13. Figure 1.13 shows that selecting either of the solar PV "S" or "T" as in the case of this study leads to great savings in the quantity of unit modules. For instance, savings of 42%, 37%, 33%, 29%, 25%, and 21% are made over modules "A" and "B", "C", "D", "E", "F", "G" and "H", respectively. When either of the modules "S" or "T" is compared with the rest of the pseudonym modules "I", "J", "K" and "L", "N" and "M", "O" and "P", "Q" and R", the results show

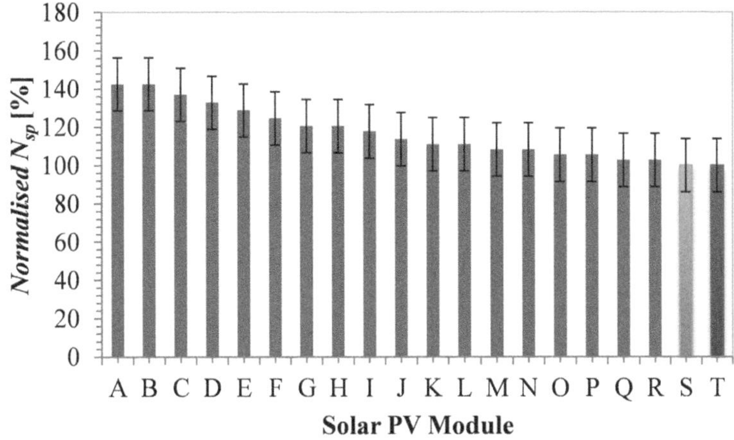

FIGURE 1.13 Normalized quantity with the different solar PV modules for the air conditioning energy use case.

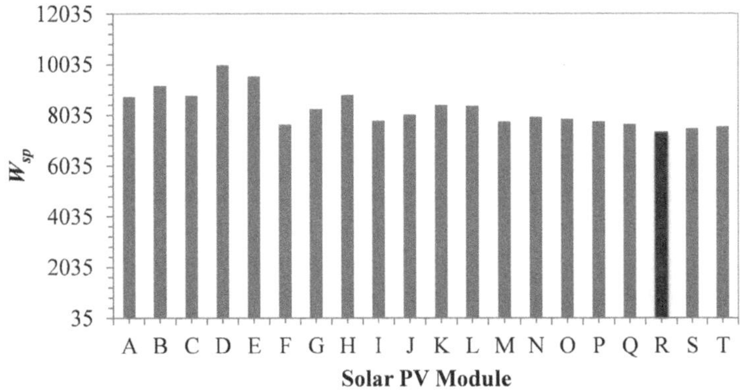

FIGURE 1.14 Total weight exerted by the different solar PV modules for the lighting case.

the accruable savings in selecting either of the modules "S" or "T" over the rest of the other number of modules in the range of 3–18%. These savings are significant when an economic variable such as cost is required for decision making as can be corroborated in the work of Iychettira [18].

1.4.3 WEIGHT OF SOLAR PV ARRAY

1.4.3.1 The Lighting Case

Figure 1.14 shows that weights of the solar PV arrays are high and thus play important role in the design and decision making in solar PV system. Figure 1.14 illustrates the total weight each pseudonym solar PV array requires for the lighting case, which is expected to be exerted on the platform or surface where the modules are installed. As shown in Figure 1.14, the results revealed that the total weight of each set of modules is high and that the weights vary significantly from one module to the other.

Having the characteristics and knowledge of the weight of the choice solar PV array to be installed, contributes significantly to decision-making and safety of the installation, particularly when the installations are planned on rooftops of lightweight buildings or structures. This is

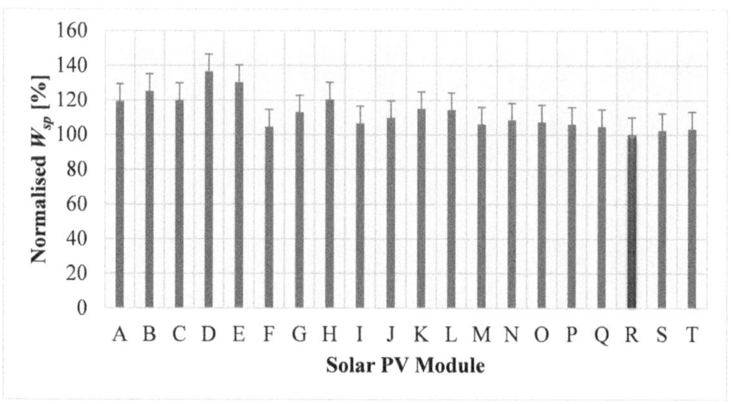

FIGURE 1.15 Normalized weight the different solar PV arrays will exert for the lighting case.

because the quantity of modules also contribute to the total weight of the solar PV infrastructure. For instance, the array that will exert the least weight is "R" with a weight of 7 338 N, while each array of "S" and "T" exert a total weight of 7 505 N and 7 571 N, respectively. Other arrays considered exert additional weight compared with array "R". The other arrays, such as "A", "B", "C", and "D" each exert a weight of 8 758 N, 9 182 N, 8 800 N, and 10 014 N, respectively. The weight of arrays "E" to "Q" ranged from 7 664 N to 9 559 N. The results show that array "D" exerts the highest weight, while array "R" exerts the least weight. So, the choice of array selected considering the least weight is "R".

The normalized weights of the different pseudonym arrays for the lighting case are illustrated in Figure 1.15 with reference to array "R". As compared with array "R", the results show that other solar PV arrays exert additional weights of varying magnitudes. Though modules "Q", "S", and "T" met the criterion of least quantity previously discussed, choosing any of these arrays exerts about 5%, 2%, and 3% weights, respectively, more than array "R". Array "D" with the highest weight exerts additional 36% weight as compared with "R". Others like "A", "F", "H", "J", "K", "O", and "P" exert additional weight of 19%, 4%, 20%, 10%, 15%, 7%, and 6%, respectively.

1.4.3.2 The Air-Conditioning Case

Figure 1.16 shows the total weight of the solar PV array for each of the module specifications considered for meeting the energy demand of the air conditioning services. As expected, the weight of each array for the air-conditioning (Figure 1.16) is significantly higher than the corresponding light case (Figure 1.14). This is because the quantity of solar PV modules required for the lighting is lower than the air-conditioning services due to significant differences in power demand and energy consumption between these two energy-consuming appliances. Since the air-conditioning system is a higher energy consumer compared with lighting, Huang, Hou, Hsu, Lin, Chen, Chen, Li, and Lee [49] advised that power generation from solar PV for meeting cooling needs to be properly designed to ameliorate loss and mismatch of power between generation and demand. This is because failure to do this could lead to the solar PV not generating enough power required to meet the cooling needs as well as influence the weight of the array.

Figure 1.16 shows that the array with the highest total weight is "D", while the array with the least weight is "S" with magnitudes of 44 153 N and 32 226 N, respectively. The closest to "S" is the array "R" with an additional magnitude of 147 N. The weights of the remaining solar PV arrays are distributed between the magnitudes of array "A" and "Q". For instance, the magnitudes of the modules that formed arrays "A", "B", "C", "E", "F", "G", and "H" are 37 953 N, 39 789 N, 38 259 N, 42 789 N, 33 209 N, 36 430 N, and 38 848 N, respectively. Arrays "I", "J", "K", "L", "M", "N",

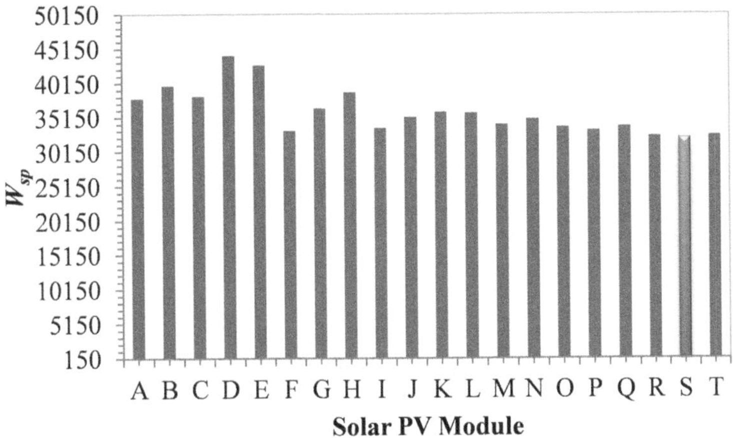

FIGURE 1.16 Total weight exerted by the different solar PV modules for the air conditioning case.

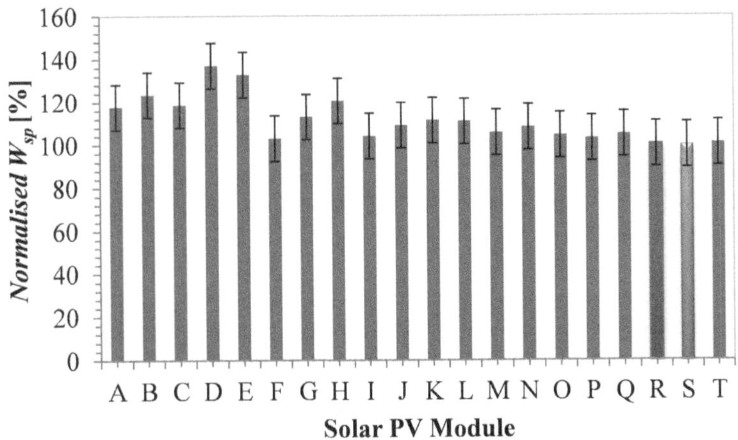

FIGURE 1.17 Normalized weight the different solar PV arrays will exert for the air conditioning case.

"O", "P", and "Q" each exert a weight of 33 578 N, 35 175 N, 35 916 N, 35 757 N, 34 100 N, 34 875 N, 33 690 N, 33 236 N, and 33 845 N, respectively. These results indicate that the weight of a solar PV array particularly when installed on the roof of a building or structure should not be ignored. These high magnitudes play significant roles in determining the pressure exerted on the installation surface and the entire building structure.

Figure 1.17 illustrates the normalized total weights of the different solar PV arrays considered for meeting the cooling energy need. This bar chart compares the weight of each solar PV array with a reference solar PV array. The reference solar PV array, in this case, is "S" because it is the one that yields the least weight. So, the weights of the other solar PV arrays were compared with this reference array to determine the weight savings. As shown in Figure 1.17, choosing another solar PV array other than "S" attracts additional weight, which in turn is exerted on the platform where the modules are installed. The solar PV arrays "R", "T" ("F" and "P"), "I" ("O" and "Q"), "M", "N", and "J" exert additional weight of 0.5%, 1% (3% each), 4%, 5%, 6%, 8%, and 9% more than that of array "S", respectively. Weight savings of 11%, 13%, 18%, 19%, 21%, 23%, 33%, and 37% are made for choosing array "S" over the remaining arrays ("L" and "K") "G", "A", "C", "H", "B", "E", and "D".

The specification and the results of this study have shown that the peak power of solar PV modules between 280 W and 400 W and other specifications like surface area as well as weight need to be given consideration because the variables play significant roles in the process of designing, analysing, and selecting PV modules to be installed on buildings facades, rooftops, platforms, or surfaces for clean energy transition.

Attention to these specifications enables existing buildings that were designed as lightweight and new lightweight buildings to be suitable for solar PV installations. Thus, it makes the buildings less external energy dependent by generating their own energy using rooftop-mounted solar PV technologies to support their power demand.

1.5 CONCLUSION

In conclusion, an energy audit was conducted on lighting and air conditioning appliances in a residential complex to enable energy users transition to clean energy and select suitable solar PV modules. The maximum daily energy consumption of each appliance was determined, and 20 commercial solar PV modules were analysed to ensure that all could produce the peak power needed to meet the maximum daily energy needs. The study then determined the quantity and weight of modules needed for the transition, selecting the modules that produced clean energy that met the daily energy consumption at the least possible quantity and weight.

The results showed that only modules culminating into arrays "Q", "R", "S", and "T" meet the condition set for quantity in the lighting case, while only modules "S" and "T" meet the condition set for quantity in the air conditioning case. Each module in the arrays "Q", "R", "S", and "T" has the needed array peak power capacity to meet the maximum daily energy demand with the least number of modules. Choosing any of these arrays leads to savings of between 2% to 36% and 0.5% to 37% in terms of quantity and space availability.

In terms of weight, only the combination of modules that form array "R" for the lighting case and "S" for the air conditioning case exerts the least weight on the building surfaces, rooftops, or platforms where the modules are installed, and this would form the overall weight to be transferred to the building structure. Arrays "S" and "T" could not be selected for the lighting case because of the propensity of exerting additional weights, and arrays "R" and "T" could not be selected for the air conditioning case because of increased weight.

In general, the modules that form array "R" considerably meet the conditions of least quantity and weight for the lighting case, while the combination of modules that form array "S" meet the set conditions for the air conditioning case as well as produce the needed yield. The study highlights the importance of considering and comparing different specifications of solar PV modules to help energy end-users and designers select a suitable combination of solar PV modules that meet the required energy needs with the least quantity and weight.

REFERENCES

[1] T. Mahachi, *Energy yield analysis and evaluation of solar irradiance models for a utility scale solar PV plant in South Africa*, Stellenbosch University, Stellenbosch, South Africa, 2016.

[2] S.M. Abolarin, A.O. Gbadegesin, M.B. Shitta, A. Yussuff, C.A. Eguma, L. Ehwerhemuepha, O. Adegbenro, A collective approach to reducing carbon dioxide emission: A case study of four University of Lagos Halls of residence, *Energy and Buildings*, 61 (2013) 318–322.

[3] R. Puertas, L. Marti, International ranking of climate change action: An analysis using the indicators from the Climate Change Performance Index, *Renewable and Sustainable Energy Reviews*, 148 (2021) 111316.

[4] M.O. Dioha, A. Kumar, D.R. Ewim, N.V. Emodi, Alternative scenarios for low-carbon transport in Nigeria: a long-range energy alternatives planning system model application, in: *Economic Effects of Natural Disasters*, Elsevier, 2021, pp. 511–527.

[5] M. Ntuli, M. Dioha, D. Ewim, A. Eloka-Eboka, Review of energy modelling, energy efficiency models improvement and carbon dioxide emissions mitigation options for the cement industry in South Africa, *Materials Today: Proceedings*, 65 (2022) 2260–2268.

[6] D.R.E. Ewim, S.M. Abolarin, T.O. Scott, C.S. Anyanwu, A survey on the understanding and viewpoints of renewable energy among South African school students, *The Journal of Engineering and Exact Sciences*, 9(2) (2023) 15375–15301e.

[7] O.I. Okoro, T.C. Madueme, Solar energy: A necessary investment in a developing economy, *Nigerian Journal of Technology*, 23(1) (2004) 58–64.

[8] O.B. Adewuyi, M.K. Kiptoo, A.F. Afolayan, T. Amara, O.I. Alawode, T. Senjyu, Challenges and prospects of Nigeria's sustainable energy transition with lessons from other countries' experiences, *Energy Reports*, 6 (2020) 993–1009.

[9] Z. Almusaied, B. Asiabanpour, S. Aslan, Optimization of solar energy harvesting: An empirical approach, *Journal of Solar Energy*, 2018 (2018) 9609735.

[10] A. Barnes, D. Marshall-Cross, B.R. Hughes, Towards a standard approach for future Vertical Axis Wind Turbine aerodynamics research and development, *Renewable and Sustainable Energy Reviews*, 148 (2021) 111221.

[11] M.A. Khan, A. Javed, S. Shakir, A.H. Syed, Optimization of a wind farm by coupled actuator disk and mesoscale models to mitigate neighboring wind farm wake interference from repowering perspective, *Applied Energy*, 298 (2021) 117229.

[12] A.S. Edun, K. Perry, J.B. Harley, C. Deline, Unsupervised azimuth estimation of solar arrays in low-resolution satellite imagery through semantic segmentation and Hough transform, *Applied Energy*, 298 (2021) 117273.

[13] D.R.E. Ewim, S.M. Abolarin, T.O. Scott, C.S. Anyanwu, A survey on the understanding and viewpoints of renewable energy among South African school students, *The Journal of Engineering and Exact Sciences*, 9(2) (2023) 1–14.

[14] R. Pokhrel, A. Walker, J.E. González, A new methodology to assess building integrated roof top photovoltaic installations at city scales: The tropical coastal city case, *ASME Journal of Engineering for Sustainable Buildings and Cities*, 1(1) (2019) 011004-14.

[15] B. Kumar Sahu, A study on global solar PV energy developments and policies with special focus on the top ten solar PV power producing countries, *Renewable and Sustainable Energy Reviews*, 43 (2015) 621–634.

[16] M.S. Chowdhury, K.S. Rahman, T. Chowdhury, N. Nuthammachot, K. Techato, M. Akhtaruzzaman, S.K. Tiong, K. Sopian, N. Amin, An overview of solar photovoltaic panels' end-of-life material recycling, *Energy Strategy Reviews*, 27 (2020) 100431.

[17] IEA, *Renewables 2020: Analysis and forecast to 2025*, International Energy Agency, Paris, France, 2020.

[18] K.K. Iychettira, Lessons for renewable integration in developing countries: The importance of cost recovery and distributional justice, *Energy Research & Social Science*, 77 (2021) 102069.

[19] I. Oghogho, O. Sulaimon, D. Egbune, V. Abanihi, Solar energy potential and its development for sustainable energy generation in Nigeria: A road map to achieving this feat, *International Journals of Engineering and Management Sciences*, 5(2) (2014) 61–67.

[20] M. Nnaji, G.O. Unachukwue, Energy efficiency and the Nigerian economy: The need for an integrated approach, *Nigerian Journal of Solar Energy*, 21 (2010) 179–181.

[21] M.O. Oseni, Improving households' access to electricity and energy consumption pattern in Nigeria: Renewable energy alternative, *Renewable and Sustainable Energy Reviews*, 16 (2012) 3967–3974.

[22] S. Abdul-Wahab, Y. Charabi, A.M. Al-Mahruqi, I. Osman, S. Osman, Selection of the best solar photovoltaic (PV) for Oman, *Solar Energy*, 188 (2019) 1156–1168.

[23] E. Fuster-Palop, C. Prades-Gil, X. Masip, J.D. Viana-Fons, J. Payá, Innovative regression-based methodology to assess the techno-economic performance of photovoltaic installations in urban areas, *Renewable and Sustainable Energy Reviews*, 149 (2021) 111357.

[24] M. Manfren, B. Nastasi, L. Tronchin, D. Groppi, D.A. Garcia, Techno-economic analysis and energy modelling as a key enablers for smart energy services and technologies in buildings, *Renewable and Sustainable Energy Reviews*, 150 (2021) 111490.

[25] B.-J. Tang, Y.-Y. Guo, B. Yu, L.D.D. Harvey, Pathways for decarbonizing China's building sector under global warming thresholds, *Applied Energy*, 298 (2021) 117213.

[26] L. Huang, R. Zheng, Energy and economic performance of solar cooling systems in the hot-summer and cold-winter zone, *Buildings*, 8(3) (2018) 37.

[27] M. Moner-Girona, A. Bender, W. Becker, K. Bódis, S. Szabó, A.G. Kararach, L.D. Anadon, A multi-dimensional high-resolution assessment approach to boost decentralised energy investments in Sub-Saharan Africa, *Renewable and Sustainable Energy Reviews*, 148 (2021) 111282.

[28] G. Ofosu-Peasah, E. Ofosu Antwi, W. Blyth, Factors characterising energy security in West Africa: An integrative review of the literature, *Renewable and Sustainable Energy Reviews*, 148 (2021) 111259.

[29] T.R. Ayodele, A.S.O. Ogunjuyigbe, O.D. Ajayi, A.A. Yusuff, T.C. Mosetlhe, Willingness to pay for green electricity derived from renewable energy sources in Nigeria, *Renewable and Sustainable Energy Reviews*, 148 (2021) 111279.

[30] Z. Usman, J. Tah, H. Abanda, C. Nche, A critical appraisal of PV-systems' performance, *Buildings*, 10(11) (2020) 192.

[31] M. Gul, Y. Kotak, T. Muneer, Review on recent trend of solar photovoltaic technology, *Energy Exploration & Exploitation*, 34(4) (2016) 485–526.

[32] W.-H. Huang, W.J. Shin, L. Wang, W.-C. Sun, M. Tao, Strategy and technology to recycle wafer-silicon solar modules, *Solar Energy*, 144 (2017) 22–31.

[33] Y. Xu, J. Li, Q. Tan, A.L. Peters, C. Yang, Global status of recycling waste solar panels: A review, *Waste Management*, 75 (2018) 450–458.

[34] S. Jung, J. Jeoung, H. Kang, T. Hong, Optimal planning of a rooftop PV system using GIS-based reinforcement learning, *Applied Energy*, 298 (2021) 117239.

[35] T. Zhong, Z. Zhang, M. Chen, K. Zhang, Z. Zhou, R. Zhu, Y. Wang, G. Lü, J. Yan, A city-scale estimation of rooftop solar photovoltaic potential based on deep learning, *Applied Energy*, 298 (2021) 117132.

[36] M. Zinaddinov, S. Mil'shtein, D. Kazmer, Design of light-weight solar panels, in: *2019 IEEE 46th Photovoltaic Specialists Conference (PVSC)*, 2019, pp. 0582–0587.

[37] S. Kim, M. Holz, S. Park, Y. Yoon, E. Cho, J. Yi, Future options for lightweight photovoltaic modules in electrical passenger cars, *Sustainability*, 13(5) (2021) 2532.

[38] B. Pillot, N. Al-Kurdi, C. Gervet, L. Linguet, Optimizing operational costs and PV production at utility scale: An optical fiber network analogy for solar park clustering, *Applied Energy*, 298 (2021) 117158.

[39] L.F. Lagrange, Energy use assessment methodology, fundamentals of energy management training (FEMT) course notes, in: *E.m. Section 6 (Ed.)*, 2016.

[40] ANSI, ASHRAE, ACCA, *Standard for commercial building energy audits*. ASHRAE, Atlanta, 2018, pp. 1–7.

[41] Leonics, *How to design solar PV system*. Leonics, 2019.

[42] M.R. Kaloop, A. Bardhan, N. Kardani, P. Samui, J.W. Hu, A. Ramzy, Novel application of adaptive swarm intelligence techniques coupled with adaptive network-based fuzzy inference system in predicting photovoltaic power, *Renewable and Sustainable Energy Reviews*, 148 (2021) 111315.

[43] S.M. Abolarin, B.M. Shitta, E.M. Aghogho, P.B. Nwosu, C.M. Aninyem, L. Lagrange, An impact of solar PV specifications on module peak power and number of modules: A case study of a five-bedroom residential duplex, *IOP Conference Series: Earth and Environmental Science*, 983(1) (2022) 012056.

[44] M.B. Shitta, S.M. Abolarin, C.A. Eguma, B. Onafeso, O. Adegbenro, Improvement of building energy performance and human occupant comfort through energy efficiency measures, *Journal of Energy Policy, Research and Development (JEPRD)*, 1(2) (2015) 14–24.

[45] J.P. Sutikno, S. Aldina, N. Sari, R. Handogo, Utilization of solar energy for air conditioning system, in: *The 24th Regional Symposium on Chemical Engineering (RSCE 2017), Processes for Energy and Environment, MATEC Web Conference*, 2018, pp. 6.

[46] D. Ürge-Vorsatz, L.F. Cabeza, S. Serrano, C. Barreneche, K. Petrichenko, Heating and cooling energy trends and drivers in buildings, *Renewable and Sustainable Energy Reviews*, 41 (2015) 85–98.

[47] L. Ning, T.Z. Taylor, J. Wei, J. Chunlian, J. Correia, L. Lai-Yung, W. Pak Chung, Climate change impacts on residential and commercial loads in the Western U.S. grid, in: *IEEE PES T&D 2010*, 2010, pp. 1–1.

[48] R. Pokhrel, N.D. Ramírez-Beltran, J.E. González, On the assessment of alternatives for building cooling load reductions for a tropical coastal city, *Energy and Buildings*, 182 (2019) 131–143.

[49] B.-J. Huang, T.-F. Hou, P.-C. Hsu, T.-H. Lin, Y.-T. Chen, C.-W. Chen, K. Li, K.Y. Lee, Design of direct solar PV driven air conditioner, *Renewable Energy*, 88 (2016) 95–101.

2 A Systematic Approach to Exploring Recent Improvements in the Sustainability of Biodiesel Production

Ochuko Felix Orikpete and Daniel Raphael Ejike Ewim

2.1 INTRODUCTION

Biodiesel is an alternative fuel that can potentially reduce greenhouse gas (GHG) emissions and dependence on fossil fuels. However, the sustainability of biodiesel production depends on various factors, including the raw materials used, production methods, and waste management strategies. Biodiesel production has gained significant attention in recent years as an alternative to fossil fuels due to its renewable nature, reduced GHG emissions, and potential for sustainable development (Ogunkunle & Ahmed, 2019; Rahpeyma & Raheb, 2019; Mahapatra et al., 2021). According to Alleman and McCormick (2016), biodiesel, also known as fatty acid methyl esters (FAMEs), owes its success to its high energy density (38–42 MJ/kg), and its compatibility with most diesel-fuelled equipment, requiring little to no modifications. In addition, biodiesel is environmentally friendly as it is non-toxic and biodegradable. Its usage also leads to a decrease in tailpipe emissions in older vehicles (Ahmad et al., 2011; Alleman & McCormick, 2016). However, concerns about the environmental and social impacts of biodiesel production have also been raised (Chisti, 2007). In response to these concerns, recent developments have been made to improve the sustainability of biodiesel production. This chapter presents a systematic review of recent improvements and innovations in biodiesel production, focusing on strategies that enhance the sustainability of this renewable energy source. The chapter will cover improvements in feedstock selection, production processes, waste management, and emerging technologies such as microbial biodiesel production, biorefinery approaches, and advanced analytical tools. This chapter offers a comprehensive review of recent advancements in the field, serving as a valuable resource for researchers, policymakers, and industry stakeholders interested in the sustainability of biodiesel production. The focus will be on examining recent enhancements in biodiesel production processes aimed at overcoming sustainability challenges, as well as exploring emerging technologies that hold the potential to further improve the sustainability of biodiesel production.

2.1.1 BACKGROUND AND SIGNIFICANCE OF BIODIESEL PRODUCTION

The increasing demand for energy and the depletion of fossil fuels has prompted the search for alternative renewable sources of energy (Konwar et al., 2014; Vasistha et al., 2021). Biodiesel is one such alternative, which is derived from renewable sources such as vegetable oils, animal fats, and algae. Biodiesel is a cleaner-burning alternative to fossil fuels and has the potential to reduce GHG

DOI: 10.1201/9781032651958-2

emissions, thereby mitigating climate change (Hasan and Rahman, 2017). Biodiesel is produced through a process called transesterification, in which the triglycerides in the feedstock are converted into FAMEs using an alcohol and a catalyst (Wang et al., 2021; Neupane, 2022). The resulting FAMEs can be used as a fuel for diesel engines, either in their pure form or blended with petroleum diesel. Biodiesel has been shown to have similar performance and efficiency to petroleum diesel, while also being biodegradable, non-toxic, and renewable (Srivastava et al., 2018). The production and use of biodiesel can also contribute to sustainable development. For example, biodiesel production can provide economic opportunities for rural communities by utilising locally available feedstocks. Additionally, biodiesel production can reduce reliance on imported fossil fuels and enhance energy security (Masudi et al., 2023). However, there are also sustainability concerns associated with biodiesel production. These include the potential for land use change, which can result in deforestation and the loss of biodiversity, as well as the use of unsustainable feedstocks such as palm oil, which can lead to environmental and social problems. Therefore, it is important to ensure that biodiesel production is sustainable and does not have negative environmental or social impacts. In light of these considerations, research and development efforts have been focused on improving the sustainability of biodiesel production.

2.1.2 Sustainability Challenges in Biodiesel Production

While biodiesel production has the potential to be a sustainable alternative to fossil fuels, there are several sustainability challenges that must be addressed to ensure that biodiesel production does not have negative environmental or social impacts (Mizik & Gyarmati, 2021).

One of the primary challenges is the selection of feedstocks. Biodiesel can be produced from a variety of feedstocks, including vegetable oils, animal fats, and algae. However, not all feedstocks are equally sustainable. For example, using feedstocks that require significant amounts of land, water, and fertiliser can result in deforestation, habitat destruction, and water pollution. Additionally, using feedstocks that compete with food production can have negative social impacts, such as food price increases and food insecurity. Another sustainability challenge is the production process itself. Biodiesel production can require significant amounts of energy and water, and can generate waste streams that require treatment and disposal. Additionally, using certain chemicals and catalysts in production can have negative environmental and health impacts (Cipolletta et al., 2022). The transportation and distribution of biodiesel also presents sustainability challenges. The transportation of biodiesel over long distances can result in significant GHG emissions, which can offset the emissions reductions achieved by using biodiesel as a fuel. Biodiesel distribution may require separate infrastructure and storage facilities, which can have additional environmental impacts. Finally, the sustainability of biodiesel production is also dependent on the policies and regulations that govern its production and use (Jangre et al., 2022). The lack of clear and consistent sustainability criteria and certification schemes can make it difficult to determine the environmental and social impacts of biodiesel production. The sustainability challenges in biodiesel production include the selection of sustainable feedstocks, the development of sustainable production processes, the transportation and distribution of biodiesel, and the establishment of clear and consistent sustainability criteria and certification schemes. Addressing these challenges is essential to ensure that biodiesel production is a sustainable alternative to fossil fuels.

2.1.3 Objectives of the Systematic Review

This systematic review aims to provide a comprehensive overview of recent improvements in biodiesel production aimed at addressing sustainability challenges. Specifically, this review aims to:

 i. Identify and evaluate recent improvements in feedstock selection for biodiesel production, including the use of sustainable and alternative feedstocks.

ii.   Evaluate recent developments in biodiesel production processes aimed at reducing energy and water use, minimising waste generation, and reducing the use of hazardous chemicals and catalysts.

iii. Assess emerging technologies that have the potential to enhance the sustainability of biodiesel production, such as microbial biodiesel production, biorefinery approaches, and advanced analytical tools.

iv.   Identify and evaluate recent improvements in waste management practices in biodiesel production, including the use of waste streams as feedstocks for other processes.

v.   Discuss the potential of these developments to enhance the sustainability of biodiesel production and reduce its negative environmental and social impacts.

This systematic review thoroughly examines recent advancements in biodiesel production, contributing to the evolution of sustainable practices and policies in the field. Moreover, the review identifies areas where additional research and development are required to improve the sustainability of biodiesel production while minimising adverse environmental and social consequences. The comprehensive review aims to facilitate the transition towards a more sustainable and renewable energy future.

2.2   METHODS

This book chapter employs a systematic review approach, similar to the method adopted by Mizik and Gyarmati (2021), to investigate recent improvements to ensure sustainability in biodiesel production. The review analysed relevant literature from various sources, including academic articles, reports, and legal regulations. The methodology followed a predetermined set of inclusion and exclusion criteria to identify relevant scientific articles. The inclusion criteria for the scientific articles were as follows: (1) published between 2000 and 2023, (2) written in English, (3) focused on recent improvements in biodiesel production aimed at ensuring sustainability, and (4) peer-reviewed. The exclusion criteria were as follows: (1) duplicates, (2) irrelevant articles, and (3) articles not meeting the inclusion criteria. A systematic search of academic databases, including Scopus, Science Direct, PubMed, ProQuest, JSTOR, and Web of Science, was conducted to identify relevant scientific articles. The search terms used were "biodiesel production", "sustainability", "feedstock selection", "production processes", "waste management", and "emerging technologies". A total of 215 articles were identified and screened, and 105 articles were selected for full-text review. The findings from the selected articles were synthesised and presented in this book chapter. The review analysed recent improvements in biodiesel production, including feedstock selection, production processes, waste management, and emerging technologies, with a focus on their potential to enhance the sustainability of biodiesel production. This review evaluates the challenges and constraints of recent advancements in biodiesel production while pinpointing areas where additional research and development could enhance sustainability. In essence, this systematic review offers a comprehensive examination of the latest improvements in biodiesel production, focusing on their potential to promote sustainability, and mitigate the negative environmental and social consequences associated with biodiesel production.

2.3   IMPROVEMENTS IN FEEDSTOCK SELECTION

The selection of appropriate feedstock is a critical aspect of sustainable biodiesel production (Ayoub et al., 2021). The use of non-food feedstocks, such as microalgae, Jatropha, and waste cooking oil, has gained significant attention as an alternative to conventional food-based feedstocks like soybean, rapeseed, and palm oil (Atabani et al., 2012). Microalgae, in particular, have shown great potential due to their high lipid content, rapid growth rate, and ability to sequester CO_2 (Chisti, 2007).

The use of feedstocks that compete with food production, such as soybean and palm oil, has raised concerns about the social and environmental impacts of biodiesel production (Mekhilef et al., 2011). To address these concerns, recent developments have focused on identifying and using non-food feedstocks such as algae, Jatropha, and waste oils. Non-edible oil crops, such as *Jatropha curcas*, *Pongamia pinnata*, and *Moringa oleifera*, have emerged as potential feedstocks that can be cultivated on marginal lands, reducing competition with food crops (Talebian-Kiakalaieh et al., 2013; Khanam, 2021). Recent studies have focused on improving the lipid productivity of microalgae through genetic engineering and optimising growth conditions (Wijffels & Barbosa, 2010). In addition, the cultivation of Jatropha on marginal lands has been suggested as a viable option to reduce the pressure on arable lands (Achten et al., 2010). These feedstocks have several advantages, including reduced land use, lower GHG emissions, and minimal impact on food security. For example, Sharmbiu et al. (2013) reported that the use of Jatropha in fuel production does not compete with food due to it being a non-edible plant, and it has a yield per hectare that is more than four times that of soybean and ten times that of corn. Moreover, it has been reported by some studies (El-Diwani et al., 2009; Moniruzzaman et al., 2017) that Jatropha seed oil or its blend with conventional diesel fuel is used as diesel in compression ignition engines without the need for engine modification. Additionally, improvements in the cultivation and processing of these feedstocks have increased their efficiency and reduced production costs. The adoption of sustainable feedstock selection practices is essential for ensuring the long-term sustainability of biodiesel production.

2.3.1 Types of Feedstock

Biodiesel can be produced from a wide range of feedstocks, including vegetable oils, animal fats, waste oils, and algae (Adewale et al., 2015; Ambaye et al., 2021). Mizik and Gyarmati (2021) classified feedstocks into generations, with the first generation comprising of rapeseed oil, sunflower oil, palm oil, soybean, and animal fat, the second generation comprised Jatropha and non-edible oils, and among the third generation are algae and seaweeds. Furthermore, a fourth generation was mentioned in Edeh (2020). The fourth generation deals with the production of biodiesel from genetically modified feedstocks, such as algae, that can be used in photobiological solar cells. The process aims to both produce biodiesel and capture CO_2 emissions through techniques like oxy-fuel combustion and geo-sequestration. The CO_2 is stored in saline aquifers, gas fields, or old oil reserves, and the production method is similar to that of second-generation biofuels. The choice of feedstock for biodiesel production can significantly affect the environmental and social sustainability of the production process. Therefore, there have been recent improvements in feedstock selection aimed at identifying more sustainable and cost-effective options for biodiesel production.

This section of the book chapter focuses on the types of feedstocks that have been used for biodiesel production and recent improvements in feedstock selection. Vegetable oils, such as soybean, rapeseed, and palm oil, are currently the most widely used feedstocks for biodiesel production. However, these feedstocks have been associated with environmental and social sustainability challenges, such as deforestation, land use change, and competition with food crops. To address these challenges, researchers and industry stakeholders have been exploring alternative feedstocks for biodiesel production. For instance, waste cooking oil, animal fats, and non-food crops, such as Jatropha, have been identified as potential feedstocks for biodiesel production. Additionally, microalgae have been identified as a promising feedstock for biodiesel production due to their high oil content and the ability to grow on non-arable land, among other advantages (Ahmad et al., 2011; Peng et al., 2020; Vasistha et al., 2021).

Recent improvements in feedstock selection have focused on identifying more sustainable and cost-effective feedstocks, such as waste cooking oil and non-food crops. Additionally, there have been efforts to promote the use of feedstocks that do not compete with food crops, such as algae and non-arable land crops (Chen et al., 2019; Ambaye et al., 2021). These efforts are aimed at enhancing the sustainability of biodiesel production and reducing its negative impacts on the environment and society. This section of the book chapter highlights the different types of feedstocks used for biodiesel production and recent improvements in feedstock selection aimed at enhancing the sustainability and cost-effectiveness of biodiesel production. The next section will focus on improvements in biodiesel production processes.

2.3.2 Sustainability Criteria for Feedstock Selection

In recent years, the sustainability of biodiesel production has become a major concern due to the negative environmental and social impacts associated with the production process (Cipolletta et al., 2022). Numerous studies have evaluated the environmental impacts of biodiesel production and usage, such as the research conducted by Abdel-Basset et al. (2021), Gonçalves et al. (2020), Liu et al. (2020), Mayer et al. (2020), and Dey et al. (2021). Falcke et al. (2017) pointed out that the best available techniques reference document (BREF) on large-volume organic chemicals addressed the environmental concerns associated with biodiesel. Jangre et al. (2022) researched the sustainability challenges of a biodiesel plant and discovered that five key factors, namely "Legal and regulatory compliance", "Political limitations", "International relations", "Health and education", and "Public safety and security", must be tackled to ensure the sustainability of the biodiesel plant. One of the key factors affecting the sustainability of biodiesel production is the selection of feedstock. Therefore, sustainability criteria for feedstock selection have been developed to ensure that biodiesel production is environmentally and socially responsible. This section of the book chapter focuses on the sustainability criteria for feedstock selection in biodiesel production. The criteria have been developed to guide feedstock selection and ensure it meets certain environmental, social, and economic standards. Some of the sustainability criteria for feedstock selection, as given by Patnaik and Mallick (2021), include the following:

- **Land use:** Feedstocks should not be sourced from land with high conservation value, high carbon stock, or land used for food production.
- **GHG emissions:** Feedstocks should have low GHG emissions throughout their lifecycle, including cultivation, harvesting, transportation, and processing.
- **Water use:** Feedstocks should use minimal water resources and not be sourced from areas with water scarcity or conflicts.
- **Biodiversity:** Feedstocks should not cause the loss of biodiversity or habitat destruction, and should not be sourced from genetically modified organisms.
- **Social impact:** Feedstocks should be sourced in a way that promotes social sustainability, including the protection of human rights, fair labour practices, and support for local communities.

These sustainability criteria are crucial for ensuring that biodiesel production is environmentally and socially sustainable. They also provide a framework for identifying sustainable feedstocks that can be used in biodiesel production.

In summary, this portion of the book chapter emphasised the significance of sustainability criteria in selecting feedstocks for biodiesel production. The following section will concentrate on how technological advancements contribute to improving the sustainability of biodiesel production.

2.3.3 Recent Developments in Feedstock Selection

Feedstock selection is a critical aspect of biodiesel production, and recent developments in this area have focused on identifying and utilising feedstocks that meet the sustainability criteria while also providing high yields and quality biodiesel. This section of the book chapter will discuss recent developments in feedstock selection for biodiesel production.

One of the recent developments in feedstock selection is the use of non-food crops as feedstock. Non-food crops such as Jatropha, camelina, and pennycress have been identified as potential feedstocks due to their high oil content and low impact on food production. These crops can be grown on marginal lands and require minimal inputs such as water and fertilisers. Additionally, research has focused on enhancing the oil yield and quality of these crops through breeding and genetic modification (Misra et al., 2016; Peng et al., 2020). Another recent development in feedstock selection is the use of waste and by-products as feedstock. Waste and by-products from various industries such as food processing, animal husbandry, and forestry can be converted into biodiesel. For example, waste cooking oil from restaurants and food processing industries can be collected and converted into biodiesel (Ambaye et al., 2021). This approach not only reduces waste but also provides a sustainable source of feedstock for biodiesel production.

Furthermore, recent developments in feedstock selection have also focused on the use of algae as a feedstock (Aggarwal & Remya, 2021). Algae have high oil content and can be grown in various environments such as saltwater, freshwater, and wastewater (Ambaye et al., 2021). Algae-based biodiesel production has the potential to be more sustainable than traditional feedstocks as it does not compete with food production or land use (Archanaa et al., 2019; Piligaev et al., 2019). Putting it briefly, recent developments in feedstock selection have focused on identifying and utilising sustainable feedstocks that provide high yields and quality biodiesel. The use of non-food crops, waste and by-products, and algae as feedstocks are promising approaches that can enhance the sustainability of biodiesel production. The next section will discuss recent technological advancements that have improved the sustainability of biodiesel production.

2.4 IMPROVEMENTS IN PRODUCTION PROCESSES

The production of biodiesel involves several steps, including transesterification, purification, and recovery. Each step can have significant impacts on the sustainability of biodiesel production, including energy consumption, waste generation, and emissions. Recent developments have focused on improving the efficiency of these processes and reducing their environmental impact. For example, new catalysts that allow for faster and more efficient transesterification reactions have been developed, reducing the energy required and improving yields. Additionally, new purification techniques, such as adsorption and membrane separation, have been developed to reduce the generation of waste, and improve the purity of the final product. The use of renewable energy sources, such as solar (León et al., 2018) and wind power, in production processes has also been explored to reduce the carbon footprint of biodiesel production. These improvements in production processes have made biodiesel production more efficient, cost-effective, and sustainable and are essential for increasing its viability as an alternative to fossil fuels.

2.4.1 TRANSESTERIFICATION PROCESS

The process of converting triglyceride in vegetable oils or animal fat into methyl or ethyl esters and a glycerol molecule through a chemical reaction with short-chain alcohol such as methanol or ethanol in the presence of a catalyst is known as transesterification (Huang et al., 2012; Ogunkunle & Ahmed, 2019; Akhihiero, 2022). The transesterification process is a critical step in the production of biodiesel, and recent improvements in this process have focused on increasing efficiency and reducing environmental impact. This section of the book chapter will discuss recent developments in the transesterification process.

One recent development in the transesterification process is the use of heterogeneous catalysts instead of homogeneous catalysts. Shah et al. (2018) reported that homogeneous catalysts such as sodium hydroxide (NaOH) and potassium hydroxide (KOH) have traditionally been used in the transesterification process because of their cost-effectiveness and high efficiency in terms of product yield, but they have several drawbacks, including the production of large amounts of waste and the need for careful handling due to their corrosive nature (Zhang et al., 2003; Konwar et al., 2014). A detailed literature summary of homogeneous-catalyzed transesterification can also be found in (Zulqarnain et al., 2021). Heterogeneous catalysts such as solid acids and bases, enzymes, and zeolites have been investigated as alternatives to homogeneous catalysts (Sakai et al., 2009; Kiss et al., 2010; Lam et al., 2010; Boey et al., 2011; Endalew et al., 2011; Mutreja et al., 2011; Birla et al., 2012; Borges & Díaz, 2012; Dias et al., 2012; Roschat et al., 2012; Farooq et al., 2013; Kazemian et al., 2013; Choudhury et al., 2014; Galadima & Muraza, 2014). These catalysts have several advantages, including easier separation from the reaction mixture, reduced waste production, and milder reaction conditions (Agarwal et al., 2012; Konwar et al., 2014).

Another recent development in the transesterification process is the use of supercritical fluid technology (Hoang et al., 2013). Supercritical fluids such as supercritical carbon dioxide have been used as a solvent and catalyst in the transesterification process. This technology has several advantages, including high reaction rates, high selectivity, and the ability to use a wide range of feedstocks. Furthermore, recent developments in the transesterification process have also focused on improving the efficiency of the process. For example, the use of ultrasound has been investigated as a means of enhancing the reaction rate and reducing the reaction time. Additionally, the use of microwave irradiation has been investigated as a means of reducing reaction time and energy consumption (Akhihiero, 2022).

Summarising, recent developments in the transesterification process have focused on increasing efficiency and reducing environmental impact. The use of heterogeneous catalysts, supercritical fluid technology, and new technologies such as ultrasound and microwave irradiation are promising approaches that can enhance the sustainability of biodiesel production. The next section will discuss recent developments in post-production processes that have improved the sustainability of biodiesel production.

2.4.2 CATALYSTS

Speeding up biodiesel production processes is an important role played by catalysts (Mahlia et al., 2020; Wang et al., 2021). Catalysts play a critical role in the production of biodiesel, and recent developments in catalysts have focused on improving the efficiency and sustainability of the production process (Konwar et al., 2014). This section of the book chapter will discuss recent developments in catalysts for biodiesel production.

One recent development in catalysts is the use of solid catalysts instead of liquid catalysts. Solid catalysts, such as heterogeneous catalysts, have several advantages over liquid catalysts, including easier separation from the reaction mixture, reduced waste production, and the ability to use a wider range of feedstocks. In addition, solid catalysts are generally less corrosive and easier to handle than liquid catalysts, reducing the risk of accidents during production. Konwar et al. (2014) have extensively explored the use of activated charcoal as a solid catalyst.

Another recent development in catalysts is the use of non-toxic catalysts. Traditional catalysts such as NaOH and KOH are corrosive and hazardous to handle, and they can produce waste that is difficult to dispose of safely (Zhang et al., 2003; Konwar et al., 2014). Non-toxic catalysts, such as enzymes and ionic liquids, have been investigated as alternatives to traditional catalysts. These catalysts are generally safer to handle and produce less waste, making them more sustainable options for biodiesel production. Furthermore, recent developments in catalysts have also focused on improving the selectivity and activity of catalysts. For example, the use of nano-catalysts has been investigated by Tahvildari et al. (2015), Akubude et al. (2019), and Khan (2021) to improve the selectivity and activity of catalysts. Nano-catalysts have a high surface area to volume ratio, which increases their activity and selectivity in the transesterification process.

To reiterate, recent developments in catalysts for biodiesel production have focused on improving the efficiency and sustainability of the production process. The use of solid catalysts, non-toxic catalysts, and nano-catalysts are promising approaches that can enhance the sustainability of biodiesel production. The next section will discuss recent developments in process optimisation that have improved the sustainability of biodiesel production.

2.4.3 Energy Efficiency

The production of biodiesel requires a significant amount of energy, which can lead to high production costs and environmental impact. Efficient production processes are crucial for the sustainable production of biodiesel. Recent developments in production processes have focused on improving energy efficiency to reduce costs and environmental impact. This section of the book chapter will discuss recent developments in energy-efficient production processes for biodiesel.

One approach to improving energy efficiency in biodiesel production is the use of alternative energy sources. Renewable energy sources, such as solar, wind, and geothermal energy, can be used to power the production process, reducing the reliance on fossil fuels. In addition, the use of waste heat recovery systems can capture and reuse waste heat generated during the production process, reducing energy consumption and costs.

Another approach to improving energy efficiency is the optimisation of process parameters. For example, the optimisation of reaction conditions, such as temperature and pressure, can improve the efficiency of the transesterification process and reduce energy consumption. Additionally, the use of continuous-flow reactors can reduce energy consumption by eliminating the need for batch processing and reducing the time required for the reaction to occur. Furthermore, the use of advanced process control techniques, such as model predictive control and adaptive control, can improve the efficiency of the production process and reduce energy consumption. These techniques use mathematical models and algorithms to optimise the process in real time, ensuring that the process operates at its maximum efficiency while meeting the desired quality specifications.

Process simulation tools can be used to model and optimise the biodiesel production process, taking into account various factors like feedstock type, production scale, and energy requirements (Nasir et al., 2013). By simulating different scenarios, it is possible to identify the most energy-efficient and environmentally friendly methods of production. Optimisation algorithms can further enhance the process by minimising resource consumption, waste generation, and GHG emissions.

In essence, recent developments in energy-efficient production processes for biodiesel have focused on the use of alternative energy sources, waste heat recovery systems, optimisation of process parameters, and advanced process control techniques. These approaches can improve the efficiency and sustainability of biodiesel production, reducing costs and environmental impact. The next section will discuss recent developments in the use of by-products and waste streams in biodiesel production.

2.4.4 Environmental Impact

Biodiesel production has the potential to reduce GHG emissions and promote sustainability. However, the production process can also have negative environmental impacts, such as water pollution, land-use change, and deforestation (Wu et al., 2018). Therefore, recent developments in biodiesel production have focused on reducing the environmental impact of the production process. This section of the book chapter will discuss recent developments in environmental sustainability in biodiesel production.

One approach to reducing the environmental impact of biodiesel production is the use of sustainable feedstocks. As discussed earlier in this chapter, sustainable feedstocks are those that meet specific sustainability criteria, such as low land-use change, low water usage, and low GHG emissions. Using sustainable feedstocks in biodiesel production can help to reduce the environmental impact of the production process. Another approach to reducing the environmental impact of biodiesel production is the implementation of sustainable production practices. This can include measures such as reducing water usage, improving waste management, and minimising the use of chemicals in the production process. Also, Agee et al. (2014) proposed using solar parabolic reflectors to minimise the amount of electricity required in the biodiesel production process. Their article also suggested using recovered biodiesel waste glycerol as a solvent system for Wolff-Kishner reduction reactions to reduce waste by-products. Furthermore, the implementation of a closed-loop production system, where waste products are recycled and reused in the production process, can reduce the environmental impact of the production process. Moreover, the use of lifecycle assessment (LCA) can help to identify the environmental impact of biodiesel production and identify areas for improvement (Jeswani et al., 2020). LCA is a tool that analyses the entire life cycle of a product, from the extraction of raw materials to the disposal of the product (Rebitzer et al., 2004; Hauschild, 2018). This analysis can help to identify areas where improvements can be made to reduce the environmental impact of biodiesel production. Several LCA studies have examined the potential of biodiesels to reduce GHG emissions over their entire lifespan and their impact on climate change. For instance, Sieverding et al. (2015) reviewed soya bean-based biodiesel, and van Eijck et al. (2014) discussed biodiesel-related issues from Jatropha. Thiruketheeswaranathan (2022) conducted a life cycle assessment of biodiesel manufactured from waste cooking oil and concluded that the transesterification phase had a larger environmental impact, accounting for 69% of the total impact, while the pre-treatment phase accounted for 29% of the overall environmental impact for the production of one metric ton of biodiesel from waste cooking oil.

Putting it in a succinct way, recent developments in environmental sustainability in biodiesel production have focused on the use of sustainable feedstocks, sustainable production practices, and the implementation of LCA. These approaches can help to reduce the environmental impact of biodiesel production and promote sustainability. The next section will discuss recent developments in the use of by-products and waste streams in biodiesel production.

2.4.5 Waste Management

Improper waste management is a critical issue in biodiesel production that can have a significant impact on the environment. The production of biodiesel generates various types of waste, including glycerol, methanol, and residual oils, which can be harmful to the environment if not disposed of properly. Therefore, improvements in waste management techniques are essential to ensure the sustainability of biodiesel production.

One of the most common waste products generated during the production of biodiesel is glycerol. Glycerol is produced during the transesterification process and can account for up to 10% of the total volume of biodiesel produced. The disposal of glycerol can be problematic since it is a high-viscosity liquid with limited applications. Therefore, it is necessary to develop efficient and cost-effective methods for glycerol disposal. Several strategies have been proposed for the utilisation of glycerol, including conversion to value-added chemicals such as propylene glycol, 1,3-propanediol, and butanol. These strategies not only help in waste reduction but also generate additional revenue streams for biodiesel production facilities. In addition to glycerol, residual oils and methanol can also pose environmental challenges. Residual oils, also known as biodiesel by-products, are typically high in free fatty acids and can be difficult to dispose of due to their high acidity. One solution is to utilise residual oils as feedstock for other industries, such as the production of soaps and detergents. Methanol is another waste product that requires careful management. Methanol is a toxic substance that can have severe impacts on the environment if not handled correctly. Several methods have been proposed for the disposal of methanol, including distillation and reuse as a feedstock for biodiesel production.

Reiterating, the proper management of waste products is crucial for the sustainability of biodiesel production. By developing effective strategies for waste management, biodiesel production can become more environmentally friendly and economically viable.

2.4.6 RECENT DEVELOPMENTS IN PRODUCTION PROCESSES

In recent years, several developments have been made in the production processes of biodiesel to improve the efficiency, sustainability, and overall quality of the fuel. One such development is the use of alternative production methods, such as enzymatic and microbial transesterification, which offer several advantages over traditional chemical processes.

Enzymatic transesterification involves the use of enzymes as catalysts, which can improve the selectivity and yield of the reaction while reducing the need for toxic chemicals and high temperatures. This method also generates fewer by-products and requires less energy input than traditional chemical processes (Ghaly et al., 2010). Enzymatic transesterification using lipases as biocatalysts have emerged as an alternative to traditional chemical processes, offering several advantages such as lower energy consumption, mild reaction conditions, and reduced waste generation. Recent progress in enzyme immobilisation, reactor design, and process integration has contributed to the commercial viability of enzymatic biodiesel production.

Microbial transesterification involves the use of microorganisms, such as bacteria and fungi, to produce biodiesel from the feedstock. This method has the potential to be more sustainable and cost-effective than traditional methods, as it uses renewable resources and produces less waste (Louhasakul et al., 2022).

Other recent developments in biodiesel production processes include the use of ultrasonic and microwave-assisted transesterification, which can reduce reaction time and increase efficiency, as well as the use of advanced catalysts and additives to improve the quality and stability of the fuel (Sara et al., 2016). In other words, these recent developments in biodiesel production processes offer promising solutions to improve the sustainability, efficiency, and quality of biodiesel production while also reducing environmental impacts and waste generation.

2.5 EMERGING TECHNOLOGIES

Emerging technologies offer promising opportunities for enhancing the sustainability of biodiesel production. Microbial biodiesel production, for example, uses microorganisms to convert feedstocks into biodiesel, offering several advantages over traditional chemical processes, including higher yields, lower energy consumption, and reduced waste generation (Uthandi, 2022).

The biorefinery concept aims to maximise the value of biomass by converting it into a range of products, such as biofuels, chemicals, and materials, in an integrated process. By co-producing valuable co-products, such as glycerol, animal feed, and biogas, biodiesel production becomes more economically and environmentally sustainable. Biorefinery approaches, which utilise multiple feedstocks to produce a range of biofuels and other value-added products, have also been explored as a way to increase the economic viability and sustainability of biodiesel production (Chisti, 2007). In addition, advanced analytical tools, such as mass spectrometry and nuclear magnetic resonance spectroscopy, are being developed to improve the efficiency and accuracy of biodiesel production monitoring and quality control. Furthermore, new materials and nanotechnologies are being explored for catalysts and membranes in production processes, leading to better yields and reduced environmental impact (Lee et al., 2015; Vasistha et al., 2021). The adoption of these emerging technologies has the potential to revolutionalise the biodiesel industry, making it more sustainable, efficient, and economically viable.

2.5.1 MICROBIAL BIODIESEL PRODUCTION

Microbial biodiesel production is an emerging technology that uses microorganisms, such as bacteria and algae, to produce biodiesel. Unlike traditional biodiesel production processes, microbial biodiesel production does not require the use of arable land, freshwater, or food crops, making it a more sustainable and environmentally friendly alternative. One of the key advantages of microbial biodiesel production is the ability to use a wide range of feedstocks, including waste materials such as agricultural and industrial residues, municipal waste, and carbon dioxide. This can significantly reduce the overall carbon footprint of biodiesel production, as well as reduce waste generation and disposal costs. Microbial biodiesel production can be achieved through several methods, including fermentation, photosynthesis, and genetic engineering (Rahul et al., 2021). Fermentation involves the use of bacteria to convert feedstock into fatty acids, which can then be converted into biodiesel through transesterification. Photosynthesis involves the use of algae to convert carbon dioxide and sunlight into lipids, which can be harvested and processed into biodiesel. Genetic engineering involves modifying microorganisms to produce specific enzymes that can catalyse the conversion of feedstock into biodiesel (Hegde et al., 2015). Despite the potential benefits of microbial biodiesel production, there are several challenges that need to be addressed, including high production costs, low yields, and technical limitations. However, ongoing research and development in this field are expected to address these challenges and improve the efficiency and viability of microbial biodiesel production.

In essence, microbial biodiesel production represents a promising technology that could greatly enhance the sustainability and efficiency of biodiesel production, as well as mitigate environmental impacts and waste generation. Additional research and development efforts are required to unlock the full potential of this technology and address the associated technical and economic challenges.

2.5.2 BIOREFINERY APPROACHES

Biorefinery approaches are emerging as a promising technology for sustainable biodiesel production. Biorefineries are facilities that convert biomass into various products such as fuels, chemicals, and materials. They integrate various processes such as biomass conversion, fractionation, and purification to maximise the value of biomass feedstocks (Mahapatra et al., 2021). The concept of biorefineries aligns with the principles of the circular economy, which seeks to minimise waste and maximise resource utilisation.

Biorefinery approaches for biodiesel production can be classified into three main categories: (1) lipid-based biorefineries, (2) sugar-based biorefineries, and (3) lignocellulose-based biorefineries.

Lipid-based biorefineries utilise lipid-rich feedstocks such as algae, waste cooking oil, and animal fat to produce biodiesel. The lipid fraction is extracted and converted to biodiesel, while the residual biomass is used for other purposes such as animal feed, fertiliser, or energy production. Lipid-based biorefineries have the advantage of producing high-quality biodiesel with a low carbon footprint.

Sugar-based biorefineries use sugar-rich feedstocks such as corn, sugarcane, and sugar beet to produce biodiesel through fermentation. The feedstock is converted into sugars, which are then fermented by microorganisms to produce biodiesel. The residual biomass is used for other purposes such as animal feed, fertiliser, or energy production. Sugar-based biorefineries have the advantage of utilising abundant feedstocks and producing high yields of biodiesel.

Lignocellulose-based biorefineries utilise lignocellulosic feedstocks such as switchgrass, wood chips, and corn stover to produce biodiesel. The feedstock is pre-treated to break down the complex lignocellulose structure into simple sugars, which are then fermented by microorganisms to produce biodiesel (Wang et al., 2021). The residual biomass is used for other purposes such as animal feed, fertiliser, or energy production. Lignocellulose-based biorefineries have the advantage of utilising non-food feedstocks and producing high yields of biodiesel (Malode et al., 2021; Sandesh & Ujwal, 2021).

Biorefinery approaches have the potential to improve the sustainability of biodiesel production by utilising renewable feedstocks and maximising the value of biomass resources. However, the technology is still in the early stages of development, and further research is needed to optimise the process and scale up the production.

2.5.3 Advanced Analytical Tools

Advanced analytical tools have gained importance in the biodiesel industry to improve the quality of biodiesel and to ensure its sustainability. They can provide valuable insights into these factors, enabling decision-makers to make informed choices that promote sustainable practices (Cherubini & Strømman, 2011). The traditional analytical methods, such as gas chromatography (GC) and high-performance liquid chromatography (HPLC), are time-consuming, labour-intensive, and require the use of harmful solvents (Knothe, 2001). Advanced analytical tools, on the other hand, provide rapid, accurate, and non-destructive analyses of biodiesel and its feedstocks.

One of the advanced analytical tools used in the biodiesel industry is near-infrared (NIR) spectroscopy. NIR spectroscopy is an advanced analytical tool that has been widely used in the biodiesel industry for quality control and process optimisation. NIR spectroscopy is a non-destructive and rapid technique that can analyse a sample in a matter of seconds, without the need for sample preparation or chemical reagents (Krzysztof et al., 2023). NIR spectroscopy can provide a rapid and non-destructive analysis of various parameters of biodiesel, such as fatty acid methyl ester (FAME) content, water content, and free glycerol content (Felizardo et al., 2007). NIR spectroscopy works by measuring the absorption of light in the NIR region of the electromagnetic spectrum, which is between 800 and 2500 nanometers (Le Pevelen & Tranter, 2017; Measurlabs, 2023). The absorption of light at different wavelengths provides information about the chemical composition of the sample, including the concentration of different components such as fatty acids, glycerol, and methanol. In the biodiesel industry, NIR spectroscopy is commonly used to analyse the quality of feedstocks, intermediate products, and final biodiesel products (López-Fernández et al., 2022). For example, NIR spectroscopy can be used to determine the moisture content, acid value, and free fatty acid content of feedstocks, which are important parameters that affect the yield and quality of the biodiesel product (Hradecká et al., 2023). NIR spectroscopy can also be used to monitor the transesterification process, which is the chemical reaction that converts the feedstock into biodiesel, and to determine the quality of the final biodiesel product, including the cetane number, density, and viscosity. One of the advantages of NIR spectroscopy is that it is a non-destructive technique that can analyse a sample in real time, without the need for sample preparation or chemical reagents. This reduces the cost and time required for analysis and allows rapid feedback to be provided to the production process. NIR spectroscopy can also be used to monitor multiple parameters simultaneously, which is important for process optimisation and control (Wolfrum, 2020).

Raman spectroscopy is another advanced analytical tool that has been increasingly used in the biodiesel industry for quality control, process optimisation, and feedstock characterisation (Hanif et al., 2018). Raman spectroscopy is a non-destructive and fast analytical technique that provides valuable information about the chemical composition of a sample, including its molecular structure and functional groups (Esmonde-White et al., 2017).

Raman spectroscopy works by measuring the inelastic scattering of light when it interacts with a sample. When a laser beam is directed onto a sample, some of the photons in the laser light interact with the molecular vibrations of the sample, causing them to shift in energy. The scattered photons that are emitted from the sample have a different energy than the incident photons, and this energy shift is used to determine the chemical composition of the sample (Jones et al., 2019). In the biodiesel industry, Raman spectroscopy is particularly useful for feedstock characterisation, process monitoring, and product analysis (da Silva et al., 2017). It can provide information about the composition of the feedstock, such as the content of fatty acids, glycerides, and other organic compounds, which can help optimise the transesterification process and improve the yield and quality of the biodiesel product. Raman spectroscopy can also be used to monitor the transesterification process in real time, allowing for rapid feedback to be provided to the production process and helping to identify potential issues before they affect the final product quality. Additionally, Raman spectroscopy can be used to analyse the quality of the final biodiesel product, such as its cetane number, density, and viscosity. One of the advantages of Raman spectroscopy is its high sensitivity and selectivity, which allows it to detect and quantify low concentrations of chemical species in a sample. Raman spectroscopy is also a non-destructive technique that can be used in situ without the need for sample preparation or chemical reagents, making it a fast and convenient tool for process monitoring and analysis (Saletnik et al., 2021).

Another emerging technology in the biodiesel industry is the use of sensors and biosensors. A biosensor is a type of analytical device that utilises biomaterials as components of the sensing system, converting biological responses into electrical signals (Hossain et al., 2010). Sensors and biosensors can provide real-time and continuous monitoring of various parameters of biodiesel production, such as temperature, pH, and enzyme activity. Sensors and biosensors are emerging technologies that have the potential to revolutionise the biodiesel industry by providing real-time monitoring of critical process parameters and enabling process optimisation and control. These technologies can help improve the quality and yield of biodiesel, reduce production costs, and enhance the sustainability of the production process (Dhar & Mazumdar, 2012). Sensors are analytical devices that detect and measure the physical, chemical, or biological properties of a sample or a system. In the biodiesel industry, sensors can be used to monitor critical process parameters, such as temperature, pressure, pH, and moisture content, which affect the yield and quality of the biodiesel product. For example, temperature sensors can be used to monitor the temperature of the transesterification reaction and ensure that it remains within the optimal range for the reaction to occur efficiently. On the other hand, biosensors are sensors that use biological components, such as enzymes or microorganisms, to detect or measure a specific chemical or biological species (Morgan et al., 2016). In the biodiesel industry, biosensors can be used to detect the presence of impurities or contaminants in the feedstock, intermediate products, or final biodiesel product. For example, biosensors can be used to detect the presence of methanol, which is a toxic and flammable substance that is used in the transesterification process and needs to be removed from the final biodiesel product. The use of sensors and biosensors in the biodiesel industry has several advantages. First, they provide real-time monitoring of critical process parameters, which allows for rapid feedback and adjustment of the process conditions to optimise the yield and quality of the biodiesel product. Second, they reduce the need for manual sampling and analysis, which can be time-consuming, labour-intensive, and costly. Third, they can detect and quantify low concentrations of chemical species, which is important for quality control and regulatory compliance. However, the use of sensors and biosensors in the biodiesel industry also has some challenges. For example, the sensitivity and selectivity of the sensors and biosensors may be affected by interfering substances in the sample matrix. In addition, the cost and complexity of the sensors and biosensors may limit their widespread use in the industry.

Other technologies that can help optimise biodiesel production processes and reduce waste and energy consumption include:

2.5.3.1 Remote Sensing and Geographic Information Systems (GIS)

Remote sensing technology and GIS tools can be employed to monitor land use changes, feedstock cultivation patterns, and the overall environmental impact of biodiesel production (Rulli et al., 2016). By integrating satellite imagery, soil and water data, and other geospatial information, it is possible to assess the sustainability of feedstock sources and identify potential areas for expansion or improvement.

2.5.3.2 Real-Time Monitoring and Control Systems

The implementation of real-time monitoring and control systems can greatly enhance the efficiency and sustainability of biodiesel production (Lopes de Sousa Jabbour et al., 2018). By continuously tracking key performance indicators (KPIs) such as energy consumption, emissions, and waste generation, plant operators can make informed decisions to optimise the production process. Advanced sensors, automation, and data analytics can also help to detect and prevent operational inefficiencies, equipment malfunctions, and potential hazards.

2.5.3.3 Waste Utilisation and Biorefinery Concepts

The development of advanced techniques for waste utilisation and the implementation of biorefinery concepts can significantly improve the sustainability of biodiesel production (Clark & Deswarte, 2008). By converting waste products, such as glycerol and biomass residues, into valuable chemicals, materials, or energy sources, it is possible to minimise waste generation and enhance the overall efficiency of the process.

2.5.4 POTENTIAL OF EMERGING TECHNOLOGIES FOR SUSTAINABLE BIODIESEL PRODUCTION

Emerging technologies hold great promise for the sustainable production of biodiesel. These technologies are being developed to address the challenges associated with traditional biodiesel production methods and to improve the efficiency and sustainability of the process. Some of the potential benefits of emerging technologies include increased yield, reduced energy consumption, and reduced environmental impact. The use of microbial biodiesel production and biorefinery approaches are examples of emerging technologies that have shown promise in sustainable biodiesel production. Microbial biodiesel production involves the use of microorganisms to convert feedstocks into biodiesel, while biorefinery approaches aim to maximise the value of all components of the feedstock by producing multiple products. Both approaches have the potential to improve the efficiency and sustainability of biodiesel production by reducing waste and increasing yield.

Advanced analytical tools are also being developed to better understand and optimise the biodiesel production process. These tools can provide valuable insights into the properties and characteristics of feedstocks, as well as the behaviour of catalysts and other production factors. This information can be used to improve the efficiency and sustainability of biodiesel production by identifying areas for optimisation and reducing waste. Overall, emerging technologies hold great potential for the sustainable production of biodiesel. However, further research and development are needed to fully realise their potential and to ensure that they are economically feasible and environmentally sustainable.

2.6 CONCLUSION

The systematic review highlights several recent improvements in biodiesel production processes that contribute to enhancing sustainability. These advancements can be categorised into different areas, as detailed below:

1. **Feedstock selection:** Sustainable feedstock selection is crucial for biodiesel production. The review emphasises that criteria for selecting sustainable feedstocks include social, environmental, and economic factors. This includes using non-food crops, waste materials, and algae to minimise competition with food production and reduce the impacts of land use change.
2. **Transesterification processes:** The review identifies advancements in transesterification processes that increase efficiency and sustainability. These include novel methods such as enzymatic transesterification, supercritical fluid technology, and ultrasound-assisted processes that can improve yield and reduce energy consumption.

3. **Catalysts:** The development of new catalysts, such as heterogeneous and enzymatic catalysts, can improve the efficiency and environmental performance of biodiesel production. These catalysts can reduce waste generation and enable the use of low-quality feedstocks.

4. **Energy efficiency:** Improvements in energy efficiency throughout the production process can significantly reduce the environmental impact of biodiesel. This includes optimising reaction conditions, heat integration, and the use of renewable energy sources.

5. **Waste management:** The review highlights advancements in waste management, such as recycling and valorisation of by-products. These approaches can reduce waste generation and create valuable co-products, further enhancing the overall sustainability of biodiesel production.

6. **Emerging technologies:** The review also identifies emerging technologies, such as bioelectrochemical systems and advanced biorefineries, that have the potential to revolutionise biodiesel production and further improve sustainability.

The findings of this review have significant implications for various stakeholders namely researchers, policy makers, and industry players:

- Researchers can use these advancements as a foundation to develop more sustainable biodiesel production methods, exploring innovative solutions to challenges within the biodiesel industry.
- Policymakers can leverage the findings to create regulations and standards that promote sustainable biodiesel production and encourage the adoption of best practices in the industry.
- Industry players can integrate the identified improvements into their production processes, leading to more sustainable practices and reducing their overall environmental impact.

Future research should build on these findings by investigating other aspects of sustainable biodiesel production, such as social sustainability and supply chain management. It is also essential to examine the environmental impacts of biodiesel production over its life cycle, from feedstock production to end-use. Additionally, exploring the use of advanced analytical tools and big data in biodiesel production can lead to more efficient and sustainable processes.

REFERENCES

Abdel-Basset, M., Gamal, A., Chakrabortty, R. K., & Ryan, M. (2021). Development of a hybrid multi-criteria decision-making approach for sustainability evaluation of bioenergy production technologies: A case study. *Journal of Cleaner Production*, 290. https://doi.org/10.1016/j.jclepro.2021.125805

Achten, W. M., Maes, W. H., Aerts, R., Verchot, L., Trabucco, A., Mathijs, E., ... & Muys, B. (2010). Jatropha: From global hype to local opportunity. *Journal of Arid Environments*, 74(1), 164–165. https://doi.org/10.1016/j.jaridenv.2009.08.010

Adewale, P., Dumont, M. J., & Ngadi, M. (2015). Recent trends of biodiesel production from animal fat wastes and associated production techniques. *Renewable and Sustainable Energy Reviews*, 45, 574–588. http://dx.doi.org/10.1016/j.rser.2015.02.039

Agee, B. M., Mullins, G., & Swartling, D. J. (2014). Use of solar energy for biodiesel production and use of biodiesel waste as a green reaction solvent. *Sustainable Chemical Processes*, 2(1), 1–10.

Aggarwal, M., & Remya, N. (2021). The state-of-the-art production of biofuel from microalgae with simultaneous wastewater treatment: Influence of process variables on biofuel yield and production cost. *BioEnergy Research*, 15, 62–76. https://doi.org/10.1007/s12155-021-10277-1

Ahmad, A. L., Yasin, N. M., Derek, C. J. C., & Lim, J. K. (2011). Microalgae as a sustainable energy source for biodiesel production: A review. *Renewable and Sustainable Energy Reviews*, 15(1), 584–593. http://dx.doi.org/10.1016/j.rser.2010.09.018

Akhihiero, E. T. (2022). Recent advances in biodiesel from plants. In M. N. Ahmed & I. Akaehomen Akii (Eds.), *Renewable Energy – Recent Advances* (Ch. 2). IntechOpen. https://doi.org/10.5772/intechopen.106924

Akubude, V. C., Nwaigwe, K. N., & Dintwa, E. (2019). Production of biodiesel from microalgae via nanocatalyzed transesterification process: A review. *Materials Science for Energy Technologies*, 2(2), 216–225. https://doi.org/10.1016/j.mset.2018.12.006

Alleman, T. L., & McCormick, R. L. (2016). *Biodiesel Handling and Use Guide*. Clean Cities – Department of Energy.

Ambaye, T. G., Vaccari, M., Bonilla-Petriciolet, A., Prasad, S., van Hullebusch, E. D., & Rtimi, S. (2021). Emerging technologies for biofuel production: A critical review on recent progress, challenges and perspectives. *Journal of Environmental Management*, 290, 112627. https://doi.org/10.1016/j.jenvman.2021.112627

Archanaa, S., Jose, S., Mukherjee, A., & Suraishkumar, G. K. (2019). Sustainable diesel feedstock: A comparison of oleaginous bacterial and microalgal model systems. *BioEnergy Research*, 12, 205–216. https://doi.org/10.1007/s12155-018-9948-6

Atabani, A. E., Silitonga, A. S., Badruddin, I. A., Mahlia, T. M. I., Masjuki, H., & Mekhilef, S. (2012). A comprehensive review on biodiesel as an alternative energy resource and its characteristics. *Renewable and Sustainable Energy Reviews*, 16(4), 2070–2093. https://doi.org/10.1016/j.rser.2012.01.003

Ayoub, M., Yusoff, M. H., Nazir, M. H., Zahid, I., Ameen, M., Sher, F., Floresyona, D., & Nursanto, E. B. (2021). A comprehensive review on oil extraction and biodiesel production technologies. *Sustainability*, 13(2), 788.

Birla, A., Singh, B., Upadhyay, S. N., & Sharma, Y. C. (2012). Kinetics studies of synthesis of biodiesel from waste frying oil using a heterogeneous catalyst derived from snail shell. *Bioresource Technology*, 106, 95–100.

Boey, P. L., Maniam, G. P., & Hamid, S. A. (2011). Performance of calcium oxide as a heterogeneous catalyst in biodiesel production: A review. *Chemical Engineering Journal*, 168(1), 15–22.

Borges, M., & Díaz, L. (2012). Recent developments on heterogeneous catalysts for biodiesel production by oil esterification and transesterification reactions: A review. *Renewable and Sustainable Energy Reviews*, 16(5), 2839–2849.

Chen, H., Li, T., & Wang, Q. (2019). Ten years of algal biofuel and bioproducts: Gains and pains. *Planta*, 249, 195–219. https://doi.org/10.1007/s00425-018-3066-8

Cherubini, F., & Strømman, A. H. (2011). Life cycle assessment of bioenergy systems: State of the art and future challenges. *Bioresource Technology*, 102(2), 437–451.

Chisti, Y. (2007). Biodiesel from microalgae. *Biotechnology Advances*, 25(3), 294–306. http://dx.doi.org/10.1016/j.biotechadv.2007.02.001

Choudhury, H. A., Chakma, S., & Moholkar, V. S. (2014). Mechanistic insight into sonochemical biodiesel synthesis using heterogeneous base catalyst. *Ultrasonics Sonochemistry*, 21, 1972–1979.

Cipolletta, M., D'Ambrosio, M., Casson Moreno, V., & Cozzani, V. (2022). Enhancing the sustainability of biodiesel fuels by inherently safer production processes. *Journal of Cleaner Production*, 344, 131075. https://doi.org/10.1016/j.jclepro.2022.131075

Clark, J. H., & Deswarte, F. E. I. (Eds.). (2008). *Introduction to Chemicals from Biomass*. John Wiley & Sons.

da Silva, J. C., Queiroz, A., Oliveira, A., & Kartnaller, V. (2017). Advances in the application of spectroscopic techniques in the biofuel area over the last few decades. In *Frontiers in Bioenergy and Biofuels* (pp. 1–25). IntechOpen. https://doi.org/10.5772/65552

Dey, S., Reang, N. M., Das, P. K., & Deb, M. (2021). A comprehensive study on prospects of economy, environment, and efficiency of palm oil biodiesel as a renewable fuel. Journal of Cleaner Production, 286 https://doi.org/10.1016/j.jclepro.2020.124981

Dhar, P., & Mazumdar, A. (2012). *Automation of Biodiesel Plant with Bio-Sensing Technologies. 2012 1st International Symposium on Physics and Technology of Sensors (ISPTS-1)* (pp. 319–325). IEEE. https://doi.org/10.1109/ISPTS.2012.6260958

Dias, J. M., Alvim-Ferraz, M., Almeida, M. F., Mendez Díaz, J. D., Polo, M. S., & Utrilla, J. R. (2012). Selection of heterogeneous catalysts for biodiesel production from animal fat. *Fuel*, 94, 418–425.

Edeh, I. (2020). Biodiesel production as a renewable resource for the potential displacement of the petroleum diesel. In *Biorefinery Concepts, Energy and Products*. IntechOpen.

El-Diwani, G., Attai, N. K., & Hawash, S. I. (2009). Development and evaluation of biodiesel fuel and byproducts from Jatropha oil. *International Journal of Environmental Science and Technology*, 6(2), 219–224.

Endalew, A. K., Kiros, Y., & Zanzi, R. (2011). Inorganic heterogeneous catalysts for biodiesel production from vegetable oils. *Biomass and Bioenergy*, 35(9), 3787–3809.

Esmonde-White, K. A., Cuellar, M., Uerpmann, C., Lenain, B., & Lewis, I. R. (2017). Raman spectroscopy as a process analytical technology for pharmaceutical manufacturing and bioprocessing. *Analytical and Bioanalytical Chemistry*, 409(3), 637–649. https://doi.org/10.1007/s00216-016-9824-1

Falcke, H., Holbrook, S., Clenahan, I., Lopez Carretero, A., Sanalan, A., Brinkmann, T., Roth, J., Zerger, B., Roudier, S., & Delgado Sancho, L. (2017). Best available techniques (BAT) reference document for the production of large volume organic chemicals. *Industrial Emissions Directive 2010/75/EU (Integrated Pollution Prevention and Control)*. https://doi.org/10.2760/77304.

Farooq, M., Ramli, A., & Subbarao, D. (2013). Biodiesel production from waste cooking oil using bifunctional heterogeneous solid catalysts. *Journal of Cleaner Production*, 59, 131–140.

Felizardo, P., Baptista, P., Uva, M. S., Menezes, J. C., & Correia, M. J. N. (2007). Monitoring biodiesel fuel quality by near infrared spectroscopy. *Journal of Near Infrared Spectroscopy*, 15(2), 97–105.

Galadima, A., & Muraza, O. (2014). Biodiesel production from algae by using heterogeneous catalysts: A critical review. *Energy*, 78, 72–83. http://dx.doi.org/10.1016/j.energy.2014.06.018

Ghaly, A. E., Dave, D., Brooks, M. S., & Budge, S. (2010). Production of biodiesel by enzymatic transesterification. *American Journal of Biochemistry and Biotechnology*, 6(2), 54–76.

Gonçalves, P. C., Monteiro, L. P. C., & Santos, L. de S. (2020). Multi-objective optimization of a biodiesel production process using process simulation. *Journal of Cleaner Production*, 270 https://doi.org/10.1016/j.jclepro.2020.122322.

Hanif, M. A., Nisar, S., Akhtar, M. N., Nisar, N., & Rashid, N. (2018). Optimized production and advanced assessment of biodiesel: A review. *International Journal of Energy Research*, 42(6), 2070–2083.

Hasan, M. M., & Rahman, M. M. (2017). Performance and emission characteristics of biodiesel-diesel blend and environmental and economic impacts of biodiesel production: A review. *Renewable and Sustainable Energy Reviews*, 74, 938–948. http://dx.doi.org/10.1016/j.rser.2017.03.045

Hauschild, M. Z. (2018). Introduction to LCA methodology. In *Life cycle assessment: Theory and Practice* (pp. 59–66). Springer.

Hegde, K., Chandra, N., Sarma, S. J., Brar, S. K., & Veeranki, V. D. (2015). Genetic engineering strategies for enhanced biodiesel production. *Molecular Biotechnology*, 57, 606–624. https://doi.org/10.1007/s12033-015-9869-y.

Hoang, D., Bensaid, S., & Saracco, G. (2013). Supercritical fluid technology in biodiesel production. *Green Processing and Synthesis*, 2(5), 407–425.

Hossain, M. Z., Shrestha, D. S., & Kleve, M. G. (2010). Biosensors for biodiesel quality sensing. *Journal of the Arkansas Academy of Science*, 64(1), 80–85.

Hradecká, I., Vráblík, A., Frątczak, J., Sharkov, N., Černý, R., & Hönig, V. (2023). Near-infrared spectroscopy as a tool for simultaneous determination of diesel fuel improvers. *ACS Omega*, 8(4), 4038–4045. https://doi.org/10.1021/acsomega.2c06845. PMID: 36743007; PMCID: PMC9893758.

Huang, D., Zhou, H., & Lin, L. (2012). Biodiesel: An alternative to conventional fuel. *Energy Procedia*, 16, 1874–1885. https://doi.org/10.1016/j.egypro.2012.01.287

Jabbour, C. J. C., Lopes de Sousa Jabbour, A. B., Godinho Filho, M., & Roubaud, D. (2018). Industry 4.0 and the circular economy: A proposed research agenda and original roadmap for sustainable operations. *Annals of Operations Research*, 270, 273–286.

Jangre, J., Prasad, K., Kaliyan, M., Kumar, D., & Patel, D. (2022). Sustainability assessment of waste cooking oil-based biodiesel plant in developing economy based on F-DEMATEL and F-ISM approaches. *Waste Management & Research*, 40(11), 1645–1659. https://doi.org/10.1177/0734242X2211001195

Jeswani, H. K., Chilvers, A., & Azapagic, A. (2020). Environmental sustainability of biofuels: A review. *Proceedings of the Royal Society A*, 476(2243), 20200351. https://doi.org/10.6084/m9.figshare.c.5208549.

Jones, R. R., Hooper, D. C., Zhang, L., Wolverson, D., & Valev, V. K. (2019). Raman techniques: Fundamentals and frontiers. *Nanoscale Research Letters*, 14(1), 1–34. https://doi.org/10.1186/s11671-019-3039-2

Kazemian, H., Turowec, B., Siddiquee, M. N., & Rohani, S. (2013). Biodiesel production using cesium modified mesoporous ordered silica as heterogeneous base catalyst. *Fuel*, 103, 719–724.

Khan, T. Y. (2021). Direct transesterification for biodiesel production and testing the engine for performance and emissions run on biodiesel-diesel-nano blends. *Nanomaterials*, 11(2), 417. https://doi.org/10.3390/nano11020417

Khanam, T., Khalid, F., Manzoor, W., Rashedi, A., Hadi, R., Ullah, F., ... & Hussain, M. (2021). Environmental sustainability assessment of biodiesel production from *Jatropha curcas* L. seeds oil in Pakistan. *PLoS ONE*, 16(11), e0258409. https://doi.org/10.1371/journal.pone.0258409

Kiss, F. E., Jovanovic, M., & Bosković, G. C. (2010). Economic and ecological aspects of biodiesel production over homogeneous and heterogeneous catalysts. *Fuel Processing Technology*, 91(10), 1316–1320.

Knothe, G. (2001). Analytical methods used in the production and fuel quality assessment of biodiesel. *Transactions of the ASAE*, 44, 193–200.

Konwar, L. J., Boro, J., & Deka, D. (2014). Review on latest developments in biodiesel production using carbon-based catalysts. *Renewable and Sustainable Energy Reviews*, 29, 546–564. http://dx.doi.org/10.1016/j.rser.2013.09.003

Lam, M. K., Lee, K. T., & Mohamed, A. R. (2010). Homogeneous, heterogeneous and enzymatic catalysis for transesterification of high free fatty acid oil (waste cooking oil) to biodiesel: A review. *Biotechnology Advances*, 28(4), 500–518.

Lee, Y. C., Lee, K., & Oh, Y. K. (2015). Recent nanoparticle engineering advances in microalgal cultivation and harvesting processes of biodiesel production: A review. *Bioresource Technology*, 184, 63–72. https://doi.org/10.1016/j.biortech.2014.10.145

León, J. A., Montero, G., Coronado, M. A., García, C., Campbell, H. E., Ayala, J. R., ... & Sagaste, C. A. (2018). Renewable energy integration: Economic assessment of solar energy to produce biodiesel at supercritical conditions. *International Journal of Photoenergy*, 2018, 1–12. https://doi.org/10.1155/2018/4658094

Le Pevelen, D. D., & Tranter, G. E. (2017). FT-IR and Raman spectroscopies, polymorphism applications. In J. C. Lindon, G. E. Tranter, & D. W. Koppenaal (Eds.), *Encyclopedia of Spectroscopy and Spectrometry* (3rd edn.). Academic Press, pp. 750–761. https://doi.org/10.1016/B978-0-12-409547-2.12161-4

Liu, W., Xu, J., Xie, X., Yan, Y., Zhou, X., & Peng, C. (2020). A new integrated framework to estimate the climate change impacts of biomass utilization for biofuel in life cycle assessment. *Journal of Cleaner Production*, 267. https://doi.org/10.1016/j.jclepro.2020.122061

López-Fernández, Josu, Moya, Desirèe, Benaiges, María Dolors, Valero, Francisco, Alcalà, Manel. (2022). Near infrared spectroscopy: A useful technique for inline monitoring of the enzyme catalyzed biosynthesis of third-generation biodiesel from waste cooking oil. *Fuel*, 319, 123794, ISSN 0016-2361, https://doi.org/10.1016/j.fuel.2022.123794

Louhasakul, Y., & Cheirsilp, B. (2022). Potential use of industrial by-products as promising feedstock for microbial lipid and lipase production and direct transesterification of wet yeast into biodiesel by lipase and acid catalysts. *Bioresource Technology*, 348, 126742. https://doi.org/10.1016/j.biortech.2022.126742

Mahapatra, S., Kumar, D., Singh, B., & Sachan, P. K. (2021). Biofuels and their sources of production: A review on cleaner sustainable alternative against conventional fuel, in the framework of the food and energy nexus. *Energy Nexus*, 4, 100036. https://doi.org/10.1016/j.nexus.2021.100036

Mahlia, T. M. I., Syazmi, Z. A. H. S., Mofijur, M., Abas, A. P., Bilad, M. R., Ong, H. C., & Silitonga, A. S. (2020). Patent landscape review on biodiesel production: Technology updates. *Renewable and Sustainable Energy Reviews*, 118, 109526. https://doi.org/10.1016/j.rser.2019.109526

Malode, S. J., Prabhu, K. K., Mascarenhas, R. J., Shetti, N. P., & Aminabhavi, T. M. (2021). Recent advances and viability in biofuel production. *Energy Conversion and Management: X*, 10, 100070. https://doi.org/10.1016/j.ecmx.2020.100070

Masudi, A., Muraza, O., Jusoh, N. W. C., & Ubaidillah, U. (2023). Improvements in the stability of biodiesel fuels: Recent progress and challenges. *Environmental Science and Pollution Research*, 30, 14104–14125. https://doi.org/10.1007/s11356-022-20375-5

Mayer, F. D., Brondani, M., Vasquez Carrillo, M. C., Hoffmann, R., Silva Lora, E. E. (2020). Revisiting energy efficiency, renewability, and sustainability indicators in biofuels life cycle: Analysis and standardization proposal. *Journal of Cleaner Production*, 252. https://doi.org/10.1016/j.jclepro.2019.119850

Mekhilef, S., Siga, S., & Saidur, R. (2011). A review on palm oil biodiesel as a source of renewable fuel. *Renewable and Sustainable Energy Reviews*, 15(4), 1937–1949. https://doi.org/10.1016/j.rser.2010.11.045

Measurlabs. (2023). *Near-Infrared Spectroscopy*. Available from: https://measurlabs.com/methods/near-infrared-spectroscopy/#:~:text=Near%2Dinfrared%20spectroscopy%20(NIRS),as%20a%20function%20of%20wavelength

Misra, N., Panda, P. K., Parida, B. K., & Mishra, B. K. (2016). Way forward to achieve sustainable and cost-effective biofuel production from microalgae: A review. *International Journal of Environmental Science and Technology*, 13, 2735–2756. https://doi.org/10.1007/s13762-016-1020-5

Mizik, T., & Gyarmati, G. (2021). Economic and sustainability of biodiesel production—A systematic literature review. *Clean Technologies*, 3(1), 1–36. https://doi.org/10.3390/cleantechnol3010002

Morgan, S. A., Nadler, D. C., Yokoo, R., & Savage, D. F. (2016). Biofuel metabolic engineering with biosensors. *Current Opinion in Chemical Biology*, 35, 150–158. https://doi.org/10.1016/j.cbpa.2016.09.020

Moniruzzaman, M., Yaakob, Z., Shahinuzzaman, M., Khatun, R., & Aminul Islam, A. K. M. (2017). Jatropha biofuel industry: The challenges. *Frontiers in Bioenergy and Biofuels*, 1(12), 23–256.

Mutreja, V., Singh, S., & Ali, A. (2011). Biodiesel from mutton fat using KOH impregnated MgO as heterogeneous catalysts. *Renewable Energy*, 36(8), 2253–2258.

Neupane, Dhurba. (2022). Biofuels from renewable sources, a potential option for biodiesel production. *Bioengineering*, 10(1), 29.

Ogunkunle, O., & Ahmed, N. A. (2019). A review of global current scenario of biodiesel adoption and combustion in vehicular diesel engines. *Energy Reports*, 5, 1560–1579. https://doi.org/10.1016/j.egyr.2019.10.028

Patnaik, R., & Mallick, N. (2021). Microalgal biodiesel production: Realizing the sustainability index. *Frontiers in Bioengineering and Biotechnology*, 9, 620777. https://doi.org/10.3389/fbioe.2021.620777

Peng, L., Fu, D., Chu, H., Wang, Z., & Qi, H. (2020). Biofuel production from microalgae: A review. *Environmental Chemistry Letters*, 18, 285–297. https://doi.org/10.1007/s10311-019-00939-0

Piligaev, A. V., Sorokina, K. N., Samoylova, Y. V., & Parmon, V. N. (2019). Production of microalgal biomass with high lipid content and their catalytic processing into biodiesel: A review. *Catalysis in Industry*, 11, 349–359. https://doi.org/10.1134/S207005041904007X

Rahul, S. M., MA, S., CS, S., & I, G. M. (2021). Insights about sustainable biodiesel production from microalgae biomass: A review. *International Journal of Energy Research*, 45(12), 17028–17056. https://doi.org/10.1002/er.6138

Rahpeyma, S. S., & Raheb, J. (2019). Microalgae biodiesel as a valuable alternative to fossil fuels. *BioEnergy Research*, 12, 958–965.

Rebitzer, G., Ekvall, T., Frischknecht, R., Hunkeler, D., Norris, G., Rydberg, T., ... & Pennington, D. W. (2004). Life cycle assessment: Part 1: Framework, goal and scope definition, inventory analysis, and applications. *Environment International*, 30(5), 701–720.

Roschat, W., Kacha, M., Yoosuk, B., Sudyoadsuk, T., & Promarak, V. (2012). Biodiesel production based on heterogeneous process catalyzed by solid waste coralfragment. *Fuel*, 98, 194–202. https://doi.org/10.1016/j.fuel.2012.04.009

Rulli, M. C., Bellomi, D., Cazzoli, A., De Carolis, G., & D'Odorico, P. (2016). The water-land-food nexus of first-generation biofuels. *Scientific Reports*, 6, 22521.

Saletnik, A., Saletnik, B., & Puchalski, C. (2021). Overview of popular techniques of Raman spectroscopy and their potential in the study of plant tissues. *Molecules*, 26(6), 1537. https://doi.org/10.3390/molecules26061537

Sakai T, Kawashima A, Koshikawa T. (2009). Economic assessment of batch biodiesel production processes using homogeneous and heterogeneous alkali catalysts. *Bioresource Technology*, 100(13), 3268–3277.

Sandesh, K., & Ujwal, P. (2021). Trends and perspectives of liquid biofuel–Process and industrial viability. *Energy Conversion and Management: X*, 10, 100075. https://doi.org/10.1016/j.ecmx.2020.100075

Sara, M., Brar, S. K., & Blais, J. F. (2016). Comparative study between microwave and ultrasonication aided in situ transesterification of microbial lipids. *RSC Advances*, 6(61), 56009–56017.

Shah, S. H., Raja, I. A., Rizwan, M., Rashid, N., Mahmood, Q., Shah, F. A., & Pervez, A. (2018). Potential of microalgal biodiesel production and its sustainability perspectives in Pakistan. *Renewable and Sustainable Energy Reviews*, 81, 76–92. http://dx.doi.org/10.1016/j.rser.2017.07.044

Sharmbiu, V. B., Bhattacharya, T. K., Nayak, I. K., Das, S. (2013). Studies on characterization of raw Jatropha oil and its biodiesel with relevance of diesel. *International Journal of Engineering Technology and Advanced Engineering*, 3(4), 48–54.

Sieverding, H. L., Bailey, L. M., Hengen, T. J., Clay, D. E., & Stone, J. J. (2015). Meta-analysis of soybean based biodiesel. *Journal of Environmental Quality*, 44, 1038–1048. https://doi:10.2134/jeq2014.07.0320

Srivastava, N., Srivastava, M., Gupta, V. K., Manikanta, A., Mishra, K., Singh, S., ... & Mishra, P. K. (2018). Recent development on sustainable biodiesel production using sewage sludge. *3 Biotech*, 8, 1–11. https://doi.org/10.1007/s13205-018-1264-5

Tahvildari, K., Anaraki, Y. N., Fazaeli, R., Mirpanji, S., & Delrish, E. (2015). The study of CaO and MgO heterogenic nano-catalyst coupling on transesterification reaction efficacy in the production of biodiesel from recycled cooking oil. *Journal of Environmental Health Science and Engineering*, 13, 1–9. https://doi.org/10.1186/s40201-015-0226-7

Talebian-Kiakalaieh, A., Amin, N. A. S., & Mazaheri, H. (2013). A review on novel processes of biodiesel production from waste cooking oil. *Applied Energy*, 104, 683–710. https://doi.org/10.1016/j.apenergy.2012.11.040

Thiruketheeswaranathan, S. (2022). Application of life cycle assessment tool on biodiesel production: Assessing the environmental impact of biodiesel. *Middle East Journal of Applied Science & Technology*, 5(1), 1–6.

Uthandi, Sivakumar, Ashokkumar Kaliyaperumal, Naganandhini Srinivasan, Kiruthika Thangavelu, Iniya Kumar Muniraj, Xinmin Zhan, Nicholas Gathergood, & Vijai Kumar Gupta. (2022). Microbial biodiesel production from lignocellulosic biomass: New insights and future challenges. *Critical Reviews in Environmental Science and Technology*, 52, no. 12, 2197–2225.

van Eijck, J., Romijn, H., Balkema, A., & Faaij, A. (2014). Global experience with Jatropha cultivation for bioenergy: An assessment of socio-economic and environmental aspects. *Renewable and Sustainable Energy Reviews*, 32, 869–889. https://doi.org/10.1016/j.rser.2014.01.028

Vasistha, S., Khanra, A., Clifford, M., & Rai, M. P. (2021). Current advances in microalgae harvesting and lipid extraction processes for improved biodiesel production: A review. *Renewable and Sustainable Energy Reviews*, 137, 110498. https://doi.org/10.1016/j.rser.2020.110498

Wang, H., Peng, X., Zhang, H., Yang, S., & Li, H. (2021). Microorganisms-promoted biodiesel production from biomass: A review. *Energy Conversion and Management: X*, 12, 100137. https://doi.org/10.1016/j.ecmx.2021.100137

Wijffels, R. H., & Barbosa, M. J. (2010). An outlook on microalgal biofuels. *Science*, 329(5993), 796–799. https://doi.org/10.1126/science.1189003

Wu, Y., Zhao, F., Liu, S., Wang, L., Qiu, L., Alexandrov, G., & Jothiprakash, V. (2018). Bioenergy production and environmental impacts. *Geoscience Letters*, 5(1), 1–9. https://doi.org/10.1186/s40562-018-0110-4

Zhang, Y., Dub, M. A., McLean, D. D., & Kates M. (2003). Biodiesel production from waste cooking oil: 1. Process design and technological assessment. *Bioresource Technology*, 89, 1–16.

Zulqarnain, Mohd Hizami Mohd Yusoff, Muhammad Ayoub, Naveed Ramzan, Muhammad Hamza Nazir, Imtisal Zahid, Nadir Abbas, Noureddine Elboughdiri, Cyrus Raza Mirza, & Tayyab Ashfaq Butt. (2021). Overview of feedstocks for sustainable biodiesel production and implementation of the biodiesel program in Pakistan. *ACS Omega*, 6 (29), 19099–19114.

3 In-Tube Condensation Heat Transfer Correlations for Horizontal Smooth Macro Tubes

S. C. Blose, D. R. E. Ewim, A. C. Eloka-Eboka, and A. O. Adelaja

NOMENCLATURE

A	area (m²)		We	Weber number
C	constant		x	vapour quality
Cp	constant pressure specific heat (J/kgK)		$X1$	experimental variable
d	diameter (mm)		Xtt	Lockhart–Martinelli parameter
D_h	hydraulic diameter (mm)		Z	experimental quantity
D_i	tube inside diameter (mm)		z	axial direction
f	function			

Fr — Froude number

g — gravitational acceleration (m/s2)

G — mass flux (kg/m²s)

h — specific enthalpy

hfg — specific latent heat of vapourization

Ja — Jakob number

J_v — dimensionless gas velocity

JVT — transition dimensionless gas velocity

k — thermal conductivity (W/mK)

L — length of the test section (m)

Φv — two-phase multiplier for Lockhart-Martinelli parameter

Nu — Nusselt number

$\dot{m}$ — mass flow rate (kg/s)

n — Akers index, number of thermocouples

Pr — Prandtl number

$\dot{q}$ — heat flux (W/m2)

$\dot{Q}$ — heat transfer rate (W)

R — thermal resistance (K/W)

Re — Reynolds number

T — temperature (°C)

$\bar{T}$ — average wall temperature (°C)

ΔT — temperature difference (°C)

V — velocity (m/s)

Greek symbols

α — heat transfer coefficient (W/m2K)

δ — uncertainty

ε — void fraction

μ — dynamic viscosity (kg/ms)

ρ — density (kg/m3)

σ — surface tension (N/m)

σ — standard deviation

Subscripts

A — ΔT independent annular flow regime

avg — average

cond — conduction

conv — convection

crit — critical

Cu — copper

cs — cross section

D — ΔT dependent flow regime/total heat transfer coefficient

e — equivalent

fri — frictional

h — hydraulic

i — inner

DOI: 10.1201/9781032651958-3

in	inlet		sat	saturation
ind	indicated		stat	static
j	measurement location		strat	fully stratified regime, stratified
l	liquid		test	test-condenser
m	mean		tot	total
post	post-condenser		tp	two-phase
pre	pre-condenser		tr	transducer
r	refrigerant		v	vapour
red	reduced		w	wall, water
s	surface			

3.1 INTRODUCTION

For decades, condensation has played a crucial role in power generation, chemical and pharmaceutical production, and refrigeration and air conditioning systems. The effectiveness of in-tube condensation is influenced by several factors, including the working fluid, tube diameter, heat flux, and mass flow rate. To optimize the design and performance of such systems, it is essential to accurately predict the heat transfer coefficients. The substantial heat transfer coefficients linked to condensation have enabled the creation of thermal management systems for many contemporary technologies requiring effective heat transfer or dispersion within confined areas. To improve condenser efficiency and consequently reduce energy consumption, it is essential to develop more accurate and comprehensive methods for predicting the thermal and fluid-dynamic behaviour of condensers. These methods should consider the use of environmentally friendly refrigerants and innovative tube designs. Previously, various condensation heat transfer correlations and models have been documented. However, most of these predictive approaches have a limited scope, as they are applicable to a specific set of fluids and operating conditions. Semi-empirical correlations [1–3], which are limited to specific fluids and operating conditions, universal correlations [4, 5], which are based on consolidated databases for a large number of fluids and a broad range of operating conditions, and analytical control-volume-based models [6] are the three types of predictive tools for annular condensation heat transfer.

Although there are several correlations for predicting heat transfer coefficients inside mini/micro-tubes in the literature, there has not been any research on the construction of a generalized correlation for condensation heat transfer coefficients inside macro-tubes that covers a wide range of working fluids. According to Qu and Mudawar [7], the strong influence of surface tension and viscosity, as well as negligible body force effects, distinguishes mini/micro-tubes from macro-tubes. Based on this, it is also important to have as many correlations as possible for predicting heat transfer coefficients inside macro-tubes in the literature.

While various correlations for condensation in macro-tubes have been proposed, the majority of them are based on relatively limited data, generally the author's own. When compared to a larger data set, such correlations fall short. However, a few general correlations have been reported that have been proven across a large range of data. Dobson and Chato [8], Thome et al. [9], Cavallini et al. [10], and Shah [11, 12] are notable. The first three are only relevant to horizontal tubes and employ the Nusselt formula for estimating heat transfer inside horizontal tubes at low flow rates when heat transfer becomes heat flux dependent. The Nusselt equation is used to compute heat transfer in the upper section of the tube, while a forced convection equation is used to calculate heat transfer in the lower part. These correlations necessitate knowledge of heat flux in order to compare them to test data. Shah [13] correlation works for both horizontal and vertical downflow, whereas the other three only work for horizontal flow.

In the review, various factors affecting in-tube condensation are discussed. These include the properties of the working fluid, such as its viscosity, density, and thermal conductivity, as well as the geometric characteristics of the tube, including its diameter, length, and orientation. Additionally, the influence of mass flow rate on the heat transfer process is examined, as it can significantly impact the performance of in-tube condensation systems. Various correlations and experimental works are presented in the chapter, offering a diverse range of approaches to predict heat transfer coefficients. Some of these correlations are based on dimensionless numbers, such as the Nusselt, Prandtl, and Reynolds numbers, which are widely used in heat transfer and fluid dynamics. These correlations can provide useful insights into the relationships between different factors affecting in-tube condensation and their impact on heat transfer coefficients. Furthermore, the chapter opens our eyes to the limitations of existing models and correlations, highlighting the need for further research and development in this area. This could involve the development of new models and correlations that are more accurate, more broadly applicable, or better suited to specific applications.

3.2 DIMENSIONLESS NUMBERS

The dimensionless parameters must be discussed in relation to flow regime maps and transitions. The most essential dimensionless parameters pertinent to condensation are discussed in the section below.

3.2.1 Reynolds Number

The Reynolds number is a dimensionless quantity used in fluid mechanics to describe the flow behaviour of a fluid within a system. It is defined as the ratio of inertial forces to viscous forces in the fluid and helps predict whether a flow is laminar or turbulent. The Reynolds number is calculated as the product of fluid density, flow velocity, and a characteristic length (such as pipe diameter or object length) divided by the fluid's dynamic viscosity. A low Reynolds number indicates that the flow is dominated by viscous forces and is likely to be laminar, while a high Reynolds number suggests that inertial forces dominate and the flow is more likely to be turbulent. The Reynolds number is widely used in engineering and fluid dynamics to analyse and design various systems involving fluid flow, such as pipes, ducts, and airfoils. It is defined mathematically as;

$$Re = \frac{\rho V d}{\mu} \tag{3.1}$$

Where ρ is the fluid density, μ is the dynamic viscosity, d is the diameter, and V is the flow velocity.

3.2.2 Nusselt's Number

The Nusselt number is a dimensionless quantity used in heat transfer calculations to quantify the ratio of convective heat transfer to conductive heat transfer across a boundary. It represents the efficiency of heat transfer in a fluid due to convection compared to the heat transfer that would occur solely by conduction. The Nusselt number is often used to assess the performance of heat exchangers and other engineering systems where heat transfer is a critical factor. It mathematically defined as follows:

$$Nu = \frac{hd}{k} \tag{3.2}$$

In this case, d represents the diameter, k denotes the thermal conductivity, and h corresponds to the convective heat transfer coefficient.

3.2.3 Dimensionless Vapour Velocity

$$J_v = \frac{xG}{\left[gd_i\rho_v\left(\rho_l-\rho_v\right)\right]^{0.5}}$$

(3.3)

Where x is the vapour quality, G is the mass flux, g is the gravitational acceleration, d_i inner diameter, ρ_v vapour density, and ρ_l is the liquid density.

3.2.4 Lockhart-Martinelli Parameter

The ratio represents the hypothetical pressure gradients that would arise if either fluid individually flowed through the pipe at its initial flow velocity.

$$X_{tt} = \left(\frac{1-x}{x}\right)^{0.9}\left(\frac{\rho_v}{\rho_l}\right)^{0.5}\left(\frac{\mu_l}{\mu_v}\right)^{0.1}$$

(3.4)

Where x is the vapour quality, ρ_v is the vapour density, ρ_l is the liquid density, μ_l is the dynamic viscosity (liquid), and μ_v is the dynamic viscosity (vapour).

$$X_{tt} = \frac{\dot{m}_l}{\dot{m}_v}\sqrt{\frac{\rho_v}{\rho_l}}$$

(3.5)

Where $\dot{m}_l$ is the liquid mass flow rate, $\dot{m}_v$ is the vapour mass flow rate, ρ_v is the vapour density, and ρ_l is the liquid density.

3.2.5 Prandtl Number

The Prandtl number is a dimensionless quantity that links a fluid's viscosity to its thermal conductivity, thereby assessing the connection between the fluid's momentum and its ability to transfer heat

$$P_r = \frac{\mu C_p}{k}$$

(3.6)

Where μ is the dynamic viscosity, C_p is the constant pressure specific heat, and k is the thermal conductivity.

3.2.6 Bond Number

A dimensionless number that characterizes the shape of bubbles or drops flowing in a fluid by comparing the relevance of gravity forces to surface tension forces.

$$Bo = \frac{g\left(\rho_l-\rho_v\right)D_h^2}{\sigma}$$

(3.7)

Where g is the gravitational acceleration, ρ_v vapour density, ρ_l is the liquid density, D_h is the hydraulic diameter, and σ is the surface tension.

3.3 CLASSIFICATION OF TUBES

Tubes are categorized according to their features and functions and are measured by their exterior diameter and wall thickness, as opposed to pipes, which are measured by their internal diameter. They come in a variety of shapes, including round, rectangular, and square. In smooth tubes, fluid generally follows a smooth path in which the particles do not interfere with each other. In the past, researchers have proposed a number of tube classifications using different methodological approaches. Cheng and Wu approach [14] is the most widely mentioned in the literature. Kandlikar [15] has one that is extensively utilized. For tube classifications, Cheng and Wu [14] utilized Bond number (Bo), while Kandlikar [15] used hydraulic equivalent diameter (D_{HYD}). Based on an investigation of the magnitudes of gravity and surface tension effects, Cheng and Wu [14] developed the following criteria:

Micro-channel: Bo < 0.5 (negligible effect of gravity)
Mini-channel: 0.5 < Bo < 3.0 (gravity and surface tension have effect)
Macro-channel: Bo > 3.0 (Surface tension has a negligible effect)

Below is the approach utilized by Kandlikar [15]

Micro-channel: 0.2 mm $\geq D_{HYD} > 0.01$ mm
Mini-channel: 3 mm $\geq D_{HYD} > 0.2$ mm
Macro-channel: $D_{HYD} > 3$ mm

The above-mentioned Kandlikar classification will be employed in this paper.

3.4 FLOW PATTERN

For condensation inside horizontal tubes, plenty of well flow regimes have been observed, including pure vapour, annular, slug, bubbling, and pure liquid, in order to decrease quality. To define dominating flow regimes, both flow regime maps and regime transition relations have been advocated [1–5]. Because of its presence throughout a substantial portion of the tube length and its capacity to produce high heat transfer coefficients, the annular regime has gotten the most attention. A thin layer sheathes the inner walls of the condensation tube in the annular regime, with shear produced by a faster-moving central vapour core. While vapour shear is the primary driving force in annular condensation, gravity can also play a role in condensation inside horizontal tubes, as evidenced by liquid stratification toward the inner surface's bottom. As a result, a thick liquid film forms near the bottom, contrasting with a very thin or no film at the top. A large influence of gravity on the amplitude and spatial variability of the condensation heat transfer coefficient is usually linked with conditions that produce pronounced stratification effects.

3.4.1 ANNULAR

Annular flow is a two-phase flow regime. It is distinguished by the presence of a liquid layer on the channel wall (creating an annular ring of liquid), while the gas travels in a continuous phase up the tube's centre. Liquid droplets can be entrained in the flow core. Annular flow regime occurs most frequently at high mass flux and high qualities. The mass flux is the key driving force impacting the heat transfer process at this point [16]. Figure 3.1 depicts this.

3.4.2 STRATIFIED

In stratified flow, a multiphase flow regime separates fluids (gas and liquid) into distinct layers, with lighter fluids flowing above heavier ones. The interface between these layers can vary in structure,

FIGURE 3.1 Annular flow regime.

FIGURE 3.2 Stratified flow regime.

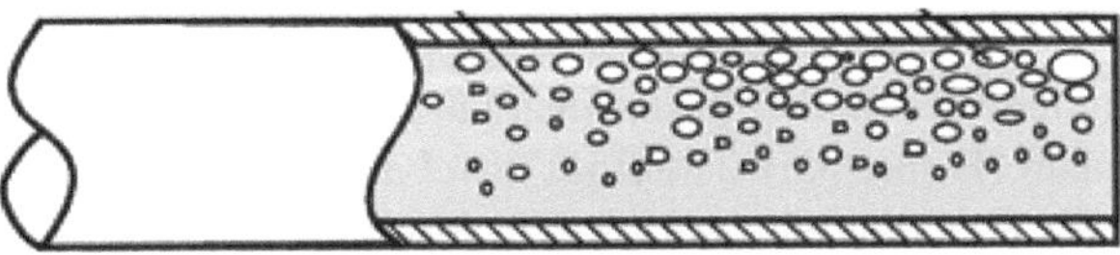

FIGURE 3.3 Bubbly flow regime.

appearing either smooth or wavy depending on the mass flux. At low mass flux, the interface tends to be flat or may exhibit small capillary waves only a few millimeters in length. Stratified flow is more common in sections of horizontal wells with low flow rates, where gravity dominates, causing the liquid to settle at the bottom of the tube. Figure 3.2 illustrates this.

3.4.3 BUBBLY

Differing from vertical tube bubbly flow, horizontal channel bubbly flow experiences significant gravitational impact. The buoyancy effect causes bubbles to scatter within the liquid, concentrating more in the tube's upper section. As gravity pulls the liquid annulus downward at lower flow velocities, this phenomenon typically occurs at increased flow rates. This is demonstrated in Figure 3.3.

3.4.4 INTERMITTENT

Predicting intermittent flow, which can be observed in horizontal or nearly horizontal pipes, is especially challenging among the various flow regimes that include annular flow. Indeed, the two-phase nature of this motion occurs irregularly in both spatial and temporal dimensions. For example, when a liquid slug reaches the top wall of the pipe (or completely blocks the section area), a gas bubble begins to flow in the liquid layer above it, this is demonstrated in Figure 3.4. Large changes of flow parameters, including local pressure, void fraction, mass flow rates of both phases, and the velocity fields in axial or radial directions, may result from the intermittent flow with constant phasic gas and liquid flow rates. Formation of slugs or plugs at high liquid flow rates as the crests of the waves intermittently reach the top of the pipe. Slugs flow occurs when the flow rate is high and the quality is low. Large amplitude waves can be characterized as the liquid slugs that separate such elongated bubbles [17].

3.4.5 SLUG AND PLUG

The waves reach the tube's peak when the speed of the gas increases. The void fraction, which causes bubble aggregation into larger plugs and slugs, determines whether the flow is a plug or slug.

FIGURE 3.4   Intermittent flow regime.

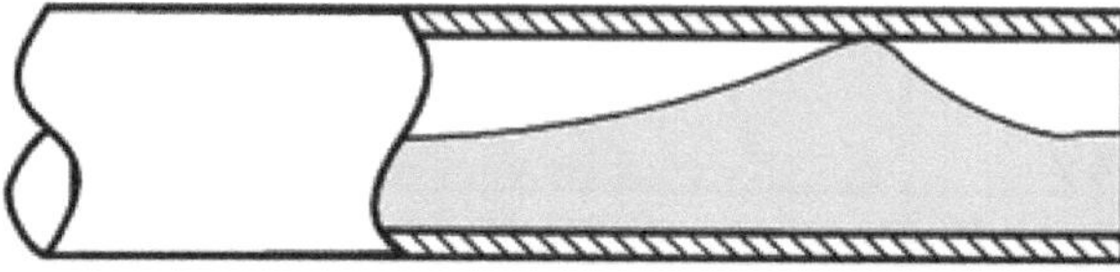

FIGURE 3.5   Slug flow regime.

The bubbles in the plug flow are smaller in size than the tube's diameter. Slugs are roughly the same size as the tube's diameter (Figure 3.5). The slugs travel at a significant fraction of the gas velocity and appear intermittently. Slug flow can be formed from stratified flow in horizontal and nearly horizontal pipes by two basic mechanisms: (i) the natural development of hydrodynamic instabilities and (ii) liquid accumulation due to a sudden imbalance between pressure and gravity forces caused by pipe undulations [18]. These waves may keep getting bigger, picking up more liquid as they go, until they reach the middle of the pipe and form slugs. If the slug fronts travel faster than the slug tails, these slugs will expand; otherwise, they will dissolve. If the front and back of the slug travel at the same speed, stable slug flow is achieved.

3.5   FLOW PATTERN MAP

Flow regime maps have been a reliable resource for researchers to characterize two-phase behaviour, leading to the development of numerous flow pattern maps for smooth tubes. Baker [19] examined air-water and oil-water flows in horizontal tubes of varying inner diameters to establish the initial two-phase flow regime map. Wang et al. [20] revised Baker's map using R22, R134a, and R407C. Taitel and Dukler [21] introduced a flow pattern map for horizontal tubes, developing a physics-based model with five dimensionless groups to predict flow regime boundaries. Their map, tested with public data across diverse operational settings, showed good agreement. Tandon et al. [22], building on previous studies, presented a flow pattern map during condensation that accurately predicted annular, semi-annular, and wavy flows. Their flow regime estimation improved by incorporating two new parameters as map coordinates. Yang and Shieh [23] explored air-water mixtures and R134a flow patterns in tubes with inner diameters from 1.0 to 3.0 mm, identifying bubble, slug, plug, stratified-wavy, scattered, and annular flow patterns. They then compared their flow maps and transition boundaries with previous research. Wojtan et al. [24] presented a flow pattern map for the boiling flow of R22 and R410A. Recent interest in the low-GWP refrigerant R1234yf has prompted further investigation into its flow regime. Padilla et al. [25] assessed the flow regimes of R134a and R1234yf, finding good agreement with Wojtan et al. [24]'s map. They visualized slug, intermittent, and annular flows using a smooth glass tube with an inner diameter of 6.70 mm and mass velocities ranging from 300 to 500 kg/m^2 s. Padilla et al. [26] analysed the flow patterns and behaviour of bubbles and vapour slugs in a vertical return curve and a rapid contraction with an area ratio of 0.49. Wang et al. [27] investigated the flow pattern of R1234yf in a 4 mm diameter tube, comparing their results with the flow regimes map created by Tandon et al. [22]. They observed some mispredictions for wavy flow and the infrequency of slug and plug regimes, despite reasonable agreement for annular flow.

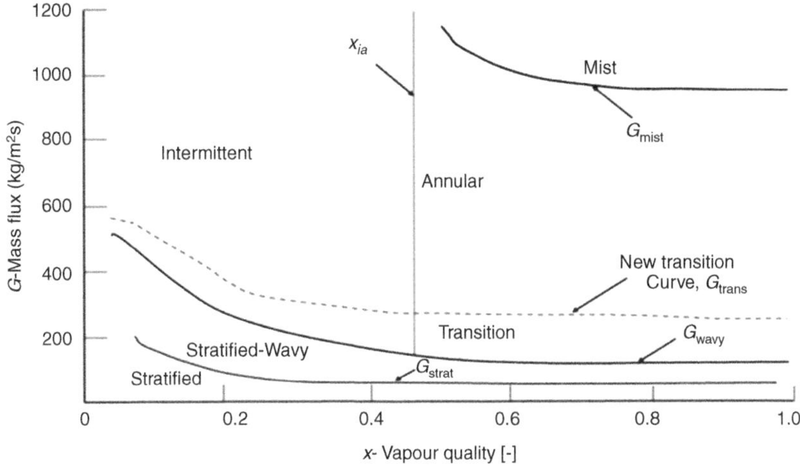

FIGURE 3.6 Flow pattern map from Suliman et al. [28].

Suliman et al. [28] proposed an updated flow pattern map for predicting heat transfer coefficients within horizontally smooth tubes, which illustrated annular, intermittent, and stratified flow regimes. Based on their findings, the El Hajal et al. [29] flow pattern map was adjusted to include a transition zone between stratified-wavy and annular or intermittent flow regimes. These flow pattern maps and the ongoing research in this field contribute to a better understanding of two-phase flow behaviour in various applications. The continuous improvements to flow pattern maps, along with the investigation of new refrigerants, aid researchers and engineers in accurately predicting and optimizing heat transfer coefficients within heat exchangers. The advances in this area will help enhance the performance and efficiency of various industrial systems that rely on condensation heat transfer.

3.6 REVIEWS

3.6.1 REVIEW OF CORRELATIONS

Table 3.1 provides an overview of selected correlations for condensation heat transfer coefficient accessible in the current technical literature. Akers and Crosser [30] proposed one of the first correlations for two-phase flow heat transfer inside horizontal tubes. Many scholars have studied two-phase condensation heat transfer inside a tube since then, using empirical and semi-empirical methods. Park and Hrnjak [31] investigated HTCs and pressure decreases for CO_2 at saturation temperatures of 15°C and 25°C. Their research used horizontal micro-channels with diameters of 0.89 mm and mass fluxes ranging from 200 to 800 kg/m² s. Shah [11, 32] proposed universal models for horizontal and vertical tubes built on 1,736 experimental data points containing 24 fluids. The models were separated into three regimes based on dimensionless vapour velocity values. Cavallini et al. [10] provided a correlation for condensation HTC in horizontal macro channels with D > 3 mm by establishing two sub-domains, namely ΔT-independent and ΔT-dependent regimes. Dorao and Fernandino [33] investigated the primary characteristics influencing the HTC at high mass velocities and proposed a model for calculating the HTC for channel diameters ranging from 0.067 mm to 14.45 mm. Dorao and Fernandino [33] employed two-phase Reynolds Prandtl numbers in their model to represent the condensation HTC inside the channels, similar to Dittus-Boelter suggestion [34].

The heat transfer coefficient correlations for the macro-tubes used in this work are shown in Table 3.1. These correlations were discovered by utilizing a variety of fluids, diameters, flow quality,

TABLE 3.1
Condensation Heat Transfer Correlations

Author(s)	Correlation	Remarks
1. Haraguchi et al. [35]	$\dfrac{h_{tp} D_h}{k_f} = 0.0152\left(1 + 0.6 Pr_f^{0.8}\right)\dfrac{\varnothing_g}{X_{tt}} Re_f^{0.77}$	$D = 8.4$ mm R22, R123, R134a $G = 90{-}400$ kg/m^2 s
2. Huang et al. [37]	$\dfrac{h_{tp} D}{k_f} = 0.0152\left(-0.33 + 0.83 Pr_f^{0.8}\right)\dfrac{\varnothing_g}{X_{tt}} Re_f^{0.77}$	R134a and R404A
3. Cavallini et al. [10]	$\alpha_{START} = 0.725\left\{1 + 0.741\left(\dfrac{1-x}{x}\right)^{0.3321}\right\}^{-1}\left[\lambda_l^3 \rho_l \left(\rho_l - \rho_v\right) g h_{lv} D\Delta T\right]^{0.25}$ $+ \left(1 - x^{-0.087}\right)\alpha_{lo}$	$D = 3.1{-}17.0$ mm R22, R134a, CO$_2$ $G = 18{-}2240$ kg/m^2 s
4. Meyer et al. [47]	$\alpha_{START} = 0.725\left\{1 + 0.741\left(\dfrac{1-x}{x}\right)^{0.332}\right\}^{-1}\left[\lambda_l^3 \rho_l \left(\rho_l - \rho_v\right) g h_{lv} D\Delta T\right]^{0.245}$ $+ \left(1 - x^{0.25}\right)\alpha_{lo}$	$D = 8.38$ mm R134a
5. Park et al. [38]	$\dfrac{h_{tp} D_h}{k_f} = 0.0055 Re_f^{0.7} Pr_f^{1.37}\dfrac{\varnothing_g}{X_{tt}}$	R134a, R236fa, R1234ze (E) $G = 50{-}260$ kg/m^2 s
6. Koyama et al. [36]	$\dfrac{h_{tp} D_h}{k_f} = 0.0152\left(1 + 0.6 Pr_f^{0.8}\right)\dfrac{\varnothing_g}{X_{tt}} Re_f^{0.77}$	R134a
7. Shah [11]	$\dfrac{h_{tp} D_h}{k_f} = 0.023 Re_{fo}^{0.8} Pr_f^{0.4}\left(\dfrac{\mu_f}{14\mu_g}\right)^n\left[\left(1-x\right)^{0.8} + \dfrac{3.8 x^{0.76}\left(1-x\right)^{0.04}}{P_R^{0.38}}\right]$	
8. Cavallini and Zecchin [48]	$\dfrac{h_{tp} D_h}{k_f} = 0.05 Re_f^{0.8} Pr_f^{0.33}\left[1 + \left(\dfrac{\rho_f}{\rho_g}\right)^{0.5}\left(\dfrac{x}{1-x}\right)\right]^{0.8}$	R12 and R22
9. Dobson and Chato [8]	$\dfrac{h_{tp} D_h}{k_f} = 0.023 Re_f^{0.8} Pr_f^{0.4}\left(1 + \dfrac{2.22}{X_{tt}^{0.89}}\right)$	$D = 3.14{-}7.04$ mm R12, R22, R134a, and R32
10. Akers et al. [49]	$\dfrac{\alpha d_i}{k_i} = C Re_e^n Pr_l^{0.333}$	$C = 0.0265$ and $n = 0.8$ for $Re_e > 50000$ $C = 5.03$ and $n = 0.333$ for $Re_e < 50000$
11. Akers and Rosson [50]	$\dfrac{h_{tp} D_h}{k_f} = 0.026 Pr_f^{\frac{1}{3}}\left\{G\left[\left(1+x\right) + x\left(\dfrac{\rho_f}{\rho_g}\right)^{0.5}\right]\dfrac{D_h}{\mu_f}\right\}^{0.8}$	R12, Propane
12. Shah [51]	$\dfrac{h_{tp} D_h}{k_f} = 0.023 Re_{fo}^{0.8} Pr_f^{0.4}\left[\left(1-x\right)^{0.8} + \dfrac{3.8 x^{0.76}\left(1-x\right)^{0.04}}{P_R^{0.38}}\right]$	$D = 7{-}40$ mm R11, R12, R22, and R113

TABLE 3.1 (Continued)
Condensation Heat Transfer Correlations

Author(s)	Correlation	Remarks
13. Ewim [17]	$$\alpha_{lo} = 0.023 Re_{lo}^{0.8} Pr_l^{0.4} \,\, k_l \big/ d$$ $$\alpha_A = \alpha_{lo}\left[1 + 1.128 x^{0.8170}\left(\rho_l/\rho_v\right)^{0.3685}\left(\mu_l/\mu_v\right)^{0.2363}\left(1-\mu_v/\mu_l\right)^{2.144} Pr_l^{-0.1}\right]$$ $$\alpha_{strat} = 0.725\left[1 + 0.741\left[\left(1-x\right)/x\right]^{0.3321}\right]^{-1}\left[k_l^3 \rho_l\left(\rho_l-\rho_v\right) g h_{lv}/\left(\mu_l d_i \Delta T\right)\right]^{T_1}$$ $$+\left(1-x^{T_2}\right)\alpha_{lo}$$	R134a
14. Moser et al. [52]	$$F = x^{0.78}\left(1-x\right)^{0.24}\left(\frac{p_l}{p_v}\right)^{0.91}\left(\frac{\mu_v}{\mu_l}\right)^{0.19}\left(1-\frac{\mu_v}{\mu_l}\right)^{0.7}$$ $$Fr_{tp} = G^2/\left(p_{tp}^2 g D\right),\ \ We_{tp} = G^2 D/\left(p_{tp}^2 \sigma\right)$$	R11, R12, R125, R22, R134a, and R410A $D = 3.14\text{--}20$ mm
15. Wang et al. [3]	$$h_{anul} = 0.0274 Pr_1 Re_1^{0.6792} x^{0.2208}\left(\frac{1,376 + 8 x_{tt}^{1.655}}{x_{tt}^2}\right)^{0.5}$$ $$h_{cov} = 0,023 Re_1^{0.8} Pr^{0.4}$$ $$f_{anul} = \left(x_{in} - x_{traw}\right)\left(x_{in} - x_{out}\right)$$	R134a $G = 75\text{--}750$ kg/m^2 s $T_{sat} = 45\text{--}63\ ^\circ$C
16. Bohdal et al. [53]	$$\frac{h_{tp} D}{k_f} = 25.084 Re_f^{0.258} Pr_f^{-0.495} P_R^{-0.288}\left(\frac{x}{1-x}\right)^{0.266}$$	
17. Shah [12]	$$h_1 = h_{LO}\left(1 + \frac{3.8}{Z^{0.95}}\right)\left(\frac{\mu_L}{14\mu_G}\right)^{\left(0.0058 + 0.557 p_r\right)}$$ $$h_{Nu} = 1.32 Re_{LO}^{-1/3}\left[\frac{\rho_{l\left(p_l - p_G\right)} g K_L^3}{\mu_L^2}\right]^{1/3}$$	40 different fluids $D = 0.08\text{--}49.0$ mm $G = 1.1\text{--}1400$ kg/m^2 s
18. Li and Norris [54]	Regime I: $h_a = \left(1 + \dfrac{1.2}{X_{tt}^{0.935}}\right) 0.023 Re_{ls}^{0.8} Pr_l^{0.4} \dfrac{k_1}{D}$ Regime II: $h_s = \dfrac{0.56}{1 + a X_{tt}^b} h_{film} + \left(1 - \theta_s/\pi\right) 0.023 Re_{ls}^{0.8} Pr_L^{0.4} \dfrac{k_1}{D}$ $D > 3$ mm; $a = 0.42$ $b = 0.786$ Regime III: $h_a^T = h_a\left(G = G_a^T\right)$ $$h_a^T = h_a\left(G = G_a^T\right)$$ $$G_a^T = \left[\frac{1.26}{14}\left(\frac{D\left(1-x\right)}{\mu_1}\right)^{1.04}\left(\frac{1 + 1.09 X_{tt}^{0.039}}{X_{tt}}\right)^{1.5}\frac{1}{Ga^{0.5}}\right]^{\frac{1}{1.59}}$$ $$G_s^T = \left[\frac{0.025}{14}\left(\frac{D\left(1-x\right)}{\mu_1}\right)^{1.59}\left(\frac{1 + 1.09 X_{tt}^{0.039}}{X_{tt}}\right)^{1.5}\frac{1}{Ga^{0.5}}\right]^{\frac{1}{1.59}}$$	CO$_2$ $D = 0.89\text{--}6.52$ mm $T_{sat} = -25$ to 30°C
19. Jassim et al. [55]	$$h_{ann} = 0.023\left(\frac{k_l}{D}\right) Re_1^{0.8} Pr_1^{0.4}\left(1 + \frac{2.22}{X_{tt}^{0.889}}\right)$$ $$X_{tt} = \left(\frac{1-x}{x}\right)^{0.9}\left(\frac{\rho_v}{\rho_l}\right)^{0.5}\left(\frac{\mu_l}{\mu_v}\right)^{0.1}$$ $$Re_1 = \frac{GD\left(1-x\right)}{\mu_1}$$	

(continued)

TABLE 3.1 (Continued)
Condensation Heat Transfer Correlations

Author(s)	Correlation	Remarks
20. Thome et al. [9]	$$\alpha_{tp} = \frac{\alpha_f \theta + (2\pi - \theta)\alpha_f}{2\pi r}$$ $$\alpha_c = 0.003 Re_l^{0.74} Pr_l^{0.5} \frac{\lambda_l}{\delta} f_i$$ $$\alpha_f = 0.728\left[\frac{\rho_l(\rho_l - \rho_v)gr\lambda_l^3}{\mu_l d" T}\right]^{0.25}$$	
21. Tandon et al. [56]	$$Nu = cPr_L^{\frac{1}{3}}\left(\frac{h_{fg}}{C_p \Delta T}\right)\left(\frac{Gx}{\mu_L}\right)^M$$ To $\dfrac{Gx}{\mu_L} > 3*10^{-4}$ $c = 0.084; M = 0.67$ To $G\dfrac{d}{\mu_L} < 3*10^{-4}$ $c = 23.1; M = 1/8$	
22. Qi et al. [57]	Annular mist intermittent Range: $J_v \geq 0.98(Z + 0.263)^{-0.62}$ and $We_v \geq 40$ Correlation: $h_{tp} = h_l = h_{LT}\left[1.37 + 1.545x^{0.817}\left(\rho_l/\rho_v\right)^{0.3685}\left(\mu_l/\mu_v\right)^{0.2363}\right.$ $\left.\left(1 - \mu_v/\mu_l\right)^{2.144} Pr_l^{-0.1}\right]$ $h_{LT} = 0.023 Re_l^{0.8} Pr_l^{0.4}\lambda_l/D$ Wavy Range: neither Regime 1 nor Regime 3 Correlation: $h_{tp} = h_l + h_{Nu}$ Stratified Range: $J_v \leq 0.7(1.254 + 2.27Z^{1.249})^{-1}$ Correlation: $h_{tp} = h_{Nu} = 0.528 Re_{lo}^{-1/3}\left[\dfrac{\rho_l(\rho_l - \rho_v)g\lambda_l^3}{\mu_l^2}\right]^{1/3}$	
23. Lee et al. [58]	$$h_l = 0.021 Re_l^{0.8} Pr_l^{0.4}\frac{k_l}{D_i}$$ $$Re_l = \frac{GD_i(1-x)}{\mu_l}$$	R404A $D = 5.6$ mm
24. Sajadi et al. [59]	$$\frac{h_{tp}h_d}{k_f} = 0.023 Re_f^{0.8} Pr_f^{0.3}\left(\frac{a}{X_{tt}^b}\right)$$ $a = 1.89$ $b = 0.87$ $$X_{tt} = \left(\frac{1-x}{x}\right)^{0.9}\left(\frac{\rho_g}{\rho_f}\right)^{0.5}\left(\frac{\mu_f}{\mu_g}\right)^{0.5}$$	

TABLE 3.1 (Continued)
Condensation Heat Transfer Correlations

Author(s)	Correlation	Remarks
25. Traviss et al. [60]	$$\frac{hD}{k_l} = \frac{0.15 Pr_l Re_l^{0.9}}{F_T}\left[\frac{1}{X_{tt}} + \frac{2.85}{X_{tt}^{0.476}}\right]$$ $$Re_l = \frac{G(1-x)D}{\mu_l}$$ $$F_T = 0.707 Pr_l Re_l^{0.5}$$	
26. Deng et al. [61]	$$h_{LO} = 0.023 Re_{LO}^{0.8} Pr_L^{0.4}\frac{\lambda_L}{D}$$	
27. Kim and Mudawar [62]	$$h = \left(0.1 + 0.06 Pr_f^{0.8}\right) Pr_f^{-1} Re_f^{-0.13} C_{fp}$$	FC-72
28. Shah [13]	$$h_1 = 0.023 Re_1^{0.8} Pr_1^{0.4}\left(\frac{\mu_1}{14\mu_v}\right)^n\left[1 + \frac{3.8}{Z^{0.95}}\right]\frac{\lambda_1}{D}$$ $$Z = \left(\frac{1}{x}-1\right)^{0.8} P_r^{0.4}$$	H_2O, CO_2, HCFCs, HCs $D = 2$–49 mm $G = 13$–820 kg/m²s $X = 0$–0.95
29. Garimella et al. [63]	$$h_A = \left[\left(\frac{\theta}{2\pi}\right) Nu_{film} + \left(1 - \frac{\theta}{2\pi}\right) Nu_{forced}\right]\frac{\lambda_1}{D}$$ $$Nu_{film} = \frac{1.93}{\theta}\left[Ra\left(\frac{1}{Ja}+1\right)\right]^{0.25}$$ $$Nu_{film} = 0.018 Re_1^{0.8} Pr_1^{1/3}\left[1+\left(\frac{x}{1-x}\right)\left(\frac{\rho_l}{\rho_v}\right)^{1.24}\right]\left(\frac{D}{0.0094}\right)^{0.34}$$	R404A and R410A $D = 0.76$–7.4 mm $G = 200$–800 kg/m² s
30. Kim and Mudawar [64]	For annular flow where $We^* > 7X_{tt}^{0.2}$ $$\frac{h_{ann} D_h}{k_f} = 0.048 Re_f^{0.69} Pr_f^{0.34}\frac{\varnothing_g}{X_{tt}}$$ For slug and bubbly flow where $We^* < 7X_{tt}^{0.2}$ $$\frac{h_{non-ann} D_h}{k_f} = \left[\left(0.048 Re_f^{0.69} Pr_f^{0.34}\frac{\varnothing_g}{X_{tt}}\right)^2 + \left(3.2 \times 10^{-7} Re_f^{-0.38} Su_{go}^{1.39}\right)^2\right]^{0.5}$$	HFCs, HCFCs, HCs, CO_2 $D = 0.242$–6.22 mm $G = 53$–1403 kg/m² s $X = 0$–1
31. Dorao and Fernandino [65]	$$h = \left(h_1^9 + h_{11}^9\right)^{1/9}$$ $$h_1 = 0.023\left(Re_v + Re_l\right)^{0.8}\left(Pr_1(1-x) + Pr_v x\right)^{0.3}\frac{\lambda_1}{D}$$ $$h_{11} = 41.5 D^{0.6}\left(Re_v + Re_l\right)^{0.4}\left(Pr_1(1-x) + Pr_v x\right)^{0.3}\frac{\lambda_1}{D}$$	R22, R125, R134a, and R410A $D = 0.067$–20 mm
32. Breber et al. [66]	$$h_l = 0.022 Re_l^{0.9} Pr_l^{0.5}\left(\varnothing_l^2\right)^{0.5}$$ $$\varnothing_l^2 = 1 + \frac{20}{X_{tt}} + \frac{1}{X_{tt}^2}$$	

(continued)

TABLE 3.1 (Continued)
Condensation Heat Transfer Correlations

Author(s)	Correlation	Remarks
33. Vu et al. [67]	$h = 0.023Re_{lo}^{0.8}Pr_l^{0.4}\dfrac{k_l}{d_i}\left[(1-x)^{0.8} + \dfrac{3.8}{Pr^{0.52}}\left(\dfrac{x}{1-x}\right)^{0.201}\right]$ Where $Re_{lo} = \dfrac{GD}{\mu_l}$; $Pr_l = \dfrac{\mu_l c_{pl}}{k_l}$; $Pr = \dfrac{P_{sat}}{P_{critical}}$	$D = 3$ mm $G = 200$–320 kg/m^2 s $X = 0.1$–0.9
34. Ribatski et al. [68]	$HTC = Nu\dfrac{k_l}{d}$ $Nu = J_h Pr_l^{1/3}$ $J_h = \begin{cases} 0.0053Re_{eq} & Re_{eq} \geq 25000 \\ 0.79Re_{eq} & Re_{eq} < 25000 \end{cases}$ $Re_{eq} = \left(\left(1-x\right)+x\left(\dfrac{\rho_l}{\rho_v}\right)^{0.5}\right)G\dfrac{d}{\mu_l}$	$G = 50$–250 kg/m^2 s R134a, R600a, R290, and R1270
35. Tang et al. [69]	$h = 0.023Re_l^{0.8}Pr_l^{0.4}\left[1+4.863\left(-ln\left(p_r\right)\dfrac{x}{1-x}\right)^{0.836}\right]$	
36. Park et al. [70]	$Nu = 0.0629.Re_{Eq}^{0.58346}Pr_l^{1.41522}\left(\dfrac{P}{P_{cr}}\right)^{-0.0615} R_x^{0.5521}\left(Fr.B_o\right)^{-0.013143}$ $R_x = \left(\dfrac{1}{cos\beta}\right)\left[\dfrac{2eN\left(1-sin\dfrac{\alpha}{2}\right)}{\pi d_f cos\dfrac{\alpha}{2}}+1\right]$	
37. Konsetov [71]	$h = 0.024Re_l^{0.8}Pr_l^{0.4}\left(\rho_l/\rho_v\right)^{0.4}\left(v_v/v_l\right)^{0.12}/\left(1-x\right)^{1.5}$	
38. Kutateladze [72]	$h = 0.04Re_f^{0.8}Pr_l^{0.4}\left(\rho_l/\rho_v\right)^{0.5}\left(\mu_v/\mu_l\right)^{0.2}$	
39. Song et al. [73]	$h_A = \dfrac{\lambda_1}{D}\dfrac{Pr_1 Re_1^{0.9}}{T^+}f\left(X_{tt}\right)$ $T^+ = 4.625Re_1^{0.149}Pr_1^{1.1945}Bd^{-0.1}$ $f\left(X_{tt}\right) = 0.15\left(\dfrac{1}{X_{tt}} + \dfrac{2.85}{X_{tt}^{0.476}}\right)$	
40. Chen et al. [74]	$h = 0.036Pr_l^{0.65}A^{0.5}\left(Re_{lo}-Re_l\right)^{0.7}Re_l^{0.2}$	
41. Jung et al. [75]	$h = 0.5152Re_1^{0.8}Pr_1^{0.4}\left[1+\dfrac{2}{X_{tt}}\right]^{0.81}Bo^{0.33}$	R12, R32, R123, R125, R134a, and R22 $D = 8.81$ mm $G = 100$–300 kg/m^2 s
42. Rahman et al. [76]	$h = Pr_1^{0.45}Re_1^{0.11}Pr_r^{0.1}\left(\dfrac{x}{1-x}\right)^{0.09}\dfrac{\lambda_1}{D}\dfrac{\phi_v}{X_{tt}}$	R134a $D = 4$ mm $G = 50$–200 kg/m^2 s

TABLE 3.1 (Continued)
Condensation Heat Transfer Correlations

Author(s)	Correlation	Remarks
43. Rifert et al. [77]	$\phi_v^2 = 1 + CX_{tt}^n + X_{tt}^2$ Where $C = 21\left[1 - e^{\left(1 - 0.2Bo^{0.5}\right)}\right]\left(1 - 0.9e^{-0.2Fr_l^{1.5}}\right)$ $Fr_l = \dfrac{Gx}{\left[gdG\left(\rho_l - \rho_v\right)\right]^{0.5}}$	
44. Shah [1]	$h_l = h_{LT}\left(\dfrac{\mu_f}{14\mu_g}\right)^n\left[\left(1-x\right)^{0.8} + \dfrac{3,8x^{0.76}\left(1-x\right)^{0.04}}{P_r^{0.38}}\right]$	$D = 2\text{--}49$ mm $G = 4\text{--}820$ kg/m² s
45. Cavallini et al. [78]	$\alpha = \alpha_{an} = \rho_L C_{pL}\left(\dfrac{\tau}{\rho_L}\right)^{0.5} / T^+$ $T^+ = 5\left\{Pr_L + ln\left(1 + 5Pr_L\right) + 0.495ln\left(\delta^+/30\right)\right\}$ $\delta^+ = 0.0504Re_l^{7/8}$	R22, R32, and R410A $D = 7\text{--}9.5$ mm
46. Wen et al. [79]	$\dfrac{hD_h}{K_f} = 0.0055Re_f^{0.7}Pr_f^{1.37}\dfrac{\varnothing_g}{X_{tt}}$	R22, R32, R125, and R134a
47. Rifert and Gorin [80]	$h_a = \left(1 + \dfrac{1.2}{X_{tt}^{0.935}}\right)0.023Re_{ls}^{0.8}Pr_l^{0.4}\dfrac{k_1}{D}$	
48. Naulboonrueng et al. [81]	$h = 0.003Re_{eq}^{0.997}Pr_l^{0.932}$ $Re_{eq} = Re_l + Re_v\left(\mu_v/\mu_l\right)\left(\rho_l/\rho_v\right)^{0.5}$	
49. Kaushik and Azer [82]	$h = 2.078Re_{eq}^{0.507}Pr_l^{0.33}\left(\Delta xd/l\right)^{0.198}P_r^{-0.14}$ $Re_{eq} = G_{eq}d / \mu_l$ $x_{av} = 0.5\left(x_{in} + x_{out}\right)$	
50. Agra and Teke [83]	$h = a^* Re_{eq}^b Pr^{1/3}$ For $Re_{eq} < 18000$ $a = 0.0289$ $b = 0.8525$ $Re_{eq} = \dfrac{G_e d}{\mu_l}$ $G_e = G\left[\left(1-x\right) + x\left(\dfrac{\rho_l}{\rho_v}\right)^{1/2}\right]$	R600a $D = 4$ mm
51. Shah [13]	$h_l = h_{LS}\left(1 + \dfrac{3.8}{Z^{0.95}}\right)\left(\dfrac{\mu_l}{14\mu_g}\right)^{\left(0.0058 + 0.557p_r\right)}$ $h_{LS} = 0.023Re_{LS}^{0.8}Pr_l^{0.4}\,k_l/D$	$D = 2\text{--}49$ mm $G = 13\text{--}820$ kg/m² s 25 different fluids

(continued)

TABLE 3.1 (Continued)
Condensation Heat Transfer Correlations

Author(s)	Correlation	Remarks
52. Shah [84]		$G = 4$–820 kg/m² s $D = 2$–49 mm R11, R12, R134a, and R404A
51. Shah [12]	$h_l = 0.023Re_{lo}^{0.8}Pr_l^{0.4}\left[1+1.128x^{0.8170}\left(\dfrac{\rho_l}{\rho_g}\right)^{0.3685}\left(\dfrac{\mu_l}{\mu_g}\right)^{0.2363}\left(1-\dfrac{\mu_g}{\mu_l}\right)^{2.144}Pr_l^{-0.100}\right]\left(\dfrac{k_l}{D}\right)$	
52. Akers et al. [85]	$h = CPr_l^{1/3}\left\{G\left[(1-x)+x\left(\dfrac{\rho_l}{\rho_v}\right)^{0.5}\right]\dfrac{D}{\mu_1}\right\}^n$	
53. Crosser [86]		
54. Moradkhani et al. [43]	$\alpha = 23.22x^{0.397}P_{red}^{0.554}Pr_{tp}^{-0.185}R^{2.928}We_{vt}^{0.894}Fr_l^{-0621}Bo^{-0.148}$	R744
55. Boyko and Kruzhilin [87]	$\alpha = cRe_{lo}^{0.8}Pr_l^{0.43}\left[1+x\left(\dfrac{\rho_l}{\rho_v}-1\right)\right]^{0.5}$	For copper tubes c = 0.032
56. Soliman [88]	$\alpha = 0.00345Re_m^{0.9}\left(\dfrac{\mu_v r}{\lambda_v\Delta T}\right)^{1/3}$ $Re_m = Gd\big/\mu_m$ $1\big/\mu_m = x\big/\mu_v + (1-x)\big/\mu_l$	
57. Iqbal and Bansal [89]	$\alpha = 0.023Re_l^{0.8}Pr_l^{0.4}\left[(1-x)^{6.43}+0.464x^{0.23}(1-x)^{-01.03}\big/Pr_r^{2.33}\right]$	

mass flux, and saturated temperature. They've all been chosen because they've been proposed for usage in macro-tubes.

Haraguchi et al. [35] developed a heat transfer coefficient correlation that included both forced and gravity-controlled convection condensation terms. To fit their experimental findings for smaller tubes and other fluids, Koyama et al. [36], Huang et al. [37], and Park et al. [38] modified Haraguchi et al. [35] correlation. Cavallini et al. [10] divided the flow condensation into two regions: ΔT-dependent and ΔT-independent, and offered correlations for each. To fit the experimental results with a mass flux of less than 100 kg/m²s, a modified Cavallini et al. [10] correlation was presented. Wang et al. [39] also conducted R134a flow condensation studies inside a 1.34 mm multi-channel and proposed a condensation correlation for two flow patterns. The vapour superficial velocity and the Weber number were used to partition the flow condensation into three zones, and correlations for each region were suggested by Shah [11]. Recently, a correlation was enhanced by Fang et al. [40], who assembled a large compound FCHT database with 5,607 data points and 30 fluids and offered an approach for establishing a novel correlation of FCHT coefficients. The database includes values for the following parameters: reduced pressure $P_r = 0.04$–0.95; liquid Prandtl number $Pr_l = 1.7$–8.5; liquid Reynolds number $Re_l = 11.6^{-5.3} \times 10^4$; gas Reynolds number $Re_g = 75.1^{-9.1} \times 10^5$. With the data and method, a new generic correlation was created with much higher accuracy; it can be used with simple channels of varying sizes and throughout a wide range of operational parameters. Its MAD for predicting the database is 14.1%, which is better than the MAD of 20.2% for the best ΔT-independent correlation currently available. The new and existing correlations (a total of 38) were evaluated for their application to different fluids. When compared to the most accurate existing correlation, the new one performs better for 8 of the 14 fluids with more than 50 data points in the whole database.

The majority of recent research has relied on "conventional intelligence procedures" or "machine learning approaches [41]," such as gaussian process regression (GPR), radial basis function (RBF), hybrid radial basis function (HRBF), and the least square fitting method (LSFM). In contrast to more traditional ways of fitting, the outcomes predicted by intelligence-based approaches tend to hold up better in a variety of engineering contexts, as evidenced by a study of the relevant literature. The most widely used form of artificial intelligence, ANN's [42] attempt to simulate the way the human brain operates. Examples of effective correlations created using machine learning techniques include those of Moradkhani et al. [43], Nie et al. [18], Li and Hrnjak [44]. Moradkhani et al. [43] performed research using machine learning approaches with the intention of enhancing a correlation for the design, scaling up, and optimization of two-phase flow heat exchangers. This study collected 11,128 experimental data points from 80 different sources, encompassing 37 condensing fluids under various operating conditions, with the aim of establishing a universal and precise correlation for predicting the HTC in both mini/micro and conventional channel heat exchangers. The new correlations were developed utilizing traditional intelligent methods such as GPR, RBF, HRBF, and the LSFM. Nie et al. [18] research is the most recent one to utilize machine learning approaches. Using a collected data set, Nie et al. [18] reported research using ML techniques to forecast flow condensing HTCs inside horizontal tubes. There were a total of 6,064 experimental data points on 28 pure fluids included in the database, which covered numerous operational parameters. Also, five ML models [41] were trained and evaluated using the database, with the K-nearest neighbors (KNN) [45], ANN [42], CNN [46], RF, and XGBoost algorithms providing the foundation. Several critical parameters governing condensation heat transfer were incorporated into a new universal correlation. With an MRD of 1.77% and an MARD of 19.21%, respectively, this correlation provided accurate predictions for each data point in the combined database.

3.6.2 REVIEW OF EXPERIMENTAL STUDIES

A significant number of experimental investigations have been carried out lately, focusing on flow condensation heat transfer within horizontal smooth tubes. In the examination of condensation heat transfer properties, parameters such as mass flux, saturation temperature, vapor quality, tube diameter, refrigerant type, and heat flux are typically considered crucial. Ewim et al. [17, 90–92] recently conducted a study in which he used a smooth copper tube with an inner diameter of 8.38 mm, the refrigerant R134a, a saturation temperature of 40°C, and mass flux values of 50, 75, 100, and 200 kg/m² s. It was found that the observed flow patterns in horizontal tubes were primarily stratified and characterized by wavy formations. Moreover, the study revealed that heat transfer coefficients depended on the temperature difference between the wall susceptible to condensation and the temperature of the condensing refrigerant. It was also found that the heat transfer coefficient diminished as the temperature difference grew. Jung et al. [75] determined the flow condensation heat transfer coefficients of R12, R22, R32, R123, R125, R134a, and R142b in a 9.52 mm inner diameter horizontal tube. The experimental results indicate that the heat transfer coefficient increases with increasing quality and mass flux. At the same mass flux, R142b, R32, and R134a have higher heat transfer coefficients than R22, while R122, R134a, and R123 have similar heat transfer coefficients to R22. Song et al. [73] conducted an experimental investigation of the flow condensation heat transfer properties of R14 in a horizontal smooth tube with an inner diameter of 4 mm and mass fluxes ranging from 200 to 650 kg/m²s. The authors noted that mass flux, vapour quality, and heat flux all had an effect on condensation heat transfer. In terms of mass flux, the heat transfer coefficient increases as the mass flux increases. Increased vapour quality resulted in increased vapour velocity and a thinner liquid film. Using an experimental setup consisting of a 4 mm diameter smooth horizontal tube, a saturation temperature of 35°C, and mass flux ranging from 100 to 400 kg/m²s, Hirose et al. [93] evaluated the condensation heat transfer coefficient characteristics for R32, R152a, and R410A refrigerants. As wetness increased, the researchers found that the heat transfer coefficient decreased as the liquid film's heat transfer resistance increased. As the refrigerant vapour velocity dropped, the amount of forced convection condensation reduced as well. Li et al. [94] showed that the heat transfer coefficient's dependence on vapour quality varied with mass flux. The heat transfer coefficient did not alter with vapour quality at lower mass fluxes. At greater mass fluxes, however, the heat transfer coefficient rose linearly with increasing vapour quality. For low vapour quality (less than 0.4), the heat transfer coefficient is almost always the same for all mass fluxes.

Experimental data from the literature covering a wide range of fluids and tube diameters for macro-tubes ranging from 3 mm to 8 mm was acquired for this study, mass flux from 50 to 1500 kg/m² s, qualities from 0 to 1. It is worth noting that in Table 3.2, experimental data using refrigerants containing oil and zeotropic mixes were not included. Because oil has a substantial impact on heat

TABLE 3.2
Experimental Data for Macro-Tubes

Author(s)	Fluid	Diameter [mm]	Shape	G [kg/m⁻² s]	T_{sat} [°C]	Points
1. Ghim et al. [104]	R245fa	7.75	Circular	150–700	50	21
2. Kondou et al. [105]	CO_2	6.10	Circular	100–200	20	47
3. Agarwal et al. [106]	R1234ze	6.10	Circular	100–300	50	37
4. Xiao et al. [107]	R134a	6.10	Circular	50–200	50	45
5. Son [108]	CO_2	4.95	Circular	400–800	20	18
6. Kim et al. [109]	CO_2	3.48	Circular	200–800	−15	16
7. Xiao et al. [110]	C1	4.00	Circular	99–255	−107 to 91	231
8. Ferreira et al. [111]	R404A	8.00	Circular	200–600	40	57
9. Cavallini et al. [101]	R134a, R125, and R32	8.00	Circular	65–750	40	68
10. Jang and Hrnjak [112]	CO_2	6.10	Circular	197–406	20	85
11. Mitra [113]	R410A	6.22	Circular	200–800	40	144
12. Agra and Teke [114]	R600a	4.00	Circular	57–118	20	50
13. Huang et al. [37]	R410A	4.18	Circular	200–600	40	35
14. Jiang et al. [115]	R410A	3.05	Circular	400–800		30
15. Kim et al. [4]	R410A	5.0	Circular	100–400		10
16. Zhuang et al. (2017)	Ethane	4.0	Circular	100–257		33
17. Zhuang et al. [116]	Methane	4.0	Circular	99–255		78
18. Li et al. [94]	CO_2	4.73	Circular	100–500		63
19. Meyer and Ewim et al. [47]	R134a	8.38	Circular	50–200		39
20. Ghorbani et al. [117]	Isobutane	8.7	Circular	110–372		29

TABLE 3.2 (Continued)
Experimental Data for Macro-Tubes

Author(s)	Fluid	Diameter [mm]	Shape	G [kg/m^{-2} s]	T_{sat} [°C]	Points
21. Keinath and Garimella [118]	R404A	3.05	Circular	200–800		30
22. Rifert and Sereda [119]	Propylene	8.8	Circular	100–300		28
23. Tang [120]	R134a	8.8	Circular	260–850	35–45	24
24. Park et al. [121]	R22	8.8	Circular	100–300	40	126
25. Laohalertdecha and Wongwises [122]	R134a	8.7	Circular	200–700	40–50	13
26. Cavallini et al. [101]	R134a/R125/R32/R410A	8.0	Circular	100–750	30–50	230
27. Sapali and Patil [123]	R134a	8.56	Circular	90–800	35–60	57
28. Cavallini et al. [124]	R22, R134a, and R125a	8.00	Circular	100–750	30–50	62
29. Pham Quang Vu et al. [125]	R410	6.61	Circular	200–320	48	37
30. Kim and Shin [126]	R22, R410A	8.70	Circular	40, 60, 80	45	22
31. Jung et al. [127]	R22, R134a, R407C, and R410A	8.82	Circular	100–300	40	13
32. Song et al. [73]	R14	4	Circular	200–650	−8.4 to (−53.17)	119
33. Long et al. [128]	R404A, R290, and R1270	4	Circular	75–800	30–40	200
34. Long et al. [129]	R32 and R410A	4	Circular	75–800	30–40	160
35. Aroonrat et al. [130]	R134a	4	Circular	300–400		14
36. Iqba and Bansal [89]	R744	6.52	Circular	50–200		87
37. Byun et al. [131]	R410A	4.32	Circular	379–758	45–55	20
38. Hossain et al. [132]	R32 and R410A	4.35	Circular	191–400		119
39. Hirose et al. [93]	R410A and R1234ze	4.35	Circular	100–400		118
40. Tang et al. [133]	R22, R134a, and R410A	8.81	Circular	250–810	35–45	31
41. Seol Sung-Hoon et al. [134]	R1234yf	3.7	Circular	100–400	40–50	72
42. Long et al. [135]	R134a, R152a, and R1234yf	4	Circular	95.4–605.3	30–40	62
43. Wang et al. [102]	R134a and R32	4	Circular	100–400	40–50	83
44. Son and Lee [136]	R22, R134a, and R410A	3.36 and 5.35	Circular	200–400	40	19
45. Lee and Son [137]	R290, R600a, and R22	6.54 and 7.73	Circular	35.5–210.4	40	22
46. Ewim et al. [17]	R134a	8.38	Circular	50–200	40	56
47. Jassim et al. [55]	R134a	8.92	Circular	100–300	24.9–25.1	32
48. Miyara et al. [138]	R410A	6.40	Circular	100–400	30–40	58
49. Shao and Granryd [139]	R12, R22, R134a, and R502	6.0	Circular	150–300	30	26
50. Jang et al. [112]	CO$_2$	6.1	Circular	200–400	−15 to (−25)	14
51. Lambrecht et al. [140]	R22 and R134a	8.1	Circular	300–800	40	6
52. Infante-Ferreira et al. [111]	R404A	8.0	Circular	200–600	40	16

(*continued*)

TABLE 3.2 (Continued)
Experimental Data for Macro-Tubes

Author(s)	Fluid	Diameter [mm]	Shape	G [kg/m^{-2} s]	T_{sat} [°C]	Points
53. Afroz and Miyara [141]	CO_2	4.35	Circular	200–500		29
54. Kaew-on et al. [142]	R134a	3.51	Circular	75–75	30	60
55. Cavallini et al. [124]	R410A	8.0	Circular	100–750	30 and 50	61
56. Zhang et al. [143]	R410A and R134a	3.78	Circular	300–600	46.85	36
57. Wu et al. [144]	R410A	3.78	Circular	99–603		45
58. Ewim et al. [92]	R134a	8.37	Circular	50–100	40	
59. Li et al. [145]	R152a	4.00	Circular	173–462	30 and 40	56
60. Mendes et al. [146]	R134a	4.80	Circular	200–300	30	71
61. Moradkhani et al. [147]	R22 and R134	7.04	Circular	300	35	72
62. Longo, Mancin, Righetti et al. [148]	R32 and R410A	3.6	Circular	200	29.9–40.1	62
63. Lee et al. [58]	R404A	5.6	Circular	80–400	45	50
64. Fu et al. [149]	R410A	4.0	Circular	307.3–998.5	40	28
65. Zilly et al. [150]	CO_2	6.1	Circular	200–400	−15 to (−25)	
66. Aroonrat et al. [151]	R134a	8.1	Circular	300–500		27
67. Macdonald [152]	Propane	7.5	Circular	150		81
68. Lee and Son [153]	R22	5.8	Circular	36–152		11
69. Nan and Infante-Ferreira [154]	Propane	8.8	Circular	150–250		6
70. Eckels and Tesene [155]	R507A	8.0	Circular	251–599		23
Total points						7682

transfer, oil-based refrigerants are not included. Because mass transfer affects zeotropic mixtures, data with zeotropic mixtures is not taken into consideration for this study.

There are several experimental studies fund in the literature, like the study of Ewim et al. [91, 92, 95] how noticed the lacking for the case of condensation in tubes at low mass fluxes. Experiments were carried out in a smooth horizontal tube with an internal diameter of 8.38 mm at five distinct mass fluxes of 50, 75, 100, 150, and 200 kg/m^2 s during the convective condensation of R134a. When mass fluxes were less than or equal to 100 kg/m^2 s, however, the dependence of heat transfer coefficients on temperature difference increased for all vapour quality. For low-mass fluxes, Ewim focuses on condensation inside horizontal, inclined, and microfin tubes [17, 90–92, 96–100]. Li et al. [94] conducted an experimental research of CO_2 flow condensation heat transfer for mass fluxes ranging from 100 to 500 kg/m^2 s within a 4.73 mm inside diameter, smooth horizontal copper tube at saturation temperatures ranging from 10 to 0°C and under a variety of vapour quality circumstances. The experimental data came from an open-loop test apparatus that released high-pressure CO_2 liquid from bottles into the atmosphere. When the test mass flux was higher than or equal to 300 kg/m^{2}s, and the vapour qualities were greater than 0.4, the rate of heat transfer rose with increasing mass flux and vapour quality, and decreased with lowering saturation temperature. Pure HFC refrigerants R134a, R236ea, R32, along with R410a, were condensed inside a smooth tube, and the heat transfer coefficients and pressure drops were evaluated experimentally by Cavallini et al. [101]. inner diameter: 8 mm. Through data analysis, the impacts of varying vapour quality, mass flux, saturation temperature, and saturation temperature differential vs tube wall on the heat transfer coefficient were studied. Wang et al. [102] observed R1234yf condensation in a horizontal tube inner diameter: 4 mm at 100–400 kg/m² s mass flux and saturation temperatures of 40–45 and 50°C. With three different mass fluxes 150, 300, and 450 kg/m² s, two different tube diameters 7.75 mm and 14.45 mm, and lowered pressures 0.25–0.95, Macdonald and Garimella [103] conducted an experimental evaluation of propane condensing inside horizontal tubes. Surprisingly, the pattern predicted by most correlations—a decrease in heat transfer coefficient with increasing tube diameter—was indeed observed for the specific tube diameter range investigated. Table 3.2 below presents a list of experimental studies that were done from the year 2000 to date.

3.7 CONCLUSION

The book chapter presented a thorough analysis of a variety of empirical and theoretical heat transfer correlations related to in-tube condensation occurring in horizontal smooth macro tubes. The chapter emphasized the importance of accurate and reliable correlations for predicting heat transfer rates in condensers, which are critical components in many industrial applications. The review identified key factors that influence heat transfer in in-tube condensation, such as mass flux, vapor quality, and tube diameter, and assessed the performance of different correlations in capturing these effects. The chapter highlighted the limitations and uncertainties associated with the existing correlations and provided insights into the strengths and weaknesses of various approaches. Overall, the chapter served as a valuable resource for researchers and engineers working in the field of heat transfer and condensation. By offering a comprehensive review of existing correlations and their limitations, the chapter laid the groundwork for the development of improved heat transfer models and more accurate predictions of heat transfer rates in industrial applications.

GLOSSARY

ANN	Artificial Neural Network
CNN	Convolutional Neural Network
FCHT	Forced Convection Heat Transfer
HFC	Hydrofluorocarbon
HTC	Heat Transfer Coefficient
MAD	Mean Absolute Deviation
MARD	Mean Absolute Relative Deviation
MRD	Mean Relative Deviation

REFERENCES

[1] M. M. Shah, "An improved and extended general correlation for heat transfer during condensation in plain tubes," *Hvac&R Research*, vol. 15, no. 5, pp. 889–913, 2009.

[2] Y.-Y. Yan and T.-F. Lin, "Condensation heat transfer and pressure drop of refrigerant R-134a in a small pipe," *International Journal of Heat and Mass Transfer*, vol. 42, no. 4, pp. 697–708, 1999.

[3] W.-W. W. Wang, T. D. Radcliff, and R. N. Christensen, "A condensation heat transfer correlation for millimeter-scale tubing with flow regime transition," *Experimental Thermal and Fluid Science*, vol. 26, no. 5, pp. 473–485, 2002.

[4] S.-M. Kim and I. Mudawar, "Universal approach to predicting heat transfer coefficient for condensing mini/micro-channel flow," *International Journal of Heat and Mass Transfer*, vol. 56, no. 1–2, pp. 238–250, 2013.

[5] S.-M. Kim and I. Mudawar, "Universal approach to predicting two-phase frictional pressure drop for adiabatic and condensing mini/micro-channel flows," *International Journal of Heat and Mass Transfer*, vol. 55, no. 11–12, pp. 3246–3261, 2012.

[6] I. Park, S.-M. Kim, and I. Mudawar, "Experimental measurement and modeling of downflow condensation in a circular tube," *International Journal of Heat and Mass Transfer*, vol. 57, no. 2, pp. 567–581, 2013.

[7] W. Qu and I. Mudawar, "Experimental and numerical study of pressure drop and heat transfer in a single-phase micro-channel heat sink," *International Journal of Heat and Mass Transfer*, vol. 45, no. 12, pp. 2549–2565, 2002.

[8] M. Dobson and J. Chato, "Condensation in smooth horizontal tubes," vol. 120, no. 1, pp. 193–213, 1998.

[9] J. R. Thome, J. El Hajal, and A. Cavallini, "Condensation in horizontal tubes, part 2: new heat transfer model based on flow regimes," *International Journal of Heat and Mass Transfer*, vol. 46, no. 18, pp. 3365–3387, 2003.

[10] A. Cavallini et al., "Condensation in horizontal smooth tubes: a new heat transfer model for heat exchanger design," *Heat Transfer Engineering*, vol. 27, no. 8, pp. 31–38, 2006.

[11] M. M. Shah, "A correlation for heat transfer during condensation in horizontal mini/micro channels," *International Journal of Refrigeration*, vol. 64, pp. 187–202, 2016.

[12] M. M. Shah, "Improved correlation for heat transfer during condensation in conventional and mini/micro channels," *International Journal of Refrigeration*, vol. 98, pp. 222–237, 2019.

[13] M. Shah, "A new flow pattern based general correlation for heat transfer during condensation in horizontal tubes," in *International Heat Transfer Conference Digital Library*. Begel House Inc., 2014.

[14] P. Cheng and H. Wu, "Mesoscale and microscale phase-change heat transfer," *Advances in Heat Transfer*, vol. 39, pp. 461–563, 2006.

[15] S. G. Kandlikar, "Fundamental issues related to flow boiling in minichannels and microchannels," *Experimental Thermal and Fluid Science*, vol. 26, no. 2–4, pp. 389–407, 2002.

[16] C. Aprea, A. Greco, and G. P. Vanoli, "Condensation heat transfer coefficients for R22 and R407C in gravity driven flow regime within a smooth horizontal tube," *International Journal of Refrigeration*, vol. 26, no. 4, pp. 393–401, 2003.

[17] D. R. E. Ewim, "Condensation inside Horizontal and Inclined Smooth Tubes at Low Mass Fluxes," University of Pretoria, 2019.

[18] F. Nie, H. Wang, Y. Zhao, Q. Song, S. Yan, and M. Gong, "A universal correlation for flow condensation heat transfer in horizontal tubes based on machine learning," *International Journal of Thermal Sciences*, vol. 184, p. 107994, 2023.

[19] O. Baker, "Design of Pipelines for Simultaneous Flow of Oil and Gas," *Oil & Gas Journal*, vol. 53, pp.185–195, 1954.

[20] C.-C. Wang, C.-S. Chiang, and D.-C. Lu, "Visual observation of two-phase flow pattern of R-22, R-134a, and R-407C in a 6.5-mm smooth tube," *Experimental Thermal and Fluid Science*, vol. 15, no. 4, pp. 395–405, 1997.

[21] Y. Taitel and A. E. Dukler, "A model for predicting flow regime transitions in horizontal and near horizontal gas-liquid flow," *AIChE Journal*, vol. 22, no. 1, pp. 47–55, 1976, January. DOI: 10.1002/aic.690220105.

[22] T. N. Tandon, H. K. Varma, and C. P. Gupta, "A new flow regimes map for condensation inside horizontal tubes," *Journal of Heat and Mass Transfer*, vol. 104, no. 4, p. 763, 1982, DOI: 10.1115/1.3245197.

[23] C. Y. Yang and C.-C. Shieh, "Flow pattern of air–water and two-phase R-134a in small circular tubes," *International Journal of Multiphase Flow*, vol. 27, 2001.

[24] L. Wojtan, T. Ursenbacher, and J. R. Thome, "Investigation of flow boiling in horizontal tubes: Part I—A new diabatic two-phase flow pattern map," *International Journal of Heat and Mass Transfer*, vol. 48, no. 14, pp. 2955–2969, July 2005. DOI: 10.1016/j.ijheatmasstransfer.2004.12.012.

[25] M. Padilla, R. Revellin, P. Haberschill, A. Bensafi, and J. Bonjour, "Flow regimes and two-phase pressure gradient in horizontal straight tubes: Experimental results for HFO-1234yf, R-134a and R-410A," *Experimental Thermal and Fluid Science*, vol. 35, no. 6, pp. 1113–1126, 2011.

[26] M. Padilla, R. Revellin, J. Wallet, and J. Bonjour, "Flow regime visualization and pressure drops of HFO-1234yf, R-134a and R-410A during downward two-phase flow in vertical return bends," *International Journal of Heat and Fluid Flow*, vol. 40, pp. 116–134, 2013.

[27] L. Wang, C. Dang, and E. Hihara, "Experimental and theoretical study on condensation heat transfer of nonazeotropic refrigerant mixtures R1234yf/R32 inside a horizontal smooth tube," International Refrigeration and Air Conditioning Conference. Paper 1308. http://docs.lib.purdue.edu/iracc/1308, 2012.

[28] R. Suliman, L. Liebenberg, and J. P. Meyer, "Improved flow pattern map for accurate prediction of the heat transfer coefficients during condensation of R-134a in smooth horizontal tubes and within the low-mass flux range," *International Journal of Heat and Mass Transfer*, vol. 52, no. 25–26, pp. 5701–5711, 2009.

[29] J. El Hajal, J. R. Thome, and A. Cavallini, "Condensation in horizontal tubes, part 1: two-phase flow pattern map," *International Journal of Heat and Mass Transfer*, vol. 46, no. 18, pp. 3349–3363, 2003.

[30] H. A. D. W.W. Akers, O.K. Crosser, "Condensation heat transfer within horizontal tubes," *Chemical Engineering Progress Symposium Series*, 55, no. 1959, 171–176, 1959.

[31] C. Y. Park and P. Hrnjak, "CO_2 flow condensation heat transfer and pressure drop in multi-port microchannels at low temperatures," *International Journal of Refrigeration*, vol. 32, no. 6, pp. 1129–1139, 2009.

[32] M. M. Shah, "Comprehensive correlations for heat transfer during condensation in conventional and mini/micro channels in all orientations," *International Journal of Refrigeration*, vol. 67, pp. 22–41, 2016.

[33] C. A. Dorao and M. Fernandino, "Dominant dimensionless groups controlling heat transfer coefficient during flow condensation inside pipes," *International Journal of Heat and Mass Transfer*, vol. 112, pp. 465–479, 2017.

[34] F. Dittus and L. Boelter, "Heat transfer in automobile radiators of the tubular type," *International Communications in Heat and Mass Transfer*, vol. 12, no. 1, pp. 3–22, 1985.

[35] H. Haraguchi, S. Koyama, and T. Fujii, "Condensation of refrigerants HCFC22, HFC134a and HCFC123 in a horizontal smooth tube. 2nd Report. Proposals of empirical expressions for local heat transfer coefficient. Reibai HCFC22, HFC134a, HCFC123 no suihei heikatsu kannai gyoshuku. 2. Kyokusho netsu dentatsu keisu ni kansuru jikkenshiki no teian," *Nippon Kikai Gakkai Ronbunshu. B Hen (Transactions of the Japan Society of Mechanical Engineers. Part B);(Japan)*, vol. 574, no. 60, 1994.

[36] S. Koyama, K. Kuwahara, K. Nakashita, and K. Yamamoto, "An experimental study on condensation of refrigerant R134a in a multi-port extruded tube," *International Journal of Refrigeration*, vol. 26, no. 4, pp. 425–432, 2003.

[37] X. Huang et al., "Influence of oil on flow condensation heat transfer of R410A inside 4.18 mm and 1.6 mm inner diameter horizontal smooth tubes," *International Journal of Refrigeration*, vol. 33, no. 1, pp. 158–169, 2010.

[38] J. E. Park, F. Vakili-Farahani, L. Consolini, and J. R. Thome, "Experimental study on condensation heat transfer in vertical minichannels for new refrigerant R1234ze (E) versus R134a and R236fa," *Experimental Thermal and Fluid Science*, vol. 35, no. 3, pp. 442–454, 2011.

[39] J. Wen, X. Gu, S. Wang, Y. Li, and J. Tu, "The comparison of condensation heat transfer and frictional pressure drop of R1234ze (E), propane and R134a in a horizontal mini-channel," *International Journal of Refrigeration*, vol. 92, pp. 208–224, 2018.

[40] X. Fang, X. Li, and Z. Luo, "Predictive method for flow condensation heat transfer in plain channels," *Physics of Fluids*, vol. 34, no. 11, p. 113321, 2022.

[41] B. Mahesh, "Machine learning algorithms-a review," *International Journal of Science and Research (IJSR).[Internet]*, vol. 9, pp. 381–386, 2020.

[42] A. Krogh, "What are artificial neural networks?," *Nature Biotechnology*, vol. 26, no. 2, pp. 195–197, 2008.

[43] M. Moradkhani, S. Hosseini, M. Mansouri, H. O. Zad, M. Karami, and G. Ahmadi, "New general models for condensation heat transfer coefficient of carbon dioxide in smooth tubes by intelligent and least square fitting approaches," *Journal of Cleaner Production*, vol. 330, p. 129762, 2022.

[44] H. Li and P. Hrnjak, "A mechanistic model in annular flow in microchannel tube for predicting heat transfer coefficient and pressure gradient," *International Journal of Heat and Mass Transfer*, vol. 203, p. 123805, 2023.

[45] S. Sun and R. Huang, "An adaptive k-nearest neighbor algorithm," in *2010 Seventh International Conference on Fuzzy Systems and Knowledge Discovery*, vol. 1: IEEE, pp. 91–94, 2010.

[46] S. Albawi, T. A. Mohammed, and S. Al-Zawi, "Understanding of a convolutional neural network," in *2017 International Conference on Engineering and Technology (ICET)*, IEEE, pp. 1–6, 2017.

[47] J. P. Meyer and D. Ewim, "Heat transfer coefficients during the condensation of low mass fluxes in smooth horizontal tubes," *International Journal of Multiphase Flow*, vol. 99, pp. 485–499, 2018.

[48] A. Cavallini and R. Zecchin, "A dimensionless correlation for heat transfer in forced convection condensation," in *International Heat Transfer Conference Digital Library*, Begel House Inc., 1974.

[49] H. A. D. W.W. Akers and O.K. Crosser, " Condensation heat transfer within horizontal tubes," *Chemical Engineering Progress Symposium Series*, vol. 55, pp. 171–176, 1959.

[50] H. F. R. W.W. Akers, "Condensation inside a horizontal tube," *Chemical Engineering Progress Symposium*, vol. 56, 145–149, 1960.

[51] M. M. Shah, "A general correlation for heat transfer during film condensation inside pipes," *International Journal of Heat and Mass Transfer*, vol. 22, no. 4, pp. 547–556, 1979.

[52] K. Moser, R. Webb, and B. Na, "A new equivalent Reynolds number model for condensation in smooth tubes," vol. 120, no. 2, pp. 410–417, 1998.

[53] T. Bohdal, H. Charun, and M. Sikora, "Comparative investigations of the condensation of R134a and R404A refrigerants in pipe minichannels," *International Journal of Heat and Mass Transfer*, vol. 54, no. 9–10, pp. 1963–1974, 2011.

[54] P. Li and S. Norris, "Heat transfer correlations for CO_2 flowing condensation in a tube at low temperatures," *Applied Thermal Engineering*, vol. 93, pp. 872–883, 2016.

[55] E. Jassim, T. Newell, and J. Chato, "Prediction of two-phase condensation in horizontal tubes using probabilistic flow regime maps," *International Journal of Heat and Mass Transfer*, vol. 51, no. 3–4, pp. 485–496, 2008.

[56] T. Tandon, H. Varma, and C. Gupta, "Heat transfer during forced convection condensation inside horizontal tube," *International Journal of Refrigeration*, vol. 18, no. 3, pp. 210–214, 1995.

[57] C. Qi, X. Chen, W. Wang, J. Miao, and H. Zhang, "Experimental investigation on flow condensation heat transfer and pressure drop of nitrogen in horizontal tubes," *International Journal of Heat and Mass Transfer*, vol. 132, pp. 985–996, 2019.

[58] B.-M. Lee, H.-H. Gook, S.-B. Lee, Y.-W. Lee, D.-H. Park, and N.-H. Kim, "Condensation heat transfer and pressure drop of low GWP R-404A alternative refrigerants (R-448A, R-449A, R-455A, R-454C) in a 5.6 mm inner diameter horizontal smooth tube," *International Journal of Refrigeration*, vol. 125, pp. 87–99, 2021.

[59] H. Fazelnia, S. Azarhazin, B. Sajadi, M. A. A. Behabadi, S. Zakeralhoseini, and M. V. Rafieinejad, "Two-phase R1234yf flow inside horizontal smooth circular tubes: Heat transfer, pressure drop, and flow pattern," *International Journal of Multiphase Flow*, vol. 135, p. 103668, 2021.

[60] D. P. Traviss, A. G. Baron, and W. M. Rohsenow, "Forced-convection condensation inside tubes," Cambridge, MA: MIT Heat Transfer Laboratory, 1971.

[61] H. Deng, M. Rossato, M. Fernandino, and D. Del Col, "A new simplified model for condensation heat transfer of zeotropic mixtures inside horizontal tubes," *Applied Thermal Engineering*, vol. 153, pp. 779–790, 2019.

[62] S.-M. Kim and I. Mudawar, "Flow condensation in parallel micro-channels–Part 2: Heat transfer results and correlation technique," *International Journal of Heat and Mass Transfer*, vol. 55, no. 4, pp. 984–994, 2012.

[63] S. Garimella, U. C. Andresen, B. Mitra, Y. Jiang, and B. M. Fronk, "Heat transfer during near-critical-pressure condensation of refrigerant blends," *Journal of Heat Transfer*, vol. 138, no. 5, p. 051503, 2016. DOI: https://doi.org/10.1115/1.4032294

[64] S.-M. Kim and I. Mudawar, "Universal approach to predicting saturated flow boiling heat transfer in mini/micro-channels – Part II. Two-phase heat transfer coefficient," *International Journal of Heat and Mass Transfer*, vol. 64, pp. 1239–1256, 2013.

[65] C. A. Dorao and M. Fernandino, "Simple and general correlation for heat transfer during flow condensation inside plain pipes," *International Journal of Heat and Mass Transfer*, vol. 122, pp. 290–305, 2018.

[66] G. Breber, J. Palen, and J. Taborek, "Prediction of horizontal tubeside condensation of pure components using flow regime criteria," *AIChE Symposium Series*, vol. 76, no. 198, pp. 342–349, 1980.

[67] P. Q. Vu et al., "An experimental investigation of condensation heat transfer coefficient using R-410a in horizontal circular tubes," *Energy Procedia*, vol. 75, pp. 3113–3118, 2015.

[68] T. A. Moreira, Z. H. Ayub, and G. Ribatski, "Convective condensation of R600a, R290, R1270 and their zeotropic binary mixtures in horizontal tubes," *International Journal of Refrigeration*, vol. 132, pp. 52–64, 2021. DOI: 10.1016/j.ijrefrig.2021.06.031

[69] L. Tang and A. T. Johnson, "Flow condensation in smooth and micro-fin with HCFC-22, HFC-134a and HFC-410 refrigerants. Part II: Design equations," *Journal of Enhanced Heat Transfer*, vol.7, no. 5, pp. 375–388, 2000.

[70]   H. L. I. Park, and I. Mudawar, "Determination of flow regimes and heat transfer coefficient for condensation in horizontal tubes," *International Journal of Heat and Mass Transfer*, vol. 80, pp. 698–716, 2015.

[71]   V. Konsetov, "Experimental investigations of the heat transfer in steam condensing in horizontal and inclined tubes (In Russian)," *Journal of Heat Power Engineering*, vol. 12, pp. 67–71, 1960.

[72]   R. Volodymyr, V. Sereda, V. Gorin, P. Barabash, and A. Solomakha. "Heat transfer during film condensation inside plain tubes. Review of experimental research," *Heat and Mass Transfer*, vol. 56, pp. 691–713, 2020.

[73]   Q. Song, G. Chen, H. Xue, Y. Zhao, and M. Gong, "R14 flow condensation heat transfer performance: Measurements and modeling based on two-phase flow patterns," *International Journal of Heat and Mass Transfer*, vol. 136, pp. 298–311, 2019.

[74]   S. Chen, F. Gerner, and C. Tien, "General film condensation correlations," *Experimental Heat Transfer An International Journal*, vol. 1, no. 2, pp. 93–107, 1987.

[75]   D. Jung, K.-h. Song, Y. Cho, and S.-j. Kim, "Flow condensation heat transfer coefficients of pure refrigerants," *International Journal of Refrigeration*, vol. 26, no. 1, pp. 4–11, 2003.

[76]   M. M. Rahman, K. Kariya, and A. Miyara, "An experimental study and development of new correlation for condensation heat transfer coefficient of refrigerant inside a multiport minichannel with and without fins," *International Journal of Heat and Mass Transfer*, vol. 116, pp. 50–60, 2018.

[77]   V. Rifert, V. Sereda, V. Gorin, P. Barabash, and A. Solomakha, "Substantiation and the range of application of a new method for heat transfer prediction in condensing inside plain tubes," *Energetika*, vol. 64, no. 3, 2018.

[78]   A. Cavallini et al., "Condensation inside and outside smooth and enhanced tubes—a review of recent research," *International Journal of Refrigeration*, vol. 26, no. 4, pp. 373–392, 2003.

[79]   X. G. J. Wen, S. Wang, Y. Li, and J. Tu, "The comparison of condensation heat transfer and frictional pressure drop of R1234ze (E), propane and R134a in a horizontal mini-channel," *International Journal of Refrigeration*, vol. 92, pp. 208–224, 2018.

[80]   V. Rifert, V. Gorin, V. Sereda, and V. Treputnev, "An improved heat transfer prediction model for film condensation inside a tube with interphacial shear effect," *World Academy of Science, Engineering and Technology International Journal of Mechanical, Aerospace, Industrial, Mechatronic and Manufacturing Engineering*, vol. 11, no. 8, pp. 1407–1412, 2017.

[81]   T. Naulboonrueng, J. Kaewon, and S. Wongwises, "Two-phase condensation heat transfer coefficients of HFC–134a at high mass flux in smooth and micro-fin tubes," *International Communications in Heat and Mass Transfer*, vol. 30, no. 4, pp. 577–590, 2003.

[82]   A. N. Kaushik, "A general heat transfer correlation for condensation inside internally finned tubes," *ASHRAE Transactions*, vol. 94, pp. 261–279, 1988.

[83]   Ö. Ağra and İ. Teke, "Experimental investigation of condensation of hydrocarbon refrigerants (R600a) in a horizontal smooth tube," *International Communications in Heat and Mass Transfer*, vol. 35, no. 9, pp. 1165–1171, 2008.

[84]   M. M. Shah, "General correlation for heat transfer during condensation in plain tubes: Further development and verification," *ASHRAE Transactions*, vol. 119, no. 2, pp. 101–118, 2013.

[85]   W. Akers, H. Deans, and O. Crosser, "Condensing heat transfer within horizontal tubes," *Chemical Engineering Progress*, vol. 54, no. 5, pp. 89–90, 1958.

[86]   O. K. Crosser, "Condensing Heat Transfer within Horizontal Tubes," Ph.D. dissertation, Rice University, Houston, TX, USA, 1955.

[87]   L. Boyko and G. Kruzhilin, "Heat transfer and hydraulic resistance during condensation of steam in a horizontal tube and in a bundle of tubes," *International Journal of Heat and Mass Transfer*, vol. 10, no. 3, pp. 361–373, 1967.

[88]   H. Soliman, "The mist-annular transition during condensation and its influence on the heat transfer mechanism," *International Journal of Multiphase Flow*, vol. 12, no. 2, pp. 277–288, 1986.

[89]   O. Iqbal and P. Bansal, "In-tube condensation heat transfer of CO_2 at low temperatures in a horizontal smooth tube," *International Journal of Refrigeration*, vol. 35, no. 2, pp. 270–277, 2012.

[90]   D. R. E. Ewim, J. P. Meyer, and S. N. R. Abadi, "Condensation heat transfer coefficients in an inclined smooth tube at low mass fluxes," *International Journal of Heat and Mass Transfer*, vol. 123, pp. 455–467, 2018.

[91] A. O. Adelaja, D. R. Ewim, J. Dirker, and J. P. Meyer, "An improved heat transfer correlation for condensation inside inclined smooth tubes," *International Communications in Heat and Mass Transfer*, vol. 117, p. 104746, 2020.

[92] D. R. Ejike Ewim, M. Mehrabi, and J. P. Meyer, "Modeling of heat transfer coefficients during condensation at low mass fluxes inside horizontal and inclined smooth tubes," *Heat Transfer Engineering*, vol. 42, no. 8, pp. 683–694, 2021.

[93] M. Hirose, J. Ichinose, and N. Inoue, "Development of the general correlation for condensation heat transfer and pressure drop inside horizontal 4 mm small-diameter smooth and microfin tubes," *International Journal of Refrigeration*, vol. 90, pp. 238–248, 2018.

[94] P. Li, J. Chen, and S. Norris, "Flow condensation heat transfer of CO_2 in a horizontal tube at low temperatures," *Applied Thermal Engineering*, vol. 130, pp. 561–570, 2018.

[95] D. R. E. Ewim, "Condensation inside Horizontal and Inclined Smooth Tubes at Low Mass Fluxes," Ph.D. dissertation, Department of Mechanical and Aeronautical Engineering, University of Pretoria, Pretoria, South Africa, 2019.

[96] D. Ewim and I. Okafor, "Condensation inside smooth and inclined smooth tubes at low mass fluxes: A quick review," in *IOP Conference Series: Earth and Environmental Science*, IOP Publishing, 2021, vol. 730, no. 1, p. 012044.

[97] D. Ewim, R. Kombo, and J. P. Meyer, "Flow pattern and experimental investigation of heat transfer coefficients during the condensation of r134a at low mass fluxes in a smooth horizontal tube," in *Proceedings of the 12th International Conference on Heat Transfer, Fluid Mechanics and Thermodynamics*, Malaga, Spain, 2016.

[98] A. O. Adelaja, D. R. Ewim, J. Dirker, and J. P. Meyer, "Heat transfer, void fraction and pressure drop during condensation inside inclined smooth and microfin tubes," *Experimental Thermal and Fluid Science,* vol. 109, p. 109905, 2019.

[99] D. Ewim, A. Adelaja, E. Onyiriuka, J. Meyer, and Z. Huan, "Modelling of heat transfer coefficients during condensation inside an enhanced inclined tube," *Journal of Thermal Analysis and Calorimetry*, vol. 146, pp. 103–115, 2021.

[100] D. Ewim and J. P. Meyer, "Pressure drop during condensation at low mass fluxes in smooth horizontal and inclined tubes," *International Journal of Heat and Mass Transfer*, vol. 133, pp. 686–701, 2019.

[101] A. Cavallini, G. Censi, D. Del Col, L. Doretti, G. Longo, and L. Rossetto, "Experimental investigation on condensation heat transfer and pressure drop of new HFC refrigerants (R134a, R125, R32, R410A, R236ea) in a horizontal smooth tube," *International Journal of Refrigeration*, vol. 24, no. 1, pp. 73–87, 2001.

[102] L. Wang, C. Dang, and E. Hihara, "Experimental study on condensation heat transfer and pressure drop of low GWP refrigerant HFO1234yf in a horizontal tube," *International Journal of Refrigeration*, vol. 35, no. 5, pp. 1418–1429, 2012.

[103] M. Macdonald and S. Garimella, "Hydrocarbon condensation in horizontal smooth tubes: Part I – Measurements," *International Journal of Heat and Mass Transfer*, vol. 93, pp. 75–85, 2016.

[104] G. Ghim and J. Lee, "Condensation heat transfer of low GWP ORC working fluids in a horizontal smooth tube," *International Journal of Heat and Mass Transfer*, vol. 104, pp. 718–728, 2017.

[105] C. Kondou and P. Hrnjak, "Heat rejection from R744 flow under uniform temperature cooling in a horizontal smooth tube around the critical point," *International Journal of Refrigeration*, vol. 34, no. 3, pp. 719–731, 2011.

[106] R. Agarwal and P. Hrnjak, "Condensation in two phase and desuperheating zone for R1234ze (E), R134a and R32 in horizontal smooth tubes," *International Journal of Refrigeration*, vol. 50, pp. 172–183, 2015.

[107] J. Xiao and P. Hrnjak, "Heat transfer and pressure drop of condensation from superheated vapor to subcooled liquid," *International Journal of Heat and Mass Transfer*, vol. 103, pp. 1327–1334, 2016.

[108] C. H. Son, H. K. Oh. "Condensation heat transfer characteristics of CO_2 in a horizontal smooth- and microfin-tube at high saturation temperatures," *Applied Thermal Engineering*, vol. 36, no. 1, pp. 51–62, 2012. DOI: 10.1016/j.applthermaleng.2011.12.017.

[109] Y. J. Kim, J. Jang, P. S. Hrnjak, and M. S. Kim, "Adiabatic horizontal and vertical pressure drop of carbon dioxide inside smooth and microfin tubes at low temperatures," *Journal of Heat Transfer*, vol. 130, no. 11, pp. 111601–111608, 2008.

[110] J. Xiao and P. Hrnjak, "A heat transfer model for condensation accounting for non-equilibrium effects," *International Journal of Heat and Mass Transfer*, vol. 111, pp. 201–210, 2017.

[111] C. I. Ferreira, T. Newell, J. Chato, and X. Nan, "R404A condensing under forced flow conditions inside smooth, microfin and cross-hatched horizontal tubes," *International Journal of Refrigeration*, vol. 26, no. 4, pp. 433–441, 2003.

[112] J. Jang and P. Hrnjak, "Condensation of CO2 at Low Temperatures," Air Conditioning and Refrigeration Center, College of Engineering, University of Illinois at Urbana-Champaign, ACRC TR-227, 2004.

[113] B. Mitra, "Supercritical Gas Cooling and Condensation of Refrigerant R410A at Near-Critical Pressures," Ph.D. dissertation, Georgia Institute of Technology, Atlanta, GA, USA, 2005.

[114] A. Dalkilic, O. Agra, I. Teke, and S. Wongwises, "Comparison of frictional pressure drop models during annular flow condensation of R600a in a horizontal tube at low mass flux and of R134a in a vertical tube at high mass flux," *International Journal of Heat and Mass Transfer*, vol. 53, no. 9–10, pp. 2052–2064, 2010.

[115] Y. Jiang, B. Mitra, S. Garimella, and C. Andresen, "Measurement of condensation heat transfer coefficients at near-critical pressures in refrigerant blends," *Transactions of the ASME Journal of Heat Transfer*, vol. 129, no. 8, pp. 958–965, 2007. https://doi.org/10.1115/1.2401618

[116] X. Zhuang, G. Chen, X. Zou, Q. Song, and M. Gong, "Experimental investigation on flow condensation of methane in a horizontal smooth tube," *International Journal of Refrigeration*, vol. 78, pp. 193–214, 2017.

[117] B. Ghorbani, M. -H Hamedi, M. Amidpour, and R. Shirmohammadi, "Implementing absorption refrigeration cycle in lieu of DMR and C3MR cycles in the integrated NGL, LNG and NRU unit," *International Journal of Refrigeration*, vol. 77, pp. 20–38, 2017, ISSN 0140-7007. https://doi.org/10.1016/j.ijrefrig.2017.02.030

[118] B. L. Keinath and S. Garimella, "High-pressure condensing refrigerant flows through microchannels, part ii: Heat transfer models," *Heat Transfer Engineering*, vol. 40, no. 9–10, pp. 830–843, 2019.

[119] V. G. Rifert and V. V. Sereda, "Condensation inside smooth horizontal tubes: Part survey of the methods of heat-exchange prediction," *Thermal Science*, vol. 19, no. 5, pp. 1769–1789, 2015.

[120] C. C.-W. Tang, "Study of Heat Transfer in Non-boiling Two-phase Gas-liquid Flow in Pipes for Horizontal, Slightly Inclined, and Vertical Orientations," Ph.D. dissertation, School of Mechanical and Aerospace Engineering, Oklahoma State University, Stillwater, OK, USA, 2011.

[121] K.-J. Park, D. Jung, and T. Seo, "Flow condensation heat transfer characteristics of hydrocarbon refrigerants and dimethyl ether inside a horizontal plain tube," *International Journal of Multiphase Flow*, vol. 34, no. 7, pp. 628–635, 2008.

[122] S. Laohalertdecha and S. Wongwises, "The effects of corrugation pitch on the condensation heat transfer coefficient and pressure drop of R-134a inside horizontal corrugated tube," *International Journal of Heat and Mass Transfer*, vol. 53, no. 13–14, pp. 2924–2931, 2010.

[123] S. Sapali and P. A. Patil, "Heat transfer during condensation of HFC-134a and R-404A inside of a horizontal smooth and micro-fin tube," *Experimental Thermal and Fluid Science*, vol. 34, no. 8, pp. 1133–1141, 2010.

[124] A. Cavallini, G. Censi, D. Del Col, L. Doretti, G. A. Longo, and L. Rossetto, "Condensation of halogenated refrigerants inside smooth tubes," *Hvac&R Research*, vol. 8, no. 4, pp. 429–451, 2002.

[125] P. Q. Vu et al., "An Experimental Investigation Of Condensation Heat Transfer Coefficient Using R-410a In Horizontal Circular Tubes," *Energy Procedia*, vol. 75, pp. 3113–3118, 2015.

[126] M.-H. Kim and J.-S. Shin, "Condensation heat transfer of R22 and R410A in horizontal smooth and microfin tubes," *International Journal of Refrigeration*, vol. 28, no. 6, pp. 949–957, 2005.

[127] D. Jung, Y. Cho, and K. Park, "Flow condensation heat transfer coefficients of R22, R134a, R407C, and R410A inside plain and microfin tubes," *International Journal of Refrigeration*, vol. 27, no. 1, pp. 25–32, 2004.

[128] G. A. Longo, S. Mancin, G. Righetti, and C. Zilio, "Saturated vapour condensation of HFC404A inside a 4 mm ID horizontal smooth tube: Comparison with the long-term low GWP substitutes HC290 (Propane) and HC1270 (Propylene)," *International Journal of Heat and Mass Transfer*, vol. 108, pp. 2088–2099, 2017.

[129] G. A. Longo, S. Mancin, G. Righetti, and C. Zilio, "Saturated vapour condensation of R410A inside a 4 mm ID horizontal smooth tube: Comparison with the low GWP substitute R32," *International Journal of Heat and Mass Transfer*, vol. 125, pp. 702–709, 2018.

[130] K. Aroonrat, L. G. Asirvatham, A. S. Dalkılıç, O. Mahian, H. S. Ahn, and S. Wongwises, "Effect of geometrical parameters on the evaporative heat transfer and pressure drop of R-134a flowing in dimpled tubes," *Heat and Mass Transfer*, vol. 57, no. 3, pp. 465–479, 2021.

[131] H.-W. BYUN, E.-J. LEE, Y.-S. SIM, J.-K. LEE, and N.-H. KIM, "Condensation heat transfer and pressure drop of R-410A in a 5.0 mm OD smooth and microfin tube," *International Journal of Air-Conditioning and Refrigeration*, vol. 21, no. 03, p. 1350018, 2013.

[132] M. A. Hossain, Y. Onaka, and A. Miyara, "Experimental study on condensation heat transfer and pressure drop in horizontal smooth tube for R1234ze (E), R32 and R410A," *International Journal of Refrigeration*, vol. 35, no. 4, pp. 927–938, 2012.

[133] L. Tang, M. M. Ohadi, and A. T. Johnson, "Flow condensation in smooth and micro-fin tubes with HCFC-22, HFC-134a and HFC-410A refrigerants. Part I: experimental results," *Journal of Enhanced Heat Transfer*, vol. 7, no. 5, 2000.

[134] S. Sung-Hoon, Y. Jung-In, L. Joon-Hyuk, C. Seung-Yun, and H. Su-Jeong, "Condensation Heat Transfer Characteristics of R-1234yf with the Variation of Tube Diameter," *International Journal of Refrigeration*, vol. 122, pp. 1–10, 2021.

[135] G. A. Longo, S. Mancin, G. Righetti, and C. Zilio, "Saturated vapour condensation of R134a inside a 4 mm ID horizontal smooth tube: Comparison with the low GWP substitutes R152a, R1234yf and R1234ze (E)," *International Journal of Heat and Mass Transfer*, vol. 133, pp. 461–473, 2019.

[136] C.-H. Son and H.-S. Lee, "Condensation heat transfer characteristics of R-22, R-134a and R-410A in small diameter tubes," *Heat and Mass Transfer*, vol. 45, no. 9, pp. 1153–1166, 2009.

[137] H.-S. Lee and C.-H. Son, "Condensation heat transfer and pressure drop characteristics of R-290, R-600a, R-134a and R-22 in horizontal tubes," *Heat and Mass Transfer*, vol. 46, no. 5, pp. 571–584, 2010.

[138] A. Miyara, K. Nonaka, and M. Taniguchi, "Condensation heat transfer and flow pattern inside a herringbone-type micro-fin tube," *International Journal of Refrigeration*, vol. 23, no. 2, pp. 141–152, 2000.

[139] D. W. Shao and E. G. Granryd, "Flow pattern, heat transfer and pressure drop in flow condensation part I: Pure and azeotropic refrigerants," *HVAC&R Research*, vol. 6, no. 2, pp. 175–195, 2000.

[140] A. Lambrechts, L. Liebenberg, A. E. Bergles, and J. P. Meyer, "Heat transfer performance during condensation inside horizontal smooth, micro-fin and herringbone tubes," *International Journal of Heat and Mass Transfer*, vol. 49, no. 1–2, pp. 431–443, 2006.

[141] H. M. Afroz, A. Miyara, and K. Tsubaki, "Heat transfer coefficients and pressure drops during in-tube condensation of CO$_2$/DME mixture refrigerant," *International Journal of Refrigeration*, vol. 31, no. 8, pp. 1458–1466, 2008.

[142] J. Kaew-On, N. Naphattharanun, R. Binmud, and S. Wongwises, "Condensation heat transfer characteristics of R134a flowing inside mini circular and flattened tubes," *International Journal of Heat and Mass Transfer*, vol. 102, pp. 86–97, 2016.

[143] J. Zhang, W. Li, and S. Sherif, "A numerical study of condensation heat transfer and pressure drop in horizontal round and flattened minichannels," *International Journal of Thermal Sciences*, vol. 106, pp. 80–93, 2016.

[144] Z. Wu, B. Sundén, L. Wang, and W. Li, "Convective condensation inside horizontal smooth and microfin tubes," *Journal of Heat Transfer*, vol. 136, no. 5, 2014.

[145] B. Li, L. Feng, L. Wang, and Y. Dai, "Experimental investigation of condensation heat transfer and pressure drop of R152a/R1234ze (E) in a smooth horizontal tube," *Heat Transfer Research*, vol. 52, no. 7, 2021.

[146] R. Mendes, D. Pottie, C. Paula, J. Pabon, and L. Machado, "Experimental study to determine the local condensation heat transfer coefficiente for R134A flowing through a 4.8 mm internal diameter smooth horizontal tube," *Revista de Engenharia Térmica*, vol. 20, no. 1, pp. 26–33, 2021.

[147] M. Moradkhani, S. Hosseini, and M. Song, "Robust and general predictive models for condensation heat transfer inside conventional and mini/micro channel heat exchangers," *Applied Thermal Engineering*, vol. 201, p. 117737, 2022.

[148] G. A. Longo, S. Mancin, G. Righetti, and C. Zilio, "Comparative analysis of microfin vs smooth tubes in R32 and R410A condensation," *International Journal of Refrigeration*, vol. 128, pp. 218–231, 2021.

[149] W. Li, J. Chen, C. Fu, and D. Xu, "Experimental investigation on convective condensation heat transfer in horizontal 4 mm diameter coated tube," *International Journal of Heat and Mass Transfer*, vol. 183, p. 122162, 2022.

[150] J. Zilly, J. Jang, and P. Hrnjak, "Condensation of CO_2 at Low Temperature Inside Horizontal Microfinned Tubes," Air Conditioning and Refrigeration Center, College of Engineering, University of Illinois at Urbana-Champaign, ACRC TR-199, 2003.

[151] K. Aroonrat and S. Wongwises, "Experimental study on two-phase condensation heat transfer and pressure drop of R-134a flowing in a dimpled tube," *International Journal of Heat and Mass Transfer*, vol. 106, pp. 437–448, 2017.

[152] M. Macdonald, "Condensation of Pure and Zeotropic Mixture of Hydrocarbons in Smooth Horizontal Tubes," Ph.D. dissertation, Department of Mechanical Engineering, University of Maryland, College Park, USA, 2015.

[153] H. Lee and C. Son, "Condensation heat transfer and pressure drop characteristics of R-290, R-600a, R-134a, and R-22 in horizontal tubes," *Heat and Mass Transfer*, vol. 46, pp. 571–584, 2010.

[154] X. H. Nan and C. A. Infante-Ferreira, "In-tube evaporation and condensation of natural refrigerant R290 (propane)," *Preliminary Proceedings of the Fourth IIR Gustav Lorentzen Conference*, 2000.

[155] S. J. Eckels, Tesene, B, "Forced convective condensation of refrigerants R-502 and R-507 in smooth and enhanced tubes," *ASHRAE Transactions*, vol. 102 (2), 627–638, 2002.

4 A Review of Numerical Tools for Solar Cells

George Chukwuebuka Enebe, Kingsley Ukoba, and Tien-Chien Jen

4.1 INTRODUCTION

Solar cells remain the building block of solar energy. Currently, solar energy is the third largest renewable energy behind hydro and wind energy. Although, solar energy holds the promise of being able to meet global energy needs with a noon-day sunlight supply. There have been attempts to research possible ways of increasing the efficiency of existing solar cells and reducing the cost [1–5]. Experimental and theoretical approaches have been used in this regard. This study aims to examine some of the theoretical tools that have been used for solar cells. This is with a view of offering the experimentalists pathway in the production of efficient and affordable solar cells and by extension improvement on solar technology.

The greatest commercially viable renewable energy source, solar energy may supply humanity with more energy in an hour than the entire planet's population uses in a year [6,4]. A considerable amount of the world's energy needs may be met by converting just 0.025% of the 165 000 terawatts of sunshine that reach the planet each year [7]. In contrast to fossil fuels, large-scale integration and high upfront expenses restrict the use of solar energy at full capacity. Research is ongoing to reduce the price to a level that is competitive with that of the existing fossil fuel power networks.

Modeling is useful for fine-tuning solar cell characteristics to increase efficiency at lower costs and with fewer resources. The creation of efficient solar cells now relies heavily on solar modeling. Modeling tools are essential for increasing solar cell efficiency and for demystifying the workings of solar cells. New varieties of vividly colored and clear solar cells can also be created via solar modeling. Integrated photovoltaic systems can make use of these colored solar cells. Unlike the electronic industry, the photovoltaic sector has not taken advantage of solar cell modeling's advantages.

The comprehension of solar cells preceding the development of solar cell modeling was based solely on empirical investigations and intuition. This is because, previous to 2008, the solar cell sector was mainly focused on increasing device scale rather than improving performance. Since 2008, solar cell modeling has attracted considerable interest, mostly because of curiosity about solar cells' increased efficiency. The main driver of the rise in the insistence on solar cell modeling points to the possibility of improving device performance by including the entire solar cell device within the optimization process.

This study is structured in a way that the introduction is followed by a background of solar cells. The background of solar cells discusses the introduction of the concept of solar cells, the history of solar cells, principles underlying operations of solar cells, and classification of the solar cells. This is followed by the experimental method of depositing solar cells. The numerical tools used for modeling solar cells are thereafter examined.

DOI: 10.1201/9781032651958-4

4.2 BACKGROUND OF SOLAR CELLS

Among these alternative energy sources, solar energy shines due to its potential for global availability. In terms of human existence, solar energy is limitless. The efficiency of solar energy is currently of great attention. One of the most promising solutions to the world's energy problems is the conversion of solar energy into economically usable energy [8–13]. Studies are focusing on low-cost, ecologically friendly, and sustainable processes and materials that will make solar technologies widely accessible [14–23].

4.2.1 Brief History of Solar Cells

As early as the 7th century BC, humans have been engaging with solar technology. Adopted to burn ants and light fires, humans first experimented with the sun by concentrating it with mirrors and glasses. These days, technology has developed into more cutting-edge technologies like solar-powered cars and planes. Horace de Saussure created the first solar collector ever. During his 1830s trip, Sir John Herschel brought this harvester to South Africa. Robert Stirling, a priest of the Church of Scotland, submitted a patent application on September 27, 1816, at Chancery in Edinburgh, Scotland, for his heat engine described as the "eEonomizer" [24, 25]. This heat engine was utilized to generate electricity using solar thermal technology to operate and generate power, concentrating the heat energy of the sun. The photovoltaic effect was not identified until 1839 [25]. It was discovered by French scientist Edmond Becquerel while performing research on an electrolytic cell [13]. Two metal rods were inserted into an electrically conductive fluid to form the electrolytic cell. The production of electricity increased when exposed to light. The first parabolic dish collector was built in the 1860s by the French mathematician August Mouchet and his helper [13]. They called it a solar-powered engine, and it served a variety of purposes.

Wilhelm Hallwachs observed in 1904 that copper and cuprous oxide are examples of photosensitive metal oxides [13]. Out of it, he created a semiconductor-junction solar cell. The presence of a barrier layer in PV devices was discovered about ten years after Albert Einstein's publication, and Robert Millikan experimentally demonstrated the photoelectric effect in 1916 [13]. The phenomenon of photovoltaic in cadmium selenide (CdSe), and PV materials in use currently, was discovered by Audobert and Stora in 1932 [24]. Gordon Teal and John Little modified the Czochralski method of crystal formation in 1948 to provide single-crystalline germanium and later silicon. Bell Labs broadcast the creation of the first usable silicon solar cell on April 25, 1954. These cells exhibited an efficiency of roughly 6%. In 1955, Western Electric granted commercial licenses for solar cell technologies [26–28]. Hoffman Electronics' Semiconductor Division produced a commercial solar cell with a 2% efficiency that year costing $25 per cell or $1,785 per watt. In 1958, the U.S. Signal Corps Laboratories of Mandelkorn invented n-on-p silicon solar cells [29, 30].

In 1988, chemists Brian O'Regan and Michael Grätzel developed the dye-sensitized solar cell [31–33]. In 2006, solar cell technology broke the "40% efficient" barrier of sunlight-to-electricity, thus setting a new record [34]. When the scientists at the National Renewable Energy Laboratory (NREL) developed a PV device with up to 40.8% conversion amount of light that struck it into power in 2008 [11, 35, 36], it set yet another new record for solar cell efficiency. Manufacturing costs for silicon PV modules dropped to roughly $1.25 per watt in 2011, thanks to China's rapidly expanding facilities, while global installations more than doubled [12]. New solar material and product development strategies, cell designs, and PV research and development are all hot topics right now.

4.2.2 Principle of Solar Cells

Photons—the energy "packets" that makeup light—hit a solar cell and are absorbed by it. This is one of the main operating principles that applies to all solar cells. When photons have enough energy, the cell releases its stored electrons [37]. Insufficient photon energy results in the conversion of photon energy to heat energy. Wires carry the freed electrons as they move around an electrical circuit. Direct current is the form of electrical current that results (DC) [38]. This current only flows in one direction and is one-way. The electrical current will be greater, and more electrons will be released each second if the light is more intense (brighter) [37–39]. The voltage in the cell is constant. As a result, solar cells transform the energy package from the sun.

4.2.3 Classification of Solar Cells

The generation of solar cells can be used to classify solar cells. Solar cells have gone through four generations thus far. Figure 4.1 illustrates this classification based on when solar cell technology and materials were discovered.

The solar panel market is dominated by the first generation. These are bigger silicon-based solar cells. Although having a high manufacturing cost, its great efficiency accounts for 86% of the global solar market. High carrier mobility is the first generation's advantage, but it is expensive and suffers from photon energy loss [40]. Large-scale monocrystalline and polycrystalline solar panels have been produced using this generation of solar cells for commercial use.

Solar cells of the second generation are thin-film solar cells. Although they have lower efficiency than first-generation cells, their production costs are lower [24,38,40–42]. Reduced materials as well as cost-effective substrates are employed, and lower-cost manufacturing techniques are utilized. The lower efficiency of second-generation solar cells experienced a significant drawback despite their flexibility and ease of integration into roofing materials. Although the second generation is less expensive than the first, it is less efficient [40]. The second generation has been made available for purchase and is currently being used in utility-scale and standalone solar power plants. The list of typical second-generation solar cells is displayed in Table 4.1.

The third generation has more efficient production costs and short-term stability. Since the third generation's efficiency is either low or their absorber material is short, they are still in the research and development stage and have not yet been marketed.

Owing to an increase in the module's efficiency, multiple junction solar cells incorporate two or more layers of several semiconductor substances. Each of the materials responds to the wavelengths

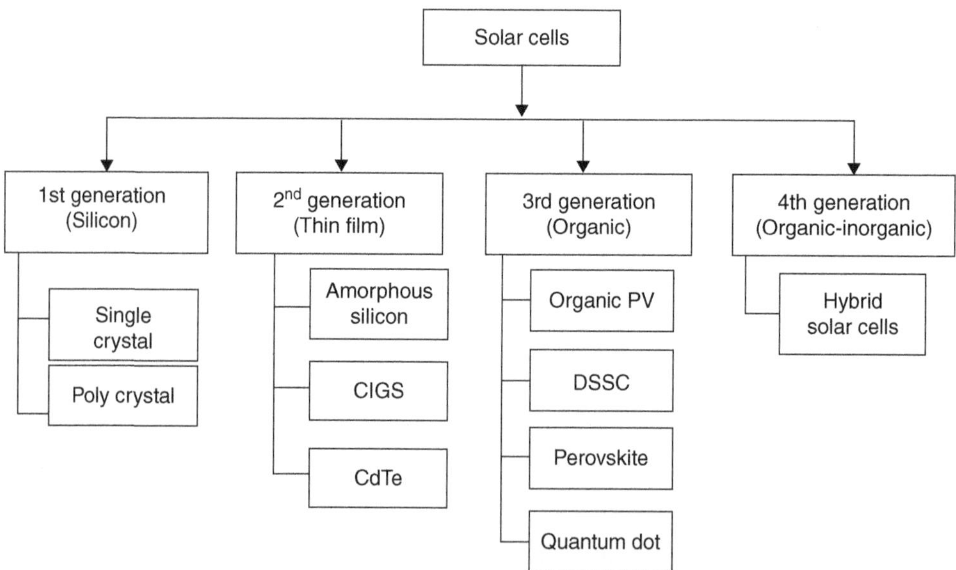

FIGURE 4.1   Classification of solar cells based on the structure of the solar cells [40].

TABLE 4.1
List of Typical Second-Generation Solar Cells

Solar Cell	Achievement
CIGS (CuInGaSe$_2$)	World record: 19.5%
	Production: 9–11%
Amorphous Si	World record: 13.2%
	not completely stable
	Production: 6–8%
CdTe	World record: 16.5%
	Production: 8–10%

of sunlight by producing electric outputs. This improves efficiency by allowing for a wider range of wavelengths to be absorbed. Its tremendous efficiency, however, comes with a considerable cost and complexity. Their influence in fields like aircraft and terrestrial applications has diminished as a result of this drawback. They have attained records of roughly 43% when illuminated by a concentrated light source and 30% when illuminated by just one light source.

4.3 METHODS OF DEPOSITING SOLAR CELLS

The deposition of solar cells is the process of fabricating the solar cells. Precursors (chemical or solution) are the primary material used for depositing solar cells. The deposition of solar cells is split into three categories based on the type of deposition and the physical or chemical processes involved, as shown in Figure 4.2.

The various methods for solution and gas-phase deposition are used in chemical processes. Atomic layer epitaxy, atomic layer deposition, and chemical vapor deposition (CVD) are examples of gas-phase techniques (ALD). Among these solution techniques for deposition are spin, sol-gel, dip-coating, and spray pyrolysis. Magnetron sputtering, pulsed laser deposition, molecular beam epitaxy, and physical vapor deposition (PVD) are examples of physical processes. Anodic oxidation, electron beam evaporation, vacuum evaporation, advanced reactive gas deposition, and chemical bath deposition are further processes.

4.4 THEORETICAL TOOLS FOR SOLAR CELLS

The successful modeling of solar cells requires a deep comprehension of the principle of solar cell operation.

As seen in Figure 4.3, the corresponding circuit diagram of the cells, solar cells are essentially P–N heterojunctions.

They display nonlinear I–V properties that are temperature and radiation intensity dependent. A solar cell can theoretically be modeled as a source of current underneath a diode under ideal circumstances. An equation can be used to express the solar cell's I–V characteristic equation (4.1)

$$I = I_{ph} - I_s \left[e^{\left(\frac{q\left(V + R_s I_{pv}\right)}{AkT_c} \right)} - 1 \right] - \frac{V + R_s I_{pv}}{R_{sh}} \tag{4.1}$$

The fill factor, open-circuit voltage, and short-circuit current are three unique factors that affect how well solar cells work. Because the fill factor depends on both variables, short circuit current along

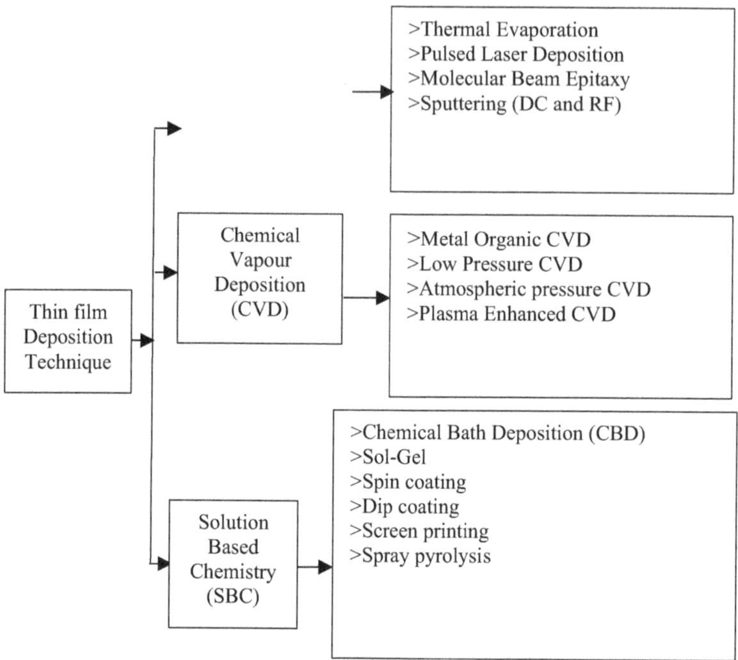

FIGURE 4.2	Classification of solar cells deposition methods [43,44].

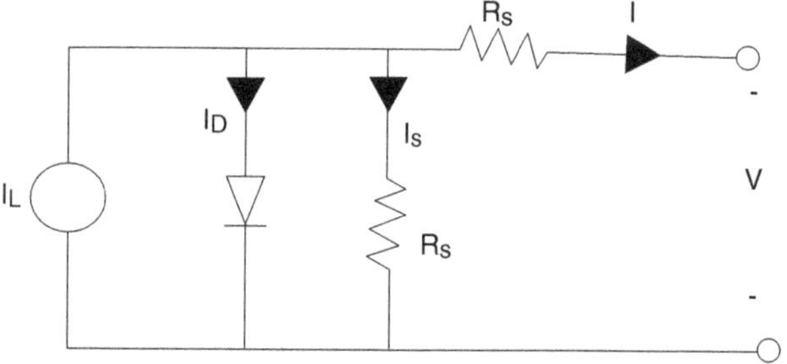

FIGURE 4.3	Equivalent circuit models of solar cells [4].

with open circuit voltage remain the principal determinants of solar cell efficiencies [30]. When incident photon energy exceeds the band gap of the cells, electrons flow within the external circuit. The solar cell's usual properties are shown in Figure 4.4. It displays how the voltage and current of solar cells respond to light exposure and temperature.

The majority of solar cells are made up of one-dimensional stacks of various semiconductor material layers. Due to the electron/hole current only flowing in one direction, the majority of thin film metal oxide solar cells may be described as one-dimensional cells. Similar modeling techniques can be used for the majority of silicon wafer solar cells. This is possible as long as the series link is not explicitly modeled. Solar cells with metal contacts implanted in a passivation layer to aid in reducing recombination, two-dimensional modeling is used [30]. In the two-dimensional simulation, the hole current and internal electron flows in two dimensions, or most situations three dimensions. The steps for modeling solar cells are shown in Figure 4.5.

The majority of solar cells are made up of one-dimensional stacks of various semiconductor material layers. Due to the electron/hole current only flowing in one direction, the majority of

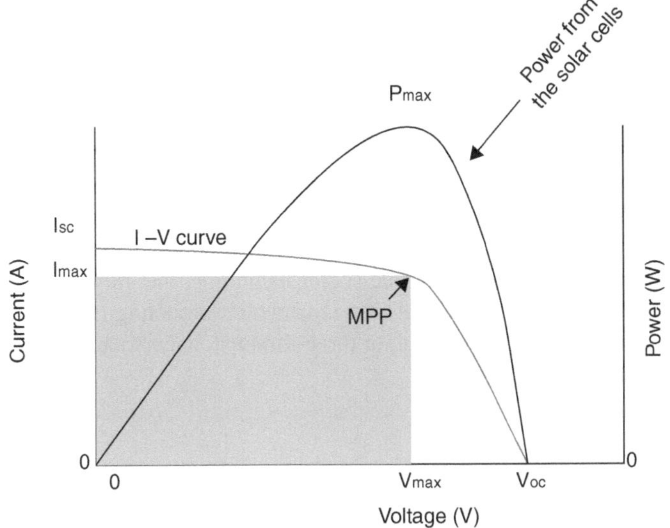

FIGURE 4.4 A schematic of solar cells characteristics [3].

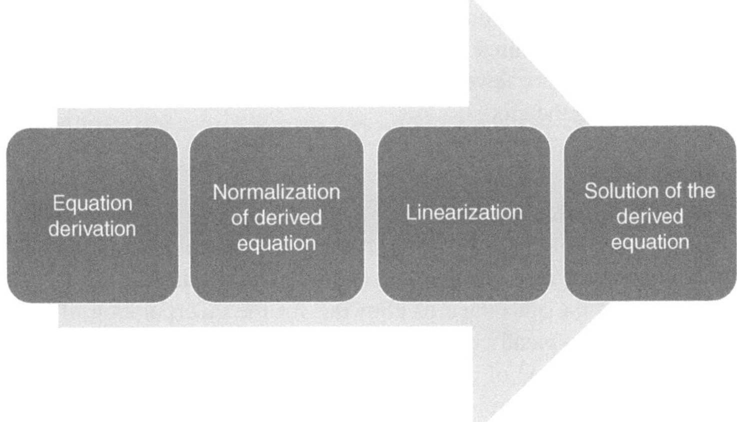

FIGURE 4.5 Step for modeling of solar cells.

metal-oxide thin film solar cells are described as one-dimensional cells. Similar modeling techniques can be used for the majority of silicon wafer solar cells. This is possible as long as the series link is not explicitly modeled.

4.5 MODELING TOOLS FOR SOLAR CELLS

For the purpose of creating and fine-tuning devices, modeling/simulations are mostly used. Numerical analysis greater knowledge of how metal oxide solar cell devices work. An early modeling tool for solar cells was created in 1980 by Mark S. Lundstrom, a Ph.D. candidate [4,45]. Thin-Film Semiconductor Simulation Program was created by Gray in 1989. (TFSSP). In 1985, Lundstrom contributed to the Solar Cell Analysis Program (SCAP). According to rumors, SCAP was created in Belgium at Ghent University. In 1989, a Ph.D. candidate in an engineering faculty applied one- and two-dimensions in SCAP for a Ph.D. thesis. Moreover, Push-Button, a set of modeling tools (one- and two-dimensional), was created by Purdue University.

4.5.1 CLASSIFICATION OF SOLAR CELL MODELING TOOLS

Current densities and the carrier can be calculated via modeling or simulation of solar cell devices. This is accomplished by resolving the Poisson's and transport equation. By implementing the necessary boundary conditions at the junctions, the overall resolution to the carrier and current densities are obtained (P–N) [46]. Yet, it is challenging to solve the current densities and carriers easily due to the nonlinear recombination. Focusing on the solver technique, the utilized modeling tool, as well as the dimension, there are three broad categories in which modeling the solar cell device can be divided (one-dimensional, two-dimensional, or three-dimensional). Solar cell modeling can be done using analytical and numerical solvers.

4.5.1.1 Analytical Solver

The analytical solver approach is the process of modeling solar cells using an analytical approach. Analytical solvers are divided into analytical and semi-analytical. In analytical, the related equations to semiconductors are resolved analytically. The equations comprise the equations of diffusion-drift current, the continuity carrier equations, and the Poisson equation. This method can calculate the important factors affecting solar cell efficiency. One of such work is the work of Kayes [47]. The device absorption is optimized using a semi-analytical method. Despite not requiring extensive computations, it is effective. This technique may evaluate the appropriate design variables for the fastest charge production over a thinner material [48,49]. Analytical models are easier to develop than numerical models but produce fewer simulation results overall. They also provide a clearer picture of the factors influencing the model.

4.5.1.2 Numerical Solver

Numerical simulation approaches for simulating solar cell devices are used. The hole density and the electron are related by Poisson's equation. Although Poisson's equation is difficult to solve using traditional mathematical methods due to its complexity, it can be done so using a numerical approach. This method has the advantage of allowing for the incorporation of important physical consequences that would not typically be taken into account. They include band gaps, lifespan, and doping. This lessens the likelihood of closed-form solutions. Tools such as finite difference, finite element [50], finite difference [51], time domain, transfer matrix, finite volume [52], and wave analysis are used for solving the differential equation for the numerical solver. With devices with tiny diameters and periodic architectures, the Transfer Matrix Technique performs well. In order to solve problems involving complex boundary conditions and complex geometry, the finite element method offers the greatest degree of flexibility. The design of most modeling tools is based on the ability to solve the fundamental equations for solar cells (Poisson together with continuity equations for electrons and holes) [53]. For the modeling or simulation of metal oxides for solar cell applications, any modeling software that can resolve fundamental semiconductor equations is suitable.

Tools such as Solar Cell Capacitance Simulator (SCAPS) [54,55], PC1D [56,57], Simulink [58], AMPS [59], Technology Computer-Aided Design (TCAD) [60], and wxAMPS [61,62] have all been utilized in the modeling of solar cells. Table 4.2 overviews the generally utilized modeling/simulation tools [63].

4.5.1.3 Analysis of Microelectronic and Photonic structures (AMPS)

This model was created for use in modeling and simulations in order to comprehend the physics and design of solar cell devices. It was created by a group at Pennsylvania State University under the direction of Fonash in 1992 [59,66]. It is a one-dimensional modeling tool that solves the Poisson

TABLE 4.2
Tools Used in the Modeling of Solar Cells Thin Films [64,65]

	AMPS	SCAPS	ASA	PC1D
Graded Bandgaps	**No**	**No**	**Yes**	**No**
Band discontinuities ΔE_c, ΔE_v All programs use the Anderson model: ΔE_c ¼ $\Delta\chi$ and ΔE_v ¼ $\Delta\chi$ þ ΔE_g				
Deep bulk states	50	3	4	No charge
Charge in deep states	Yes	Yes	Yes	No
Deep interface states	No	Yes	No	No
Maximum number of layers	30	7	Unlimited	5
Simulation of non-routine measurements	No	C–V C–f	C–V	Transients
Friendliness to User, Interactivity	++	++	−	++
Speed	−	+	++	++
Numerical robustness (Convergence)	++	OK	++	OK

and continuity equations of solar cells using finite differences and the Newton-Raphson approach [67]. It has the capacity to work on multiple models at once. The phrase "device case" is used. It is simple to describe the model because of the user-friendly interface. There are around 30 layers that can be given to any case or device, each with its own set of established parameters. Around 50 deep levels of donors and acceptors can be given to these layers, allowing for the creation of any density of state distributions. These deep levels can have uniform, discontinuous, or Gaussian distributions. The temperature has no bearing on characteristics like bandgap, electron/hole mobility, and others. Because it can add different layers with varied settings and simulate in both light and dark, it can imitate a graded junction. The main parameters for device modeling in AMPS (2010) are shown in Table 4.3.

One of AMPS' drawbacks is its inability to handle more than 3000 discretization nodes in the most recent version. Manually entering the absorption coefficients, wavelength, and spectrum intensity into AMPs has another drawback. Numerous studies have been conducted using this simulation tool [18,59,66–68]. The explicit definition of interface recombination is also challenging. Due to the processing speed being slower than other modeling tools, batch-mode processing is preferred. Yet, it provides a fantastic plotting capability for the analysis and design of two terminal structures. P-n, schottky barrier devices, single/heterojunction p-i-n, as well as similar devices can all be modeled using AMPS [30]. Since it can operate in both light and darkness, it can replicate a variety of optoelectronic components, such as solar cells and diodes.

4.5 SIMULINK

This is a graphical block diagram and modeling tool of MatLab that is used for different applications. It is used for simulating, and analyzing multidomain dynamical systems and modeling. Tsai et al. [69] produced a generalized model of photovoltaics using Simulink. Kumar Kundu and Ghosh [58,70] used Simulink to model solar cells. Other researchers have also deployed the versatility of Simulink to understand by modeling and simulating different solar cells [71,58,72,73]. Several studies have been carried out using this simulation tool [74–80].

4.5.1 wxAMPS

This is more stable than AMPS and an enhancement. Tunneling currents, greater visibility, faster speed, and convergence are the benefits [81]. This is because the Newton and Gummel technique

TABLE 4.3

AMPS-1D Parameters for Use in the Simulation of Heterojunction Solar Cells [67]

Parameter and Units e, h for Electrons and Holes, Respectively	a-Si:H(i)	a-Si:H(p+)	a-Si:H(n+)	c-Si(n)	Layer at the a-Si:H/c-Si Interface
Thickness (nm)	5	10	10	300000	10
Electron affinity (eV)	3.8	3.8	3.8	4.05	4.05
Band gap (eV)	1.72	1.72	1.72	1.12	1.12
Effective conduction band density (cm^{-3})	2.50×10^{20}	2.50×10^{20}	2.50×10^{20}	2.80×10^{19}	2.80×10^{19}
Effective valence band density (cm^{-3})	2.50×10^{20}	2.50×10^{20}	2.50×10^{20}	1.04×10^{19}	1.04×10^{19}
Electron mobility ($cm^2\ V^{-1}\ s^{-1}$)	20	10	10	1350	1350
Hole mobility ($cm^2\ V^{-1}\ s^{-1}$)	2	1	1	450	450
Doping concentration of acceptors (cm^{-3})	0	1×10^{20}	0	0	0
Doping concentration of donors (cm^{-3})	0	0	1×10^{20}	3×10^{15}	3×10^{15}
Band tail density of states ($cm^{-3}\ eV^{-1}$)	2×10^{21}	2×10^{21}	2×10^{21}	1×10^{14}	1×10^{14}
Characteristic energy (eV) donors, acceptors	0.06, 0.03	0.06, 0.03	0.06, 0.03	0.01	0.01
Capture cross section for donor states, e, h, (cm^2)	$1 \times 10^{-15}, 1 \times 10^{-17}$	$1 \times 10^{-15}, 1 \times 10^{-17}$	$1 \times 10^{-15}, 1 \times 10^{-17}$	$1 \times 10^{-15}, 1 \times 10^{-17}$	$1 \times 10^{-15}, 1 \times 10^{-17}$
Capture cross section for acceptor states, e, h, (cm^2)	$1 \times 10^{-17}, 1 \times 10^{-15}$	$1 \times 10^{-17}, 1 \times 10^{-15}$	$1 \times 10^{-17}, 1 \times 10^{-15}$	$1 \times 10^{-17}, 1 \times 10^{-15}$	$1 \times 10^{-17}, 1 \times 10^{-15}$
Gaussian density of states (cm^{-3})	$8 \times 10^{15}–8 \times 10^{17}$	$8 \times 10^{17}–8 \times 10^{20}$	$8 \times 10^{17}–8 \times 10^{20}$	0	0
Gaussian peak energy (eV) donors, acceptors	1.22, 0.70	1.22, 0.70	1.22, 0.70	0	0
Standard deviation (eV)	0.23	0.23	0.23	0	0
Capture cross section for donor states, e, h, (cm^2)	$1 \times 10^{-14}, 1 \times 10^{-15}$	$1 \times 10^{-14}, 1 \times 10^{-15}$	$1 \times 10^{-14}, 1 \times 10^{-15}$	0	0
Capture cross section for acceptor states, e, h, (cm^2)	$1 \times 10^{-15}, 1 \times 10^{-14}$	$1 \times 10^{-15}, 1 \times 10^{-14}$	$1 \times 10^{-15}, 1 \times 10^{-14}$	0	0
Midgap density of states ($cm^{-3}\ eV^{-1}$)	0	0	0	1×10^{11}	$4 \times 10^{15}–4 \times 10^{19}$
Switch over energy (eV)	0	0	0	0.56	0.56
Capture cross section for donor states, e, h, (cm^2)	0	0	0	$1 \times 10^{-15}, 1 \times 10^{-17}$	$1 \times 10^{-15}, 1 \times 10^{-17}$
Capture cross section for acceptor states, e, h, (cm^2)	0	0	0	$1 \times 10^{-17}, 1 \times 10^{-15}$	$1 \times 10^{-17}, 1 \times 10^{-15}$

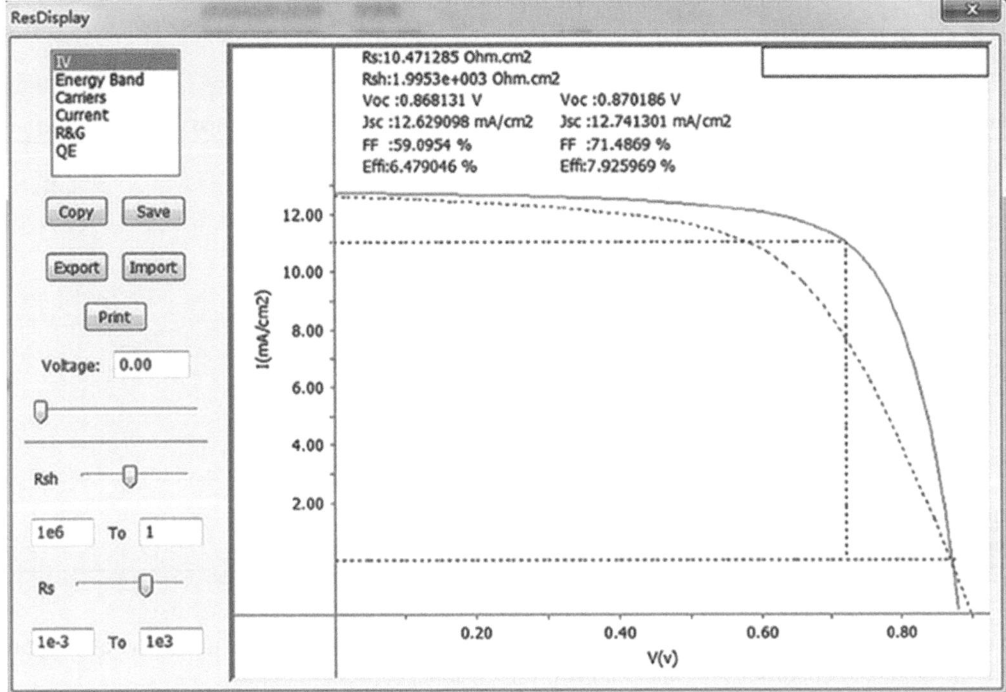

FIGURE 4.6 The interface of wxAMPS [61,86].

was used to solve the enhanced algorithm. A team at Nankai University and the University of Illinois created it [81,82]. wxAMPS's user interface uses a cross-platform library. Moreover, it offers improved data entering [18,81,83]. It is an open-source utility that performs well when compared to SCAP and other programs. Materials that possess high defect concentrations and band tails provide greater modeling capacity. Both tandem and graded solar cells can be used. Sharing device parameters is made possible by the WIKI function. The interface is shown in Figure 4.6. It has been used in some research for cell tunning and modifications [84,85].

4.5.2 TCAD

This tool was created using the Klaassen concentration-dependent SHR model, Poisson equation, drift-diffusion equation, Auger model, and Klassen's low-field mobility model [60]. A photo-generation model simulates the transmission, refraction, reflection, and absorption of light at the interface. It utilizes solar spectrum AM 1.5 G with a 100 mW/cm^2 incident power density. Table 4.4 displays the parameters for a typical optoelectronic device [87].

4.5.3 SCAPS

SCAPS is an acronym for the Ghent University-developed SCAP in One and Two Dimensions (SCAP1D and SCAP2D). Opto-electrical simulation of the 1D and/or 2D structures of semiconductor layers is done using its solar cell simulation application [54,55,89]. SCAPS was initially created for the CuInSe2 and CdTe family of cell architectures. The differential equations, along with several laws from the physics of semiconductors, show that mathematically, solar cell operation can be solved by SCAPS using finite difference techniques. The continuity equations together with

TABLE 4.4
TCAD Optoelectronic Device Parameters [60,88]

Parameters	n-cSi	n^+-cSi	Front Interface	Rear Interface
Electron affinity (eV)	4.17	4.17	-	-
Permittivity ε_r (eV)	11.7	11.7	-	-
Energy band gap E_g (eV)	Doping dependent	1.12	-	-
Interface recombination velocity (cm/s)	-	-	1×10^5	1×10^5
Effective DOS in conduction band N_c (cm^{-3})	2.77×10^{19}	2.77×10^{19}	-	-
Effective DOS in valence band N_V (cm^{-3})	1.03×10^{19}	1.03×10^{19}	-	-
SRH Electron lifetime τ_n (sec)	1×10^{-4}	1×10^{-4}	-	-
SRH Hole lifetime τ_p (sec)	1×10^{-3}	1×10^{-3}	-	-
SRH concentration for electrons N_{SRHn} (cm^{-3})	5×10^{16}	5×10^{16}	-	-
SRH concentration for holes N_{SRHp} (cm^{-3})	5×10^{16}	5×10^{16}	-	-
Trap energy for SRH recombination E_{trap} (eV)	0	0	-	-
Electron Auger coefficient C_n (cm^6/s)	2.8×10^{-31}	2.8×10^{-31}	-	-
Hole Auger coefficient C_p (cm^6/s)	9.9×10^{-32}	9.9×10^{-32}	-	-
Optical recombination rate C_{opt} (cm^3/s)	1.1×10^{-14}	1.1×10^{-14}	-	-

Poisson's equation are fully simultaneously solved numerically by SCAPS under the assumption that the boundary conditions for one- and two-dimensional cells are met [90].

$$\nabla^2 v = -\frac{q}{\epsilon}\left(p - n + N_D - N_A\right) \tag{4.2}$$

$$\nabla.J_p = q\left(G - R\right) \tag{4.3}$$

$$\nabla.J_n = q\left(R - G\right) \tag{4.4}$$

Represented in Eq. (4.5) are the general terms of Eqs. (4.3) and (4.4):

$$G(x) = \int_0^\infty \phi a e^{-ax} d\lambda \tag{4.5}$$

The hole and the electron current densities are shown in Eqs. (4.3) and (4.4) are stated as:

$$J_p = -q\mu_p p\nabla V_p - kT\mu_p\nabla_p \tag{4.6}$$

$$J_n = -q\mu_n n\nabla V_n + kT\mu_n\nabla_n \tag{4.7}$$

$$V_p = V - \left(1 - \gamma\right)\frac{\Delta_G}{q} \tag{4.8}$$

$$V_n = V + \gamma\frac{\Delta_G}{q} \tag{4.9}$$

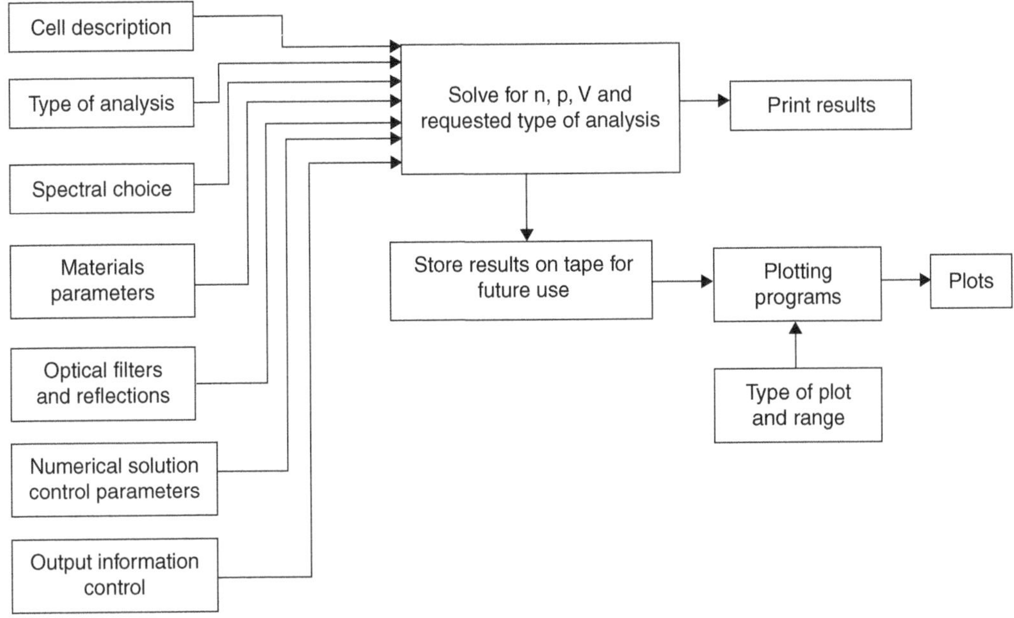

FIGURE 4.7 Structures of SCAP1D and SCAP2D block diagram [3].

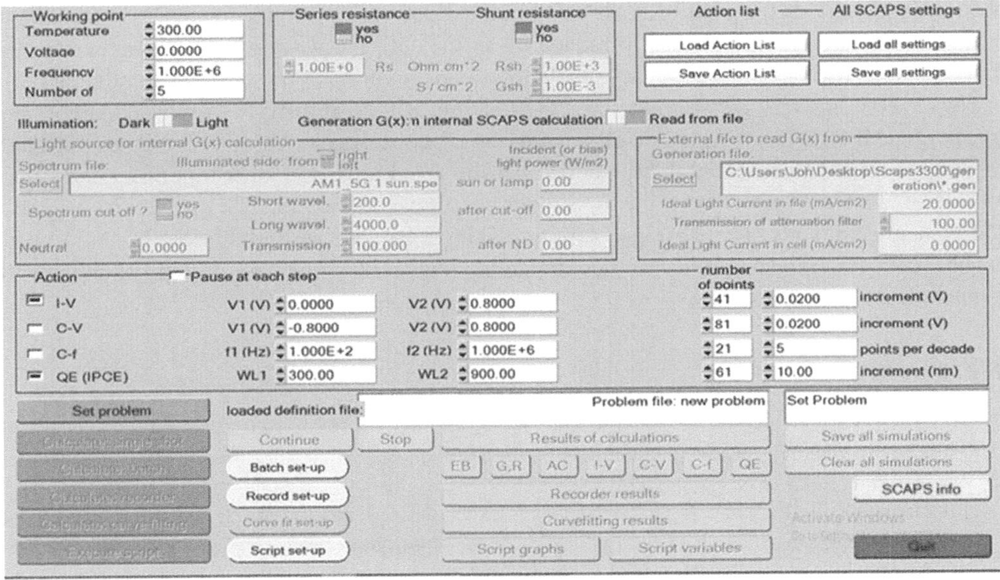

FIGURE 4.8 The parameters utilized in modeling solar cells [45].

where v_p and v_n are the effective potentials illustrated in Eqs. (4.8) and (4.9). Δ_G and γ represent the changes in the band structure, which include the density of states, band gap, and the account for Fermi-Dirac statistics.

The block diagram of the working principle of SCAPs is presented in Figure 4.7.

SCAP graphic user interface showing input parameters is depicted in Figure 4.8. SCAPS has been used by researchers to model and simulate different types of solar cells [1,5,3,91–93].

4.5.4 OTHER TOOLS

A 1D modeling tool for solar cells is called the "Numerical Solar Cell Simulation Program (NSSP)." This program was used by Amin, Sopian, and Konagai to model the architecture of CdTe solar cells from various angles [68,94]. It concentrated on applying theoretical research to reduce the thickness of the CdTe absorber. More innovative works have also been done using this simulation tool for solar cell analysis and improvements [95,96].

Silvaco [97–100], Crosslight, Advanced Semiconductor Analysis (ASA), AFORS-HET (Automat For Simulating Hetero-structures), Synopsis, and Sunshine [101,102] are a few more modeling programs that have become more popular. Silvaco simulates using cutting-edge technology known as Thin Film Transistors (TFT). Thin film transistors and solar cells are simulated using both physical models and numerical techniques. Advanced Physical Models of Semiconductor Devices is a technology that Crosslight employs (APSYS) [103]. Modeling the optoelectronic characteristics of two-dimensional thin film solar cell systems is done here using the finite element method. Numerous works have been done on solar cells and their modifications using APSYS [103–109].

4.6 CONCLUSION

A detailed discussion of the modeling and simulation techniques used for solar cells was provided in this work. Research and development of solar technologies especially in emergent nations can be greatly aided through mathematical models and the theoretical approval of solar cells. Today, a wide variety of commercial and laboratory-scale solar cell modeling tools are utilized all around the world. These solar cell simulation tools have recorded significant contributions to the body of knowledge and in the area of solar cell modifications and optimizations. The contributions of these simulation tools have led to reductions in the cost and wastage of trials and errors initially incurred without the use of these tools to provide a guide. Some of these simulation tools, like wxAMPS, are very consistent solar cell modeling tools that provide improved stability and speed. wxAMPS performs similarly to previous simulations, such as SCAPS, but it has a stronger capacity to mimic materials with high defect density, band tails, and other thin film photovoltaic properties. wxAMPS provides all the tools required to create robust and fast simulations of graded and tandem solar cells. The main problem is striking a balance and modifying the tool to meet the necessary requirements. Most of these modeling tools rely on the ability to solve fundamental semiconductor equations.

REFERENCES

[1] Enebe, G. C., Lukong, V. T., Mouchou, R. T., Ukoba, K. O., & Jen, T. C. (2022). Optimizing nanostructured TiO_2/Cu_2O pn heterojunction solar cells using SCAPS for fourth industrial revolution. *Materials Today: Proceedings*, *62*, S145–S150.

[2] Lukong, V. T., Mouchou, R. T., Enebe, G. C., Ukoba, K., & Jen, T. C. (2022). Deposition and characterization of self-cleaning TiO_2 thin films for photovoltaic application. *Materials Today: Proceedings*, *62*, S63–S72.

[3] Ukoba, K. O., & Inambao, F. L. (2018). Modeling of fabricated NiO/TiO_2 PN heterojunction solar cells. *International Journal of Applied Engineering Research*, *13*(11), 9701–9705. ISSN 0973-4562.

[4] Enebe, G. C., Ukoba, K., & Jen, T. C. (2019). Numerical modeling of effect of annealing on nanostructured CuO/TiO 2 pn heterojunction solar cells using SCAPS. *AIMS Energy*, *7*(4), 527–538.

[5] Ukoba, K., Imoisili, P. E., & Jen, T. C. (2021). Numerical analysis and performance improvement of nanostructured Cu_2O/TiO_2 pn heterojunction solar cells using SCAPS. *Materials Today: Proceedings*, *38*, 887–892.

[6] Shahsavari, A., & Akbari, M. (2018). Potential of solar energy in developing countries for reducing energy-related emissions. *Renewable and Sustainable Energy Reviews*, *90*, 275–291.

[7] Seba, T. (2010). *Solar trillions: 7 market and investment opportunities in the emerging clean-energy economy*. Tony Seba.

[8] Hamakawa, Y. (2004). Background and motivation for thin-film solar-cell development. *Thin-film solar cells: Next generation photovoltaics and its applications*, pp. 1–14.

[9] Luque, A. (1989). Solar cells and optics for photovoltaic concentration. CRC Press/Taylor and Francis Group.

[10] McEvoy, A., Castaner, L., & Markvart, T. (2012). *Solar cells: Materials, manufacture and operation*. Academic Press.

[11] Tao, M. (2008). Inorganic photovoltaic solar cells: Silicon and beyond. *The Electrochemical Society Interface*, *17*(4), 30.

[12] Hegedus, S., & Luque, A. (Eds.). (2011). *Handbook of photovoltaic science and engineering*. John Wiley & Sons.

[13] Demming, A. (2010). Solar harvest. *Nanotechnology*, *21*(49), 490201.

[14] Pavan, M., Rühle, S., Ginsburg, A., Keller, D. A., Barad, H. N., Sberna, P. M., ... & Fortunato, E. (2015). TiO_2/Cu_2O all-oxide heterojunction solar cells produced by spray pyrolysis. *Solar Energy Materials and Solar Cells*, *132*, 549–556.

[15] Shi, E., Zhang, L., Li, Z., Li, P., Shang, Y., Jia, Y., ... & Cao, A. (2012). TiO2-coated carbon nanotube-silicon solar cells with efficiency of 15%. *Scientific reports*, *2*(1), 884.

[16] Iqbal, K., Ikram, M., Afzal, M., & Ali, S. (2018). Efficient, low-dimensional nanocomposite bilayer CuO/ZnO solar cell at various annealing temperatures. *Materials for Renewable and Sustainable Energy*, *7*, 1–7.

[17] Yaqoob, U., Ayub, A. R., Rafiq, S., Khalid, M., El-Badry, Y. A., El-Bahy, Z. M., & Iqbal, J. (2021). Structural, optical and photovoltaic properties of unfused Non-Fullerene acceptors for efficient solution processable organic solar cell (Estimated PCE greater than 12.4%): A DFT approach. *Journal of Molecular Liquids*, *341*, 117428.

[18] Smucker, J., & Gong, J. (2021). A comparative study on the band diagrams and efficiencies of silicon and perovskite solar cells using wxAMPS and AMPS-1D. *Solar Energy*, *228*, 187–199.

[19] Hossain, M. A., Mondal, J., Ali, M. F., & Humayun, M. A. A. (2014). Design of high efficient InN quantum dot based solar cell. *International Journal of Scientific Engineering and Technology*, *3*(4), 346–349.

[20] Guo, Q., Ford, G. M., Yang, W. C., Walker, B. C., Stach, E. A., Hillhouse, H. W., & Agrawal, R. (2010). Fabrication of 7.2% efficient CZTSSe solar cells using CZTS nanocrystals. *Journal of the American Chemical Society*, *132*(49), 17384–17386.

[21] Green, M., Dunlop, E., Hohl-Ebinger, J., Yoshita, M., Kopidakis, N., & Hao, X. (2021). Solar cell efficiency tables (version 57). *Progress in Photovoltaics: Research and Applications*, *29*(1), 3–15.

[22] Beek, W. J., Wienk, M. M., & Janssen, R. A. (2004). Efficient hybrid solar cells from zinc oxide nanoparticles and a conjugated polymer. *Advanced Materials*, *16*(12), 1009–1013.

[23] Yin, X., Chen, P., Que, M., Xing, Y., Que, W., Niu, C., & Shao, J. (2016). Highly efficient flexible perovskite solar cells using solution-derived NiO x hole contacts. *ACS Nano*, *10*(3), 3630–3636.

[24] Goetzberger, A., Luther, J., & Willeke, G. (2002). Solar cells: Past, present, future. *Solar Energy Materials and Solar Cells*, *74*(1–4), 1–11.

[25] Fraas, L. M., & Fraas, L. M. (2014). History of solar cell development. *Low-Cost Solar Electric Power*, 1–12.

[26] Hegedus, S., & Luque, A. (2011). Achievements and challenges of solar electricity from photovoltaics. *Handbook of photovoltaic science and engineering* (pp. 1–38). John Wiley & Sons, Ltd.

[27] Sun, S. S., & Sariciftci, N. S. (Eds.). (2017). *Organic photovoltaics: mechanisms, materials, and devices*. CRC Press.

[28] Perlin, J. (1999). *From space to earth: the story of solar electricity*. Earthscan.

[29] Fraas, L. M. (2014). *Low-cost solar electric power* (p. 97). Springer.

[30] Enebe, G. C. (2019). *Modeling and Simulation of Nanostructured Copper Oxides Solar Cells for Photovoltaic Application*. University of Johannesburg (South Africa).

[31] Grätzel, M. (2003). Dye-sensitized solar cells. *Journal of photochemistry and Photobiology C: Photochemistry Reviews*, *4*(2), 145–153.

[32] O'regan, B., & Grätzel, M. (1991). A low-cost, high-efficiency solar cell based on dye-sensitized colloidal TiO2 films. *Nature*, *353*(6346), 737–740.

[33] Lee, C. P., Li, C. T., & Ho, K. C. (2017). Use of organic materials in dye-sensitized solar cells. *Materials Today*, *20*(5), 267–283.

[34] Chen, F. F., & Chen, F. F. (2011). The future of energy II: Renewable energy. *An Indispensable Truth: How Fusion Power Can Save the Planet*, Springer, New York, NY, 75–175.

[35] El Chaar, L., & El Zein, N. (2011). Review of photovoltaic technologies. *Renewable and sustainable energy reviews*, *15*(5), 2165–2175.

[36] Oghogho, I. (2014). Solar energy potential and its development fortainable energy generation in Nigeria: A road map to achieving this feat. *International Journal of Engineering and Management Sciences*, *5*(2), 61–67.

[37] Markvart, T., & Castañer, L. (2018). Principles of solar cell operation. In *McEvoy's Handbook of Photovoltaics* (pp. 3–28). Academic Press.

[38] Green, M. A. (1982). *Solar cells: Operating principles, technology, and system applications*. Prentice-Hall, Inc.

[39] Würfel, P., & Würfel, U. (2016). *Physics of solar cells: From basic principles to advanced concepts*. John Wiley & Sons.

[40] Pastuszak, J., & Węgierek, P. (2022). Photovoltaic Cell Generations and Current Research Directions for Their Development. *Materials*, *15*(16), 5542.

[41] Kingsley O. Ukoba, freddie I. Inambao, and E. A. Adeoye. (2018). "Modeling and Simulation of Metal Oxide Solar Cells: An Overview," *Proceeedings of the 16th industrial and commercial use of energy conference*, 2018.

[42] Chopra, K. L., Paulson, P. D., & Dutta, V. (2004). Thin-film solar cells: An overview. *Progress in Photovoltaics: Research and applications*, *12*(2–3), 69–92.

[43] Ukoba, K. O., Inambao, F. L., & Eloka-Eboka, A. C. (2018). Fabrication of affordable and sustainable solar cells using NiO/TiO$_2$ PN heterojunction. *International Journal of Photoenergy*, 2018(1), 1–7.

[44] Ukoba, K. O., Eloka-Eboka, A. C., & Inambao, F. L. (2018). Review of nanostructured NiO thin film deposition using the spray pyrolysis technique. *Renewable and Sustainable Energy Reviews*, 82, 2900–2915.

[45] Burgelman, M., Decock, K., Niemegeers, A., Verschraegen, J., & Degrave, S. (2016, February). *SCAPS manual*.

[46] Xiao, W., Dunford, W. G., & Capel, A. (2004, June). A novel modeling method for photovoltaic cells. In *2004 IEEE 35th Annual Power Electronics Specialists Conference (IEEE Cat. No. 04CH37551)* (vol. 3, pp. 1950–1956). IEEE.

[47] Kayes, B. M., Atwater, H. A., & Lewis, N. S. (2005). Comparison of the device physics principles of planar and radial p-n junction nanorod solar cells. *Journal of Applied Physics*, *97*(11), 114302.

[48] Ali, N. M., Allam, N. K., Abdel Haleem, A. M., & Rafat, N. H. (2014). Analytical modeling of the radial pn junction nanowire solar cells. *Journal of Applied Physics*, *116*(2), 024308.

[49] Petrosyan, S., Yesayan, A., & Nersesyan, S. (2012). Theory of nanowire radial pn-junction. *World Acad. Sci. Eng. Technol*, *71*, 1065–1070.

[50] Strang, G., Fix, G. J., & Griffin, D. S. (1974). *An analysis of the finite-element method*.

[51] Smith, G. D., & Smith, G. D. (1985). *Numerical solution of partial differential equations: finite difference methods*. Oxford University Press.

[52] Kabir, E., Kumar, P., Kumar, S., Adelodun, A. A., & Kim, K. H. (2018). Solar energy: Potential and future prospects. *Renewable and Sustainable Energy Reviews*, *82*, 894–900.

[53] Ukoba, K. O., Inambao, F. I., & Adeoye, E. A. (2018). "Modeling and Simulation of Metal Oxide Solar Cells: An Overview," *Proceeedings of the 16th industrial and commercial use of energy conference*, 2018.

[54] Sawicka-Chudy, P., Starowicz, Z., Wisz, G., Yavorskyi, R., Zapukhlyak, Z., Bester, M., ... & Cholewa, M. (2019). Simulation of TiO$_2$/CuO solar cells with SCAPS-1D software. *Materials Research Express*, *6*(8), 085918.

[55] Burgelman, M., Decock, K., Niemegeers, A., Verschraegen, J., & Degrave, S. (2016, February). *SCAPS manual*.

[56] Assal, S., Yadir, S., Amiry, H., Sidki, M., & Benhmida, M. (2014, October). Analysis of technological factors's influence on solar cell's performance by PC1D simulation. In *2014 International renewable and sustainable energy conference (IRSEC)* (pp. 1–5). IEEE.

[57] Clugston, D. A., & Basore, P. A. (1997, September). PC1D version 5: 32-bit solar cell modeling on personal computers. In *Conference record of the twenty sixth IEEE photovoltaic specialists conference-1997* (pp. 207–210). IEEE.

[58] Kumar Kundu, P., & Ghosh, P. (2017). Modeling and simulation of solar cell using embedded MATLAB simulink tool. *International Journal of Electronics, Electrical and Computational System IJEECS*, 6(7), 101–114.

[59] Zhu, H., Kalkan, A. K., Hou, J., & Fonash, S. J. (1999, March). Applications of AMPS-1D for solar cell simulation. In *AIP conference proceedings* (vol. 462, no. 1, pp. 309–314). American Institute of Physics.

[60] Nasser, H., Kökbudak, G., Mehmood, H., & Turan, R. (2017). Dependence of n-cSi/MoOx heterojunction performance on cSi doping concentration. *Energy Procedia*, *124*, 418–424.

[61] Liu, Y., Heinzel, D., & Rockett, A. (2011, June). A new solar cell simulator: WxAMPS. In *2011 37th IEEE photovoltaic specialists conference* (pp. 002753–002756). IEEE.

[62] Liu, Y., Sun, Y., & Rockett, A. (2012). A new simulation software of solar cells – WxAMPS. *Solar Energy Materials and Solar Cells*, *98*, 124–128, DOI: 10.1016/j.solmat.2011.10.010.

[63] Burgelman, M., Verschraegen, J., Degrave, S., & Nollet, P. (2004). Modeling thin-film PV devices. *Progress in Photovoltaics: Research and Applications*, *12*(23), pp. 143–153. DOI: 10.1002/pip.524.

[64] Burgelman, M., Verschraegen, J., Degrave, S., & Nollet, P. (2004). Modeling thin-film PV devices. *Progress in Photovoltaics: Research and Applications*, *12*(2–3), 143–153.

[65] Burgelman, M., Nollet, P., & Degrave, S. (2000). Modelling polycrystalline semiconductor solar cells. *Thin Solid Films*, *361*, 527–532.

[66] Fonash, S., Arch, J., Cuiffi, J., Hou, J., Howland, W., McElheny, P., ... & Rubinelli, F. (1997). *A manual for AMPS-1D for Windows 95/NT a one-dimensional device simulation program for the analysis of microelectronic and photonic structures.* The Pennsylvania State University.

[67] Hernández-Como, N., & Morales-Acevedo, A. (2010). Simulation of hetero-junction silicon solar cells with AMPS-1D. *Solar Energy Materials and Solar Cells*, *94*(1), 62–67.

[68] Amin, N., Tang, M., & Sopian, K. (2007, December). Numerical modeling of the copper-indium-selenium (CIS) based solar cell performance by AMPS-1D. In *2007 5th Student Conference on Research and Development* (pp. 1–6). IEEE.

[69] Tsai, H. L., Tu, C. S., & Su, Y. J. (2008, October). Development of generalized photovoltaic model using MATLAB/SIMULINK. In *Proceedings of the world congress on Engineering and computer science* (vol. 2008, pp. 1–6).

[70] Salmi, T., Bouzguenda, M., Gastli, A., & Masmoudi, A. (2012). Matlab/simulink based modeling of photovoltaic cell. *International Journal of Renewable Energy Research*, *2*(2), 213–218.

[71] Altas, I. H., & Sharaf, A. M. (2007, May). A photovoltaic array simulation model for Matlab-Simulink GUI environment. In *2007 International conference on clean electrical power* (pp. 341–345). IEEE.

[72] Hussain, A. B., Abdalla, A. S., Mukhtar, A. S., Elamin, M., Alammari, R., & Iqbal, A. (2017). Modelling and simulation of single-and triple-junction solar cells using MATLAB/SIMULINK. *International Journal of Ambient Energy*, *38*(6), 613–621.

[73] Dehghanzadeh, A., Farahani, G., & Maboodi, M. (2017). A novel approximate explicit double-diode model of solar cells for use in simulation studies. *Renewable Energy*, *103*, 468–477.

[74] Dessaint, L.-A., Al-Haddad, K., Le-Huy, H., Sybille, G., & Brunelle, P. A power system simulation tool based on Simulink. *IEEE Transactions on Industrial Electronics*, *46*(6), 1252–1254.

[75] Altas, I. H., & Sharaf, A. M. (2007, May). A photovoltaic array simulation model for Matlab-Simulink GUI environment. In *2007 International conference on clean electrical power* (pp. 341–345). IEEE.

[76] Keles, C., Alagoz, B. B., Akcin, M., Kaygusuz, A., & Karabiber, A. (2013, May). A photovoltaic system model for Matlab/Simulink simulations. In *4th International conference on power engineering, energy and electrical drives* (pp. 1643–1647). IEEE.

[77] Ishaque, K., Salam, Z., & Taheri, H. (2011). Accurate MATLAB simulink PV system simulator based on a two-diode model. *Journal of Power Electronics*, *11*(2), 179–187.

[78] Xue, D., & Chen, Y. (2013). *System simulation techniques with MATLAB and Simulink.* John Wiley & Sons.

[79] Klee, H., & Allen, R. (2016). *Simulation of dynamic systems with MATLAB and Simulink.* CRC Press.

[80] Chaturvedi, D. K. (2017). *Modeling and simulation of systems using MATLAB and Simulink.* CRC Press.

[81] Liu, Y., Sun, Y., & Rockett, A. (2012). A new simulation software of solar cells—wxAMPS. *Solar Energy Materials and Solar Cells, 98,* 124–128.

[82] Liu, Y., Sun, Y., & Rockett, A. (2012, June). Batch simulation of solar cells by using Matlab and wxAMPS. In *2012 38th IEEE photovoltaic specialists conference* (pp. 000902–000905). IEEE.

[83] Hossain, M. K., Rubel, M. H. K., Toki, G. I., Alam, I., Rahman, M. F., & Bencherif, H. (2022). Effect of various electron and hole transport layers on the performance of CsPbI3-based perovskite solar cells: A numerical investigation in DFT, SCAPS-1D, and wxAMPS frameworks. *ACS Omega, 7*(47), 43210–43230.

[84] Messei, N. (2016). *Study of the effect of grading in composition on the performance of thin film solar cells based on AlGaAs and CZTSSe, a numerical simulation approach.* Doctoral Dissertation, Thèse de doctorat, Freres Mentouri Constantine1 Faculty of Sciences Exactes.

[85] Sayeed, M. A., & Salem, I. (2017). *Design and Simulation of Efficient Perovskite Solar Cells.* Doctoral Dissertation, East West University.

[86] Liu, Y., Sun, Y., & Rockett, A. (2012). A new simulation software of solar cells—wxAMPS. *Solar Energy Materials and Solar Cells, 98,* 124–128.

[87] Sheng, Y., Xia, C. S., Simon Li, Z. M., & Cheng, L. W. (2013). Simulation of InGaN/GaN light-emitting diodes with patterned sapphire substrate. *Optical and Quantum Electronics, 45,* 605–610.

[88] Piprek, J. (2013). *Semiconductor optoelectronic devices: introduction to physics and simulation.* Elsevier.

[89] Burgelman, M., & Marlein, J. (2008, September). Analysis of graded band gap solar cells with SCAPS. In *Proceedings of the 23rd European photovoltaic solar energy conference, Valencia* (pp. 2151–2155).

[90] Schwartz, R. J., Gray, J. L., & Lundstrom, M. S. (1985, May). Current status of one-and two-dimensional numerical models: Successes and limitations. In *JPL Proc. of the flat-plate solar array Proj. Res. forum on high-efficiency crystalline silicon solar cells.*

[91] Mouchou, R. T., Jen, T. C., Laseinde, O. T., & Ukoba, K. O. (2021). Numerical simulation and optimization of p-NiO/n-TiO$_2$ solar cell system using SCAPS. *Materials Today: Proceedings, 38,* 835–841.

[92] Hussain, S. S., Riaz, S., Nowsherwan, G. A., Jahangir, K., Raza, A., Iqbal, M. J., ... & Naseem, S. (2021). Numerical modeling and optimization of lead-free hybrid double perovskite solar cell by using SCAPS-1D. *Journal of Renewable Energy, 2021,* 1–12.

[93] Rahman, M. A. (2021). Design and simulation of a high-performance Cd-free Cu 2 SnSe 3 solar cells with SnS electron-blocking hole transport layer and TiO 2 electron transport layer by SCAPS-1D. *SN Applied Sciences, 3,* 1–15.

[94] Amin, N., Chelvanathan, P., Hossain, M. I., & Sopian, K. (2012). Numerical modelling of ultra thin Cu (In, Ga) Se2 solar cells. *Energy Procedia, 15,* 291–298.

[95] Okamoto, C., Mitsuda, N., Minemoto, T., Azuma, Y., Omae, S., Takakura, H., & Hamakawa, Y. (2003, May). A new type of spherical silicon solar cells with semi-concentrated reflector system. In *Proceedings of 3rd world conference onphotovoltaic energy conversion, 2003* (vol. 2, pp. 1324–1327). IEEE.

[96] Amin, N., Sopian, K., & Konagai, M. (2007). Numerical modeling of CdS/CdTe and CdS/CdTe/ZnTe solar cells as a function of CdTe thickness. *Solar Energy Materials and Solar Cells, 91*(13), 1202–1208.

[97] Kotov, I. V., Humanic, T. J., Nouais, D., Randel, J., Rashevsky, A., & Alice-Its Collaboration. (2006). Electric fields in nonhomogeneously doped silicon. Summary of simulations. *Nuclear Instruments and Methods in Physics Research Section A: Accelerators, Spectrometers, Detectors and Associated Equipment, 568*(1), 41–45.

[98] Levinshtein, M. E., & Mnatsakanov, T. T. (2002). On the transport equations in popular commercial device simulators. *IEEE Transactions on Electron Devices, 49*(4), 702–703.

[99] Kersys, T., Anilionis, R., & Eidukas, D. (2008). Simulation of doped Si oxidation in nano-dimension scale. *Elektronika ir Elektrotechnika, 84*(4), 43–46.

[100] Garcia Jr, B. (2007). *Indium gallium nitride multijunction solar cell simulation using silvaco atlas.* PhD diss., Monterey California. Naval Postgraduate School.

[101] Zeman, M., Willemen, J. A., Vosteen, L. L. A., Tao, G., & Metselaar, J. W. (1997). Computer modelling of current matching in a-Si: H/a-Si: H tandem solar cells on textured TCO substrates. *Solar Energy Materials and Solar Cells*, *46*(2), 81–99.

[102] Zeman, M., Isabella, O., Jäger, K., Babal, P., Solntsev, S., & Santbergen, R. (2011). Modeling of advanced light trapping approaches in thin-film silicon solar cells. *MRS Online Proceedings Library (OPL)*, *1321* .

[103] BhagyaRajasekaraiah, U., Saini, A., Krishnan, S., & Vatedka, R. (2023). Studying the effect of space radiation induced defects in multijunction solar cell using APSYS simulation software and comparison with the experimental data. *Nuclear Instruments and Methods in Physics Research Section B: Beam Interactions with Materials and Atoms*, *535*, 74–87.

[104] Xiao, Y. G., Li, Z. Q., & Li, Z. S. (2008, September). Modeling of GaInP/GaAs/Ge and the inverted-grown metamorphic GaInP/GaAs/GaInAs triple-junction solar cells. In *High and Low Concentration for Solar Electric Applications III* (vol. 7043, pp. 49–60). SPIE.

[105] He, Y., & Yan, W. (2021). Simulation of 0.67 eV-InGaAs solar cell with InP window-layers of various thickness on virtual InP/AlGaInAs/GaAs substrates. *Engineering Research Express*, *3*(4), 045007.

[106] Zhang, R., Hollars, D. R., & Kanicki, J. (2012, July). CIGS solar cell on flexible stainless steel substrate fabricated by sputtering method: Simulation and experimental results. In *2012 19th International workshop on active-matrix flatpanel displays and devices (AM-FPD)* (pp. 289–292). IEEE.

[107] Xiao, Y. G., Lestrade, M., Li, Z. Q., & Li, Z. S. (2007, October). Modeling of Si-based solar cells with V-grooved surface texture by Crosslight APSYS. In *Photovoltaic Cell and Module Technologies* (vol. 6651, pp. 143–150). SPIE.

[108] Li, Z. Q., Xiao, Y. G., & Li, Z. S. (2007). Two-dimensional simulation of GaInP/GaAs/Ge triple junction solar cell. *Physica Status Solidi C*, *4*(5), 1637–1640.

[109] Li, Z. Q., Xiao, Y. G., & Li, Z. S. (2006, September). Modeling of multi-junction solar cells by Crosslight APSYS. In *High and low concentration for solar electric applications* (vol. 6339, pp. 29–36). SPIE.

5 Performance Evaluation of a Batch

Reactor for Catalytic Depolymerization of Polymeric Waste

O. L. Rominiyi, M. A. Akintunde, E. I Bello, L. Lajide, and O. M. Ikumapayi

5.1 INTRODUCTION

Municipal solid waste (MSW) is fast growing all over the world as a result of consumption patterns and increasing urbanization coupled with bad management practices [1]. It is estimated that within the next 34 years (from 2016 to 2050) the MSW will increase from 2.01 to 3.40 billion tons [2]. Plastic waste accounts for a large part of MSW due to its wide range of applications. Moreover, waste polyethylene (WPE) accounts for 40% of plastics in MSW [3]. About 70% of the total produced plastics have been directly deposited into the environment. Mismanagement of plastic waste has aggravated environmental pollution and endangered human health.

Therefore, there is a need to recycle waste plastics, especially WPE. Depolymerization has been found to be the best alternative in the management of polymeric waste. This involved heating the polymeric waste in an inert environment at an elevated temperature to produce the condensable gas (gaseous fuel), non-condensable gas (liquid fuel), and solid residue (Char). These products have a reasonable calorific value that makes them suitable as alternative energy resources. The efficiency of polymeric waste to liquid fuel in a batch reactor can be improved with the addition of a catalyst [4]. Rominiyi et al. [5] developed a gasifier for the production of liquid fuel from *Spondias mombin* an agro waste, which is a common pollutant found in the mechanic workshop.

Apart from the prevention of air pollution as one of the benefits of pyrolysis, it is safer and more environmental friendly than incineration, landfill, and other means of energy recovery like gasification. The waste is converted into new products such as useful liquid fuel, carbon black, and combustible gases [6,7].

The temperature and heating rates can be controlled to produce desired solid, gas, and liquid products because they have considerable influence on the pyrolysis process [8].

Pyrolysis of waste plastic is one of the most feasible large-scale methods of energy regeneration. This is because waste plastic is a valuable source of liquid and gas fuels as well as chemicals [8].

Using a straightforward batch pyrolysis process, research by Low et al. [9] demonstrated that it is possible to convert a mixture of waste plastics, notably polyolefins and polystyrene, into a usable liquid fuel with a yield of at least 70%. It was established in their initial test that it is promising and practical to create liquid products utilizing waste plastic that has not been sorted. The liquid product's chemical compositions resemble those of current hydrocarbon fuels, including diesel, gasoline, and kerosene. As a result, the liquid product can either be processed in a refinery to further

DOI: 10.1201/9781032651958-5

raise its quality or used directly as fuel. Their report's findings also point to the potential benefit of employing a small-scale, straightforward pyrolysis technique for dealing with plastic wastes, but they also point out that more research is needed to determine how to separate PVC and polyethylene terephthalate (PET) from mixed waste plastics and determine how closely the patterns of thermal decomposition of plastic mixtures conform to those of the individual polymers present [9].

The use of incinerator generates some pollutants to the air, which also cause environmental issues. Therefore, recycling and recovering methods have been used to minimize the environmental impacts and to reduce the damage of plastic waste. Chemical recycling via the pyrolysis process is one of the promising methods to recycle waste plastics, which involves thermo-chemical decomposition of organic and synthetic materials at elevated temperatures in the absence of oxygen to produce fuels. The process is usually conducted at temperatures between 400°C and 800°C [10].

Pyrolysis has been shown to be a viable method of processing this type of trash by Lopez et al. study on the recycling of packing and packaging plastic wastes using the approach, as the achieved transformation to liquid + gas can reach more than 95%. The amount of temperature, which is a key factor in the procedure, does, nevertheless, have a significant impact on the pyrolysis product. Although a highly aromatic liquid and more short-chain hydrocarbons in the gases can increase at a higher working temperature, a small temperature operation can produce a significant proportion of wax-like liquids. Certain chars are thought to have formed as a result of subsequent repolymerization reactions. A laboratory-scale pyrolysis reactor using polyethylene, polypropylene, and polystyrene underwent theoretical and experimental research.

According to reports, the pyrolysis cracking temperature for polypropylene and polyethylene is 450°C. However, the value for polystyrene is dramatically lower at 320°C. Relatively high temperatures and heating rates can have a substantial effect on the manufacturing of light hydrocarbons. The generation of light hydrocarbon compounds can also be aided by prolonged residence times. In their research study, they looked into a variety of parameters, including the catalyst, the kind of reactor, the reflux rate, and the pressure. Polypropylene pyrolysis product with high reflux rates has 84.2% liquid products, 15.7% non-condensable gases, and less than 0.025% char. Polystyrene was pyrolyzed to generate an end result that is 93% liquid, 4% non-condensable gases, and 3% char [11].

High-density polyethylene (HDPE) and polystyrene (PS) waste were used as raw material in a semi-batch reactor with 9.1 wt% catalysts at a temperature of 400°C; it was discovered that the quantity of liquid products increases clearly with an increase in the mixing proportions of polystyrene against HDPE. Increased Polystyrene content in the PS and HDPE blend results in elevated liquid yields and rapid biodegradability. The non-catalytic pyrolysis process has also been studied using waste PE, PP, and PS. The results showed that waste PS produced higher liquid while waste PE and PP produced higher gaseous products [12].

This suggests that in the aforementioned admixture, polystyrene pyrolysis predominates over that of HDPE. Even without the support of knowledgeable trading companies, foreign manufacturers and companies that produce these conversion technologies may find it difficult to export their products to countries like the Philippines, or they may refuse to sell their waste plastic conversion technologies due to business and intellectual property concerns. It can appear unfeasible to import machinery for waste plastic conversion technologies from other nations given the expense of the technology, equipment, and transportation [13].

A waste plastic oil converter was conceived and built by Elmo and Rapsing [14]. It was a model that was intentionally created as a prototype in an effort to establish an eco-friendly way to recycle garbage. Electricity is used to operate the machinery, which uses a non-catalytic pyrolysis method to turn waste plastic into oil. It runs in batches and has a fixed-bed reactor. Using polystyrene waste plastic as a feedstock, it was demonstrated to be functional and operational. Other waste plastics were not used to test the apparatus, though. So, this study was carried out to learn how well locally produced machinery performed at turning various forms of waste plastics into oil. This investigation is seen to be important in solving the growing issue of trash disposal. The amount of plastic that is burned and dumped in a landfill or dumpsite will be kept to a minimum, lowering the number of carbon emissions and carbon sinks in the environment. As a result of the apparatus, only the converted oil and char were gathered [14].

Due to the lack of a vapor recovery facility, uncondensed vapor or gas was not collected. The economic aspects of the equipment were left out of the study because they are still in the prototype stage and are solely designed for laboratory experiments. The study did not provide include information on the physical and chemical characteristics of the gathered oil [14].

Waste plastic can be successfully converted to fuel through pyrolysis, but it takes a lot of energy to maintain the severe temperatures inside the reactor reliably and continuously. The pyrolysis process has a significant energy requirement, which might make it challenging to allow for industrial and financial viability. Yet, there are ways to lower or regulate the temperature and energy requirements. The temperature and energy requirements can be considerably reduced by lowering the pressure inside the vessel, which inevitably results in a change in design that gives far higher industrial viability and eventually makes it possible to scale up operations. Also, the resulting decrease in reactor energy need suggests that using renewable energy sources and recycling streams might satisfy the entire energy demand [15].

Nevertheless, reduced ambient temperature and slow pyrolysis could both minimize the thermal lag phenomena and increase liquid biofuel [16,17]. Slow pyrolysis facilitates homogenous temperature distributions inside the reactor and the reactants [18] favoring heat and mass transfer for liquid fuel synthesis [19,16]. The proportion of the products, according to Sharuddin et al. [20] depends on the reactor type, the presence of catalysts, and the operational parameters such as temperature, residence time, and carrier gas flow rate. Semi-batch reactors have been used in numerous investigations on the thermal pyrolysis of polyethylene (PE). High liquid fuel outputs have been claimed to be possible [21–26].

When it comes to operating circumstances, temperature is the key factor in the pyrolysis of plastics. At temperatures exceeding 425°C, the virgin low-density PE was totally pyrolyzed, according to Onwudili et al. [3]. At 425°C, the liquid fuel output was 89.5 weight percent. In addition, they discovered that the output of liquid oil sharply decreased as the temperature rose above 425°C. According to Quesada et al. [23], 450°C was the ideal temperature for the pyrolysis of WPE to produce liquid oil.

A 500°C temperature was said to be ideal for producing liquid gasoline [23]. In addition, Sharuddin et al. [20] found that plastic pyrolysis for the creation of liquid fuel was best done at temperatures under 500°C. The residence time was connected to the amount of time the study took place at the target temperature, according to Onwudili et al. [3]. According to Quesada et al. [23], residence duration is also an important consideration when figuring out the makeup of pyrolysis products.

In a 200 mL bench-scale semi-batch reactor using nitrogen as the carrier gas. Quesada et al. [23], further studied the effect of the carrier gas flow rate on PE's pyrolysis. Using gas flow rates between 0 and 60 mL/min, the formation of liquid fuel and char was achieved. The concentration of PE's thermal pyrolysis products might also be affected by the flow rate of the carrier gas, and a higher gas carrier flow rate would result in a higher yield of liquid fuel and a lower yield of char.

The temperature of the biomass-fed reactor is quite difficult to control because there are many factors affecting the combustion efficiency of the biomass stove; these factors are the flow rate of air into the combustion chamber, air-fuel ratio, air feeding distance to the combustion chamber, and air feeding conduit diameter [27,28].

The temperature range during the operation can be as low as 195°C (because of blower operation gets interrupted by an electrical power outage) to the maximum temperature of 650°C [29].

The thermal degradation behavior of plastics has been investigated by Aboulkas et al. [30]. The activation energy and the reaction model of the pyrolysis of polyethylene (PE) and polypropylene

(PP) have been estimated for non-isothermal kinetic results. Several reactor systems have been developed and used, such as batch/semi-batch [31], fixed bed, fluidized bed, spouted bed, microwave [32], and screw kiln. Many researchers have used batch or semi-batch reactors because of their simple design and easy operation.

Akinola [33] utilized thermodynamic principles to obtain the reactor efficiency at varying temperatures. The results of linear regression equations for pyrolytic temperatures and the calculated temperatures showed a correlation coefficient of 95.50% for product temperature, 91.00% for reactor efficiency at generating gas, and 71.60% for reactor thermal efficiency.

A mean efficiency of reactor 86.00% was established. A full factorial experimental approach was adopted by Ogedengbe et al. [34] to conduct the pyrolysis experiments on used tyres. The values of the process parameters used for the experiments varied at three levels and were combined to give nine experiments. Each of the nine experiments was repeated thrice to arrive at a total of 27 experimental sets, which were carried out and assessed using analysis of variance (ANOVA).

Subsequently, models were formulated and validated for estimating the pyrolysis product yield with respect to the operating parameter considered. The result shows that the main effect of change in pyrolysis temperature and feedstock size on all product yield was statistically significant, while the effect resulting from the interaction between the process parameters was not. The models formulated for predicting the product yield were able to adequately estimate the percentage yield obtained from experiments, using the values of the pyrolysis temperature and feedstock size, with a correlation coefficient of 90.00%, 78.63%, and 95.11% for char, pyro-oil, and pyro-gas yields, respectively. The mean fuel conversion efficiency was 87.91%. A fuel conversion efficiency of 80% was obtained in a reactor by Ibrahim [35].

In summary, a lot of studies have been conducted to examine how the process parameters (temperature, residence time, and carrier gas flow rate) affect the amount of liquid fuel produced from the pyrolysis of PE. On the other hand, the majority of these studies examined these effects by altering the operational requirements one at a time.

In this research, the depolymerization process was utilized to recycle the polymeric waste for liquid and gaseous fuel production in a batch reactor using local content, and the performance evaluation of the energy system was carried out.

5.2 MATERIALS AND METHOD

The materials used for the performance evaluation of the polymeric liquid fuel include but not limited to the following: spring balance; digital scale; calcium oxide; activated carbon; reactor assembly; thermal sensor; Flowmeter; pressure guage; pyrometer; electricity; heating element; controller; LCD board; stop watch, raw HDPE liquid fuel sample; activated carbon-catalyzed HDPE liquid fuel sample; calcium oxide-catalyzed HDPE liquid fuel sample; raw low-density polyethylene (LDPE) liquid fuel sample; activated carbon-catalyzed LDPE liquid fuel sample; calcium oxide-catalyzed LDPE liquid fuel sample; raw HDPE:LDPE mixture (1:1); liquid fuel sample; HDPE:LDPE-activated carbon-catalyzed mixture (1:1) liquid fuel sample; HDPE:LDPE calcium oxide-catalyzed mixture (1:1) liquid fuel sample; HDPE:conventional diesel blended (1:1) liquid fuel sample; HDPE:conventional diesel blended (1:2) liquid fuel sample; HDPE/conventional diesel blended (2:1) liquid fuel sample; LDPE:conventional diesel blended (1:1) liquid fuel sample; LDPE:conventional diesel blended (1:2) liquid fuel sample; LDPE:conventional diesel blended (2:1) liquid fuel sample; conventional diesel; kerosine and petrol/gasoline.

The polymeric waste, which consists of: HDPE in the form of garbage containers, toilet buckets, and kegs; LDPEas used in low-grade plastic bags, nylons; Polystyrene (PS), in the form of disposable cutlery, take away packs and cups; polypropylene (PP) as used waste plastic containers and

TABLE 5.1
Melting Point of Some Selected Polymeric Wastes

Polymeric Wastes	Melting Point (°C)
High-density polyethylene (HDPE)	180
Low-density polyethylene (LDPE)	160
Polyethylene terephthalate (PET)	260
Polypropylene (PP)	190
Polystyrene (PS)	270
Polyvinyl chloride (PVC)	260
Polyamide	220
Acrylic	140

Source: Extracted from [36].

PET as used plastic bottles, were collected from various strategic places in Ado-Ekiti Metropolis, Ekiti State, Nigeria. The waste plastic materials were segregated and categorized. PET, HDPE, and LDPE plastic materials were thoroughly washed with water, sun-dried, and shredded to increase the surface area of the materials available for the reaction to proceed and in contact with the catalyst during the pyrolysis.

The melting point of polymeric waste was predetermined before shredding from the literature as indicated in Table 5.1, and waste plastics of 2 kg were fed into the batch reactor set up as in plate three without catalyst, flushed with nitrogen gas for 30 seconds to both create an inert environment and also act as a gas carrier, and fired with tungsten heating element of 15 kW capacity for 3 h and 30 min. The batch reactor outlet was connected to a condenser and also linked with the sub-cooler to improve the efficiency of the condensation process of the vaporized product from the reactor during an anaerobic thermal cracking of the polymeric waste obtained after melting and superheating the polymeric waste. The temperature was measured with the K type of thermocouple controlled by the thermal sensor. Fluid parameters like flow rate, temperature, and pressure of the products coming out of the system during the depolymerization process were measured by the multipurpose controller incorporated into the energy conversion system at intervals of 10 min.

Similarly, the retention time as well as the mass of the liquid fuel, gaseous fuel, and bio-char produced were measured with the digital weighing scale and spring balance. Also, 10 g of activated charcoal and 10 g of calcium oxide were singly added to the feedstock before being electrically fired to effect the depolymerization process of PET, HDPE, and LDPE polymeric wastes, respectively. The temperature, pressure, and flow rate of the depolymerized product readings were similarly taken at 10-min interval for the raw (un-catalyzed) PET, PET with activated carbon, PET with calcium oxide, HDPE raw (un-catalyzed), HDPE-activated carbon, HDPE mixed with calcium oxide, LDPE raw, LDPE with activated carbon and LDPE mixed with calcium oxide, the admixture is depicted in Figures 5.1.

The temperature inside the reactor was maintained between 100°C and 350°C for 3 h and 30 min. The retention time of the substrate for the reaction to proceed was measured by a stopwatch. The gaseous fuel was tapped off periodically on top of the collected oil due to the buoyancy effect. Figure 5.1 shows the gaseous fuel samples and liquid fuel samples.

The prepared oil was fractionated to obtain the pure diesel from the sludge accompanied by the liquid fuel produced, as in Figure 5.2. Table 5.1 presents the melting point of some polymeric wastes.

FIGURE 5.1   The liquid fuel (Bio-diesel) produced through depolymerization process in the batch reactor.

FIGURE 5.2   The fractionating sludge obtained during the depolymerization process.

5.3   RESULTS AND DISCUSSIONS

5.3.1   PET Depolymerized Without Catalyst

It was observed in Figure 5.3 that at the temperature of 140°C, there was a noticeable flow of the vaporized product from the reactor with a flow rate of 0.1317 L/min and a pressure of 26.1 kPa within 50 min of firing the reactor. The maximum flow rate of the gaseous product in the energy conversion system was observed to be 0.1790 L/min after 90 min of starting up with a pressure of 35.4 kPa and a temperature of 170°C. The progress of the polymeric cracking was disrupted due to the clogging of pipe. It was also shown in Figure 5.4 that gaseous fuel, liquid fuel, and residue produced are: (346 g) 17.3%, (20 g) 1%, and (588 g) 29.4%, respectively, while the unaccountable loss is (1046 g) 52.3%. The maximum temperature within 210 min of firing the system is 270°C at a pressure and flow rate of 35.8 kPa and 0.1368 L/min.

5.3.1.2   Assembly During Depolymerization of PET Without Catalyst
5.3.2   PET and Calcium Oxide-Catalyzed Depolymerization Process

Figure 5.5 shows the results of the observations of the batch reactor system when calcium oxide (CaO) was used as a catalyst with PET (polymeric waste) substrate. The retention time for this process was recorded as 45 min. It was shown that the temperature inside the reactor steadily increased

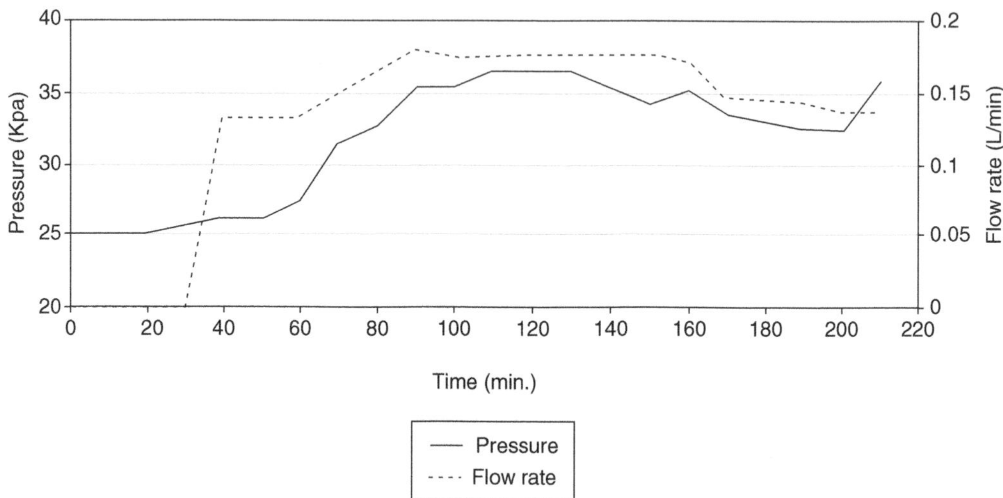

FIGURE 5.3 Variation of flow rate and pressure with time in batch reactor assembly during depolymerization of PET without catalyst.

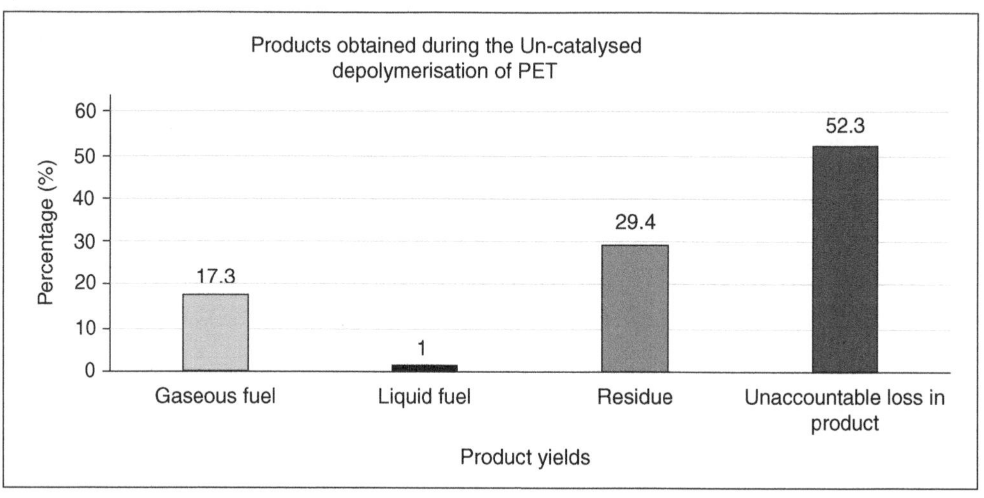

FIGURE 5.4 Product obtained during depolymerization of PET without catalyst.

from the ambient temperature of 30–211°C, which is higher than the melting point of polyethylene therephthalate during 210 min (3 h and 30 min) of firing. The maximum flow rate of the vaporized polymeric waste and pressure was obtained as 36.7 kPa and 0.1768 L/min, respectively, at the temperature of 179°C within 2 h of firing the reactor. It was also shown that there was no flow within the next 35 min of the firing but at a pressure of 29.3 kPa and temperature of 130°C there was an appreciable flow of 0.1301 L/min of the vaporized product from the reactor. Similarly, it was observed in the bar chart presented in Figure 5.6 also that, there was a clogging (yellowish thick fluid) that solidified along the tube, which prevented the reasonable mass of liquid fuel production of 60 g (3%) during the depolymerization process. The mass of gaseous fuel produced is 445 g (22.25%), while 827 g (44.35%) was the residue obtained and the unaccountable loss is 668 g (33.4%), 60 g (3%) of liquid fuel was produced.

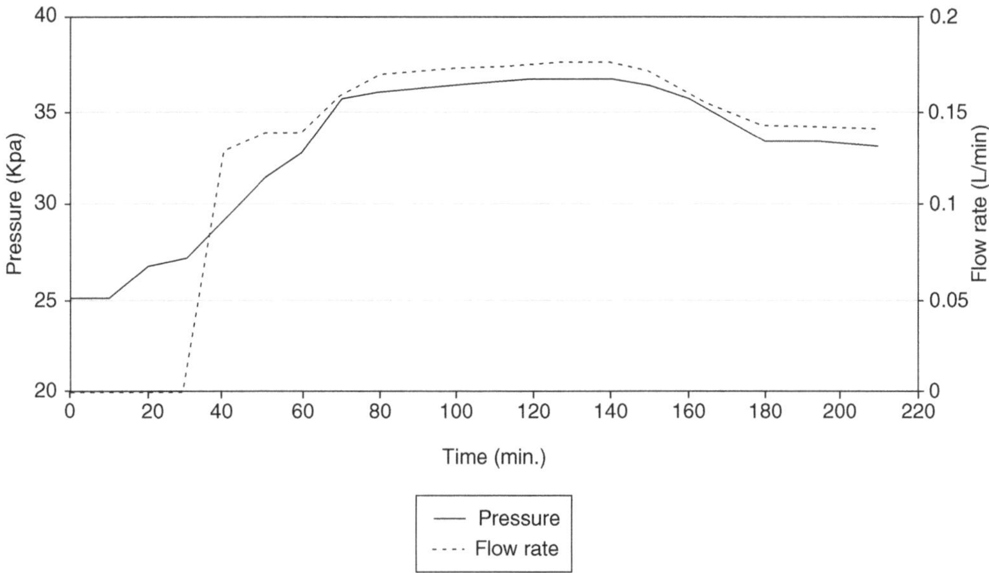

FIGURE 5.5 Variation of flow rate and pressure with time in batch reactor assembly during depolymerization of PET with calcium oxide as catalyst.

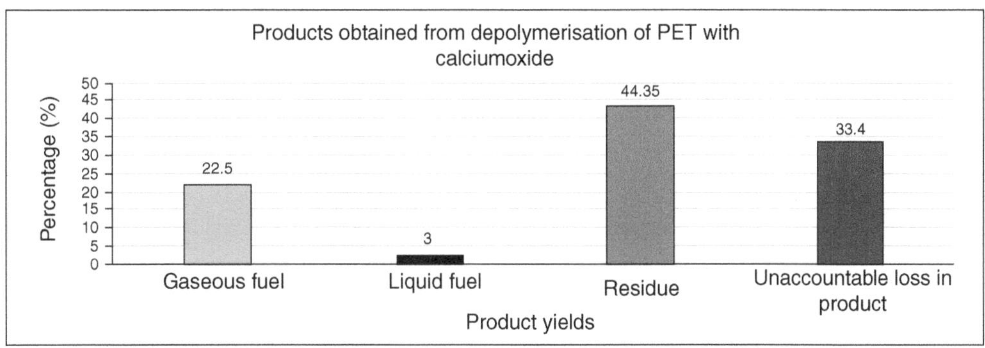

FIGURE 5.6 Product obtained from depolymerization of PET with calcium oxide as catalyst.

5.3.3 Depolymerized PET and Activated Carbon in the Batch Reactor System

Figure 5.7 shows that the retention time when the activated carbon was used was 38 min while the maximum flow rate of the depolymerized product coming out from the batch reactor system was 0.1987 L/min at a pressure of 37.6 kPa and 170°C within 100 min of heating the polymeric waste in the system. However, there is a time lag of 30 min for the wastes to acquire enough heat energy for the pressure of 26.8 kPa to build up in the system before there is an appreciable flow of the vaporized product at 0.1382 L/min during the process. It was observed from Figure 5.8 that out of the 2000 g of the feedstock fed into the reactor 600 g (30%) was converted to gaseous fuel, 40 g (2%) liquid fuel, the residual biochar was 960 g (48%) while the unaccountable loss in the product during the process was 400 g (20%) Figure 5.9 indicated that un-catalyzed depolymerized PET has the highest retention time of 97 min while the least of 38 min was obtained when PET was catalyzed with calcium oxide.

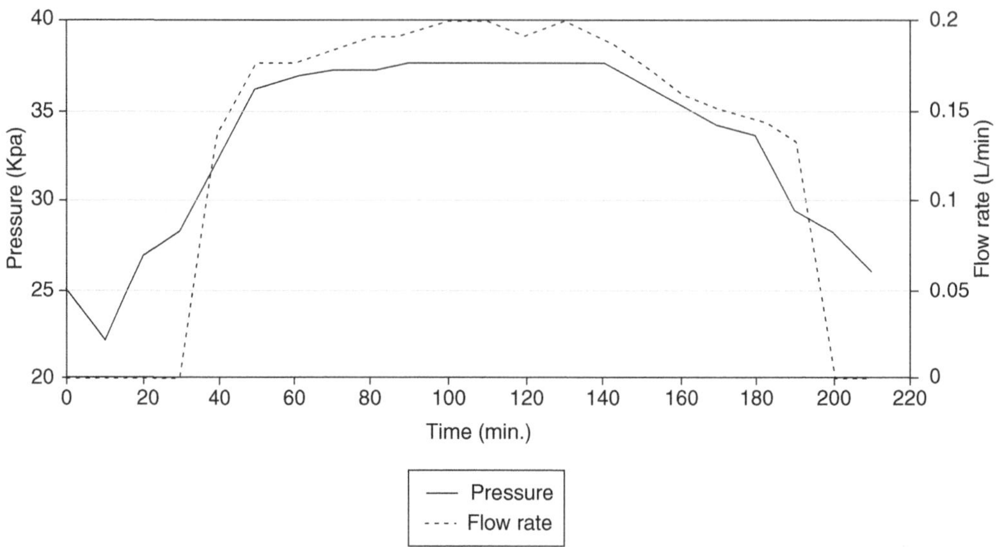

FIGURE 5.7　Variation of flow rate and pressure with time in batch reactor assembly during depolymerization of PET with 10 g of activated carbon as catalyst.

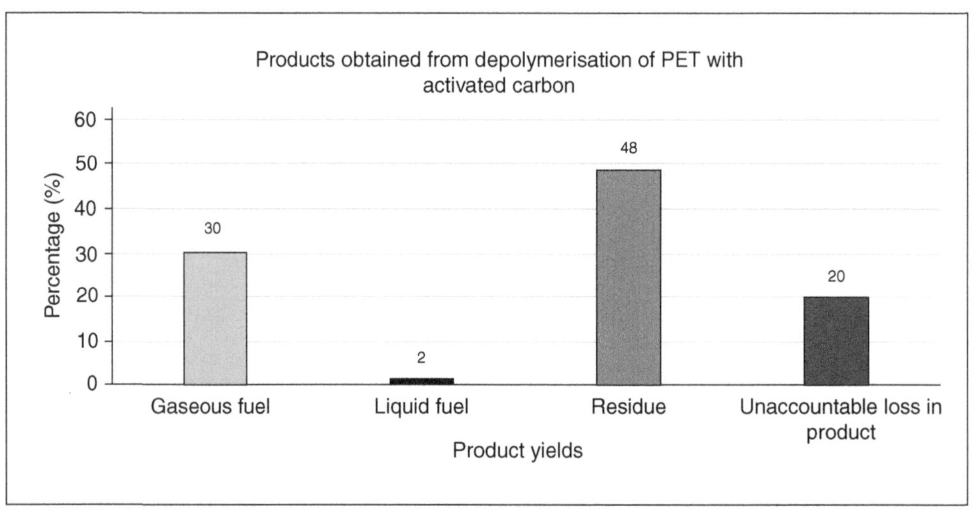

FIGURE 5.8　Product obtained from depolymerization of PET with activated carbon.

5.3.4　Depolymerization of Un-Catalyzed HDPE in the Batch Reactor

It was indicated in Figures 5.10 and 5.11, respectively, that when polymeric waste of HDPE whose melting point is 125°C was pyrolyzed without the addition of catalyst in the batch reactor energy system the time for the first drop of oil was 80 min at a temperature of 161°C and pressure of 31.1 kPa and flow rate of 0.1371 L/min. There was no appreciable flow of the vaporized HDPE between 10 min and 50 min because the feedstock has to gain enough heat energy for it to be melted before it became vaporized along the copper tube from the outlet of the reactor to the shell in the tube condenser inlet where the hot vaporized product losses heat to the water used as the cooling medium.

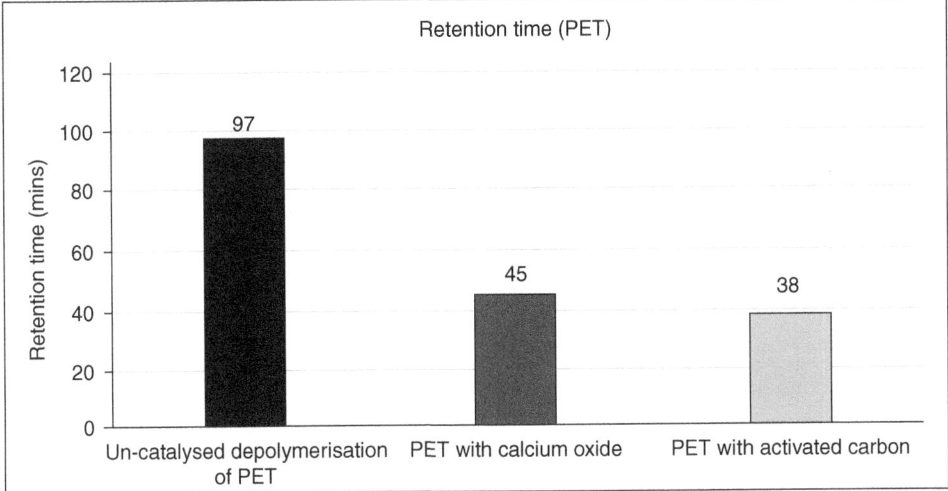

FIGURE 5.9   Retention time of PET during depolymerization process in batch reactor system.

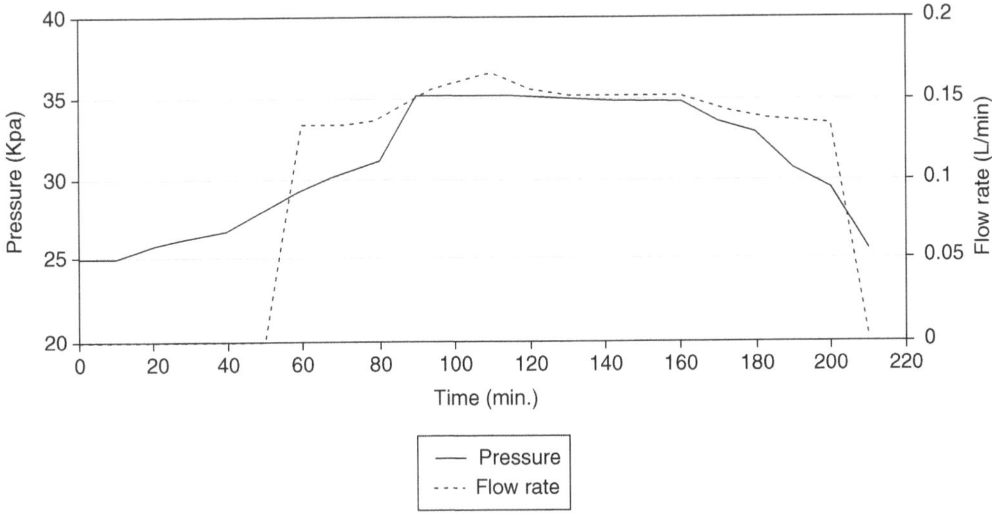

FIGURE 5.10   Variation of flow rate and pressure with time in batch reactor assembly during assembly during un-catalyzed depolymerization of HDPE.

The maximum flow rate of 1 product coming out from the reactor to the condenser was observed as 0.1661 L/min at a temperature of 175°C and pressure of 35.3 kPa within 110 min of carrying out the performance evaluation of the system. The percentage of non-condensable combustible gaseous fuel obtained during the process was 3.75%, while the liquid fuel produced was 50.06% with a residue of 4.21% when 2 kg of un-catalyzed HDPE polymeric waste sample was fed into the reactor.

5.3.5   DEPOLYMERIZATION OF HDPE IN THE BATCH REACTOR BY USING ACTIVATED CARBON AS A CATALYST

Figures 5.12 and 5.13 indicate that the retention time when activated carbon is used for the depolymerization of HDPE polymeric waste is 45 min, which is less than when the process is not catalyzed.

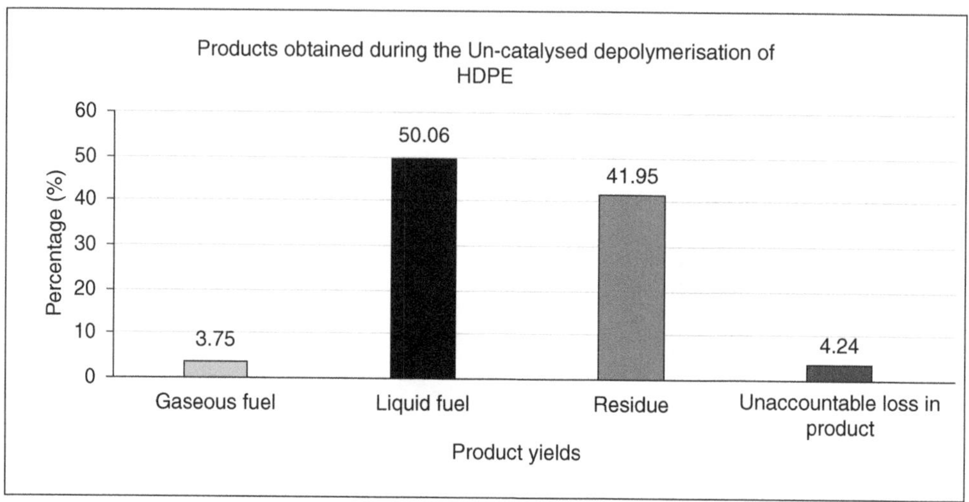

FIGURE 5.11   Product obtained from un-catalyzed depolymerization of HDPE.

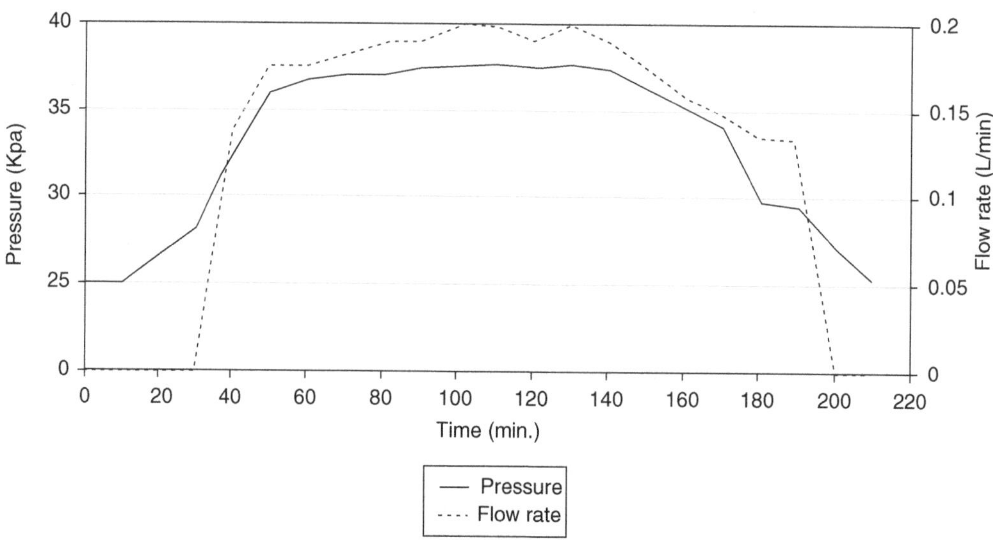

FIGURE 5.12   Variation of flow rate and pressure with time in batch reactor assembly during depolymerization of HDPE with 10 g of activated carbon as catalyst.

There is a noticeable increase in the temperature from the commencement of the catalytic depolymerization process throughout the 3 h and 30 min of firing the system. At 100 min of heating the reactor, the maximum flow rate of the end product coming out from the reactor via its outlet at 171°C was 0.1985 L/min at a pressure of 37.6 kPa. The percentage of non-condensable combustible fuel was measured and recorded as 5.25%, the liquid fuel produced was 53.60%, while the residual sludge was measured as 39.65% from the 2 kg feedstock (HDPE).

5.3.6   Depolymerization of HDPE in the Batch Reactor by Using Calcium Oxide as a Catalyst

The flow rate of the vaporized polymeric waste in the reactor was obtained to be maximum at 0.1768 L/min. When HDPE was catalyzed with calcium oxide (CaO) as shown in Figure 5.14 after

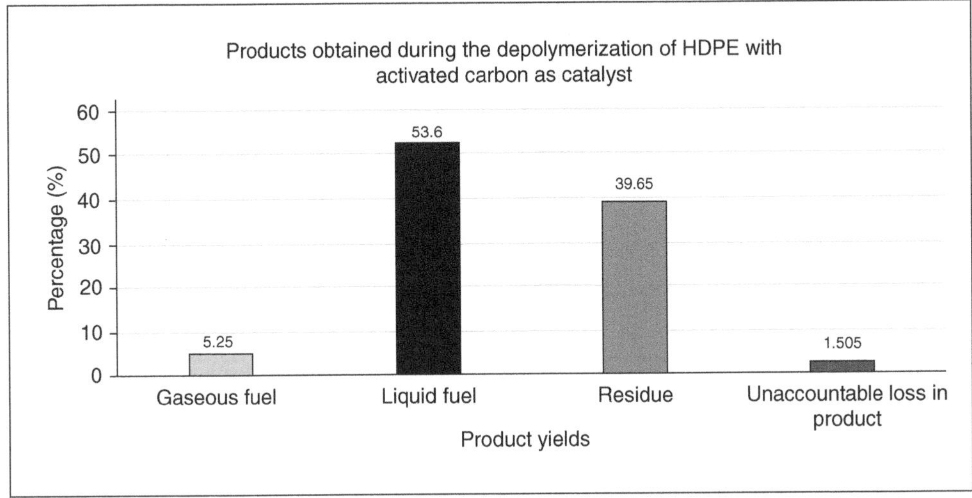

FIGURE 5.13 Product obtained from depolymerization of HDPE with activated carbon as a catalyast.

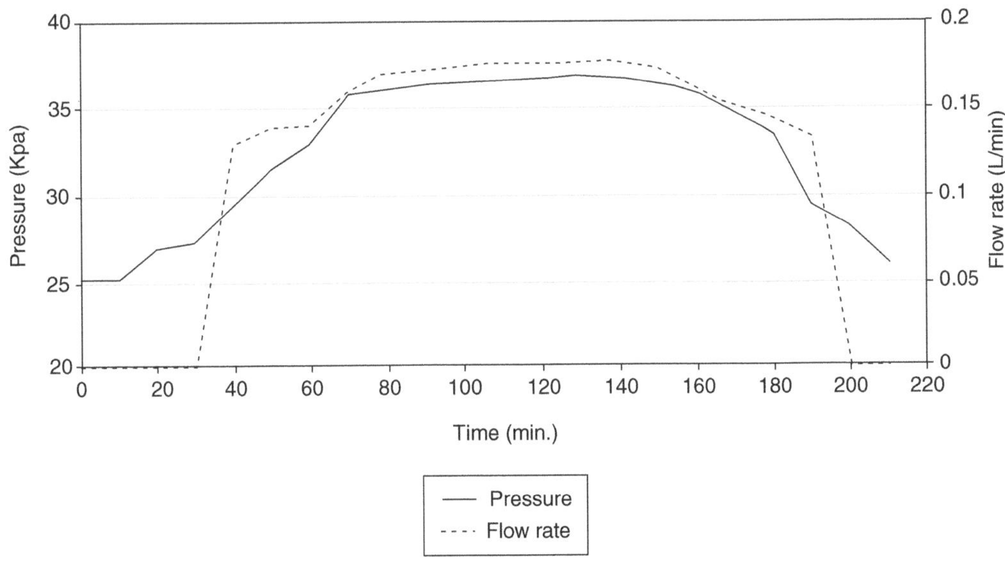

FIGURE 5.14 Variation of flow rate and pressure with time in batch reactor assembly during depolymerization of HDPE with 10 g of calcium oxide as catalyst.

being melted consequence of firing the system for 130 min at a temperature and pressure of 182 and 36.80 kPa, respectively. The retention time when the first drop of the liquid fuel was observed to have dropped is 55 min. Figure 5.15 indicates that the quantity of gaseous fuel, liquid fuel, and residue that was measured and recorded was 82.50 g, 1022.90 g, and 801 g, respectively, when 2 kg feedstock was used. Figure 5.16 indicates the highest retention time of 80 min was obtained in un-catalyzed HDPE, while the least was obtained with HDPE catalyzed with activated carbon during the depolymerization process, as shown in Figure 5.16.

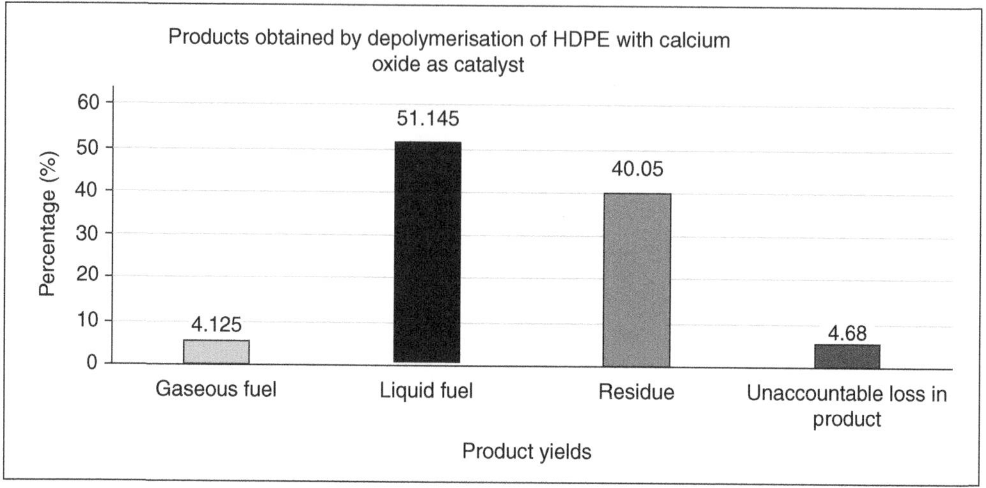

FIGURE 5.15 Products obtained by depolymerization of HDPE with calcium oxide as catalyst.

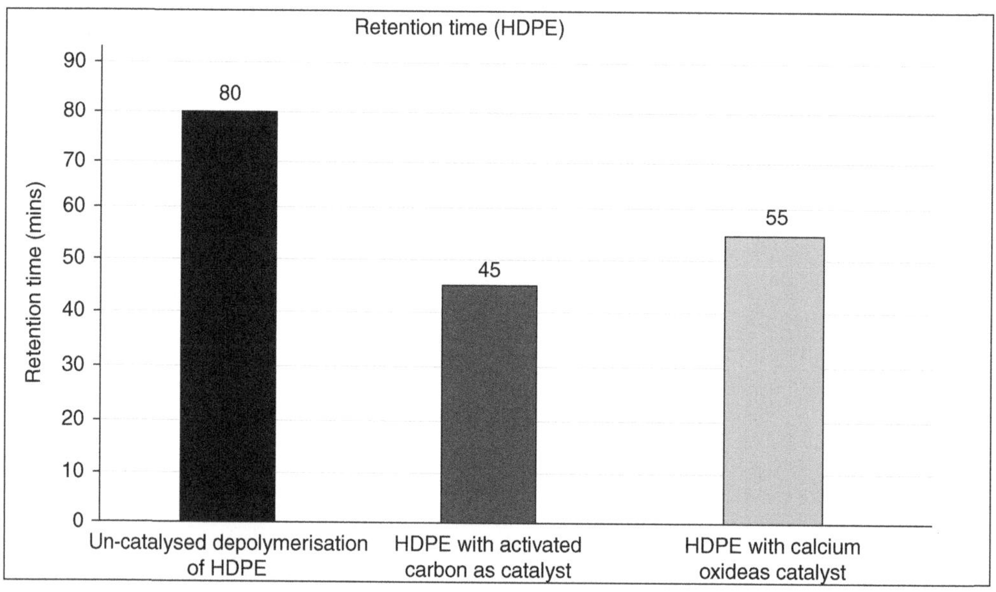

FIGURE 5.16 Retention time of HDPE during depolymerization process in batch reactor.

5.3.7 Depolymerization of Un-Catalyzed LDPE in the Batch

It was shown in Figure 5.17 that the flow remain still in the first 27 min of firing the reactor but there was a manifestation of the flowing of vaporized product from the reactor at 30 min of operating the energy conversion system with the flow rate of 0.1331 L/min recorded at a pressure of 30.1 kPa and temperature of 118°C. The retention time was measured as 35 min and the maximum flow rate of vaporized content coming out from the reactor heating chamber to the condenser was recorded as 0.1895 L/min, and this was attained at a temperature and pressure of 173°C and 37.3 kPa, respectively, within 100 min of operating the system while at 200 min of operation, there was no noticeable flow of vaporized content. Figure 5.18 indicates that 8.25% of gaseous fuel, 9.05% of liquid fuel,

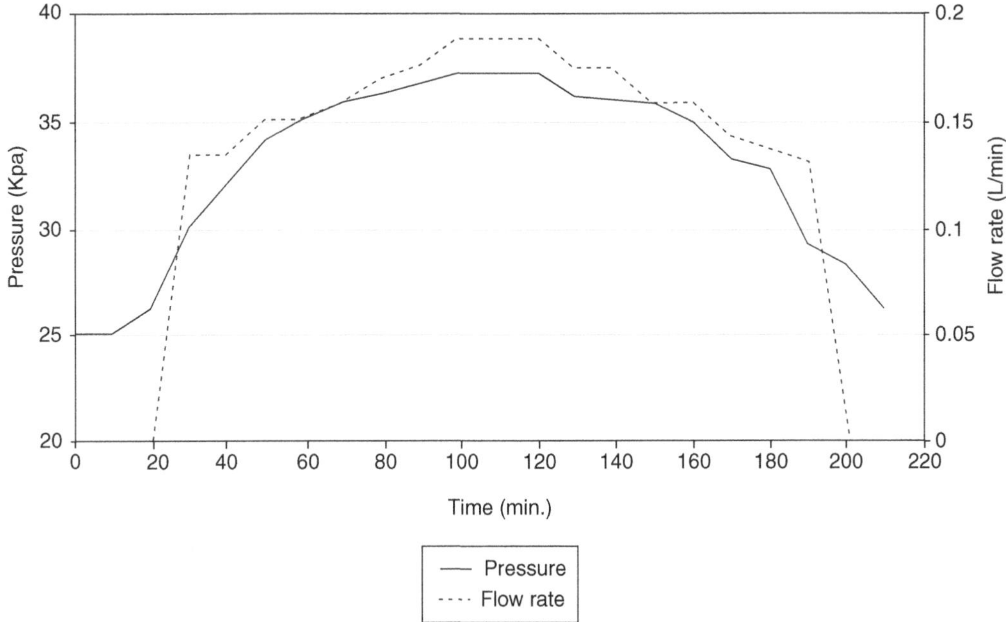

FIGURE 5.17   Variation of flow rate and pressure with time in batch reactor assembly during depolymerization of un-catalyzed LDPE.

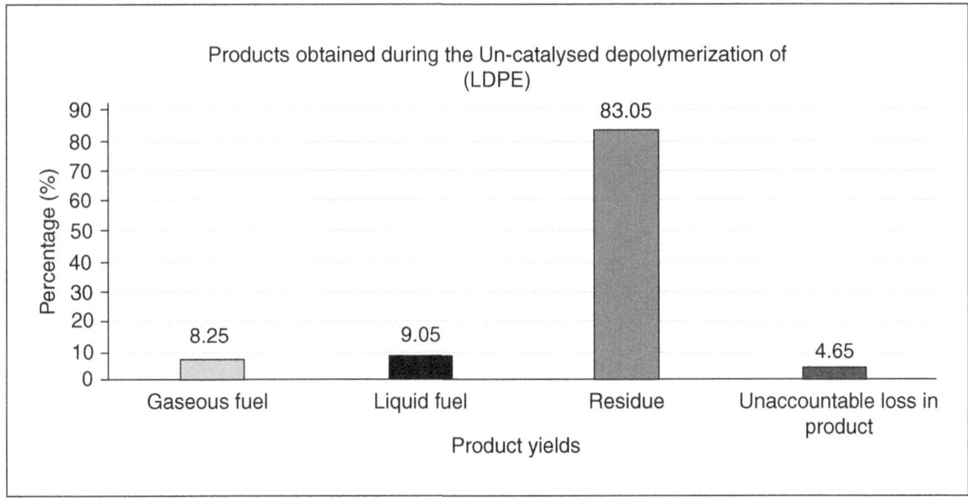

FIGURE 5.18   Products obtained during the un-catalyzed depolymerization of LDPE.

and 83.05% of biochar were the products obtained, while the unaccountable losses amounted to 4.65% during the un-catalyzed depolymerization of LDPE in this batch reactor.

5.3.8   DEPOLYMERIZATION OF LDPE WITH ACTIVATED CARBON AS CATALYST IN THE BATCH REACTOR

Figure 5.19 shows that at 100 min, the maximum flow rate of the depolymerized product from the reactor going to the condenser is 0.1899 L/min at a pressure and temperature of 37.4 kPa and 174°C,

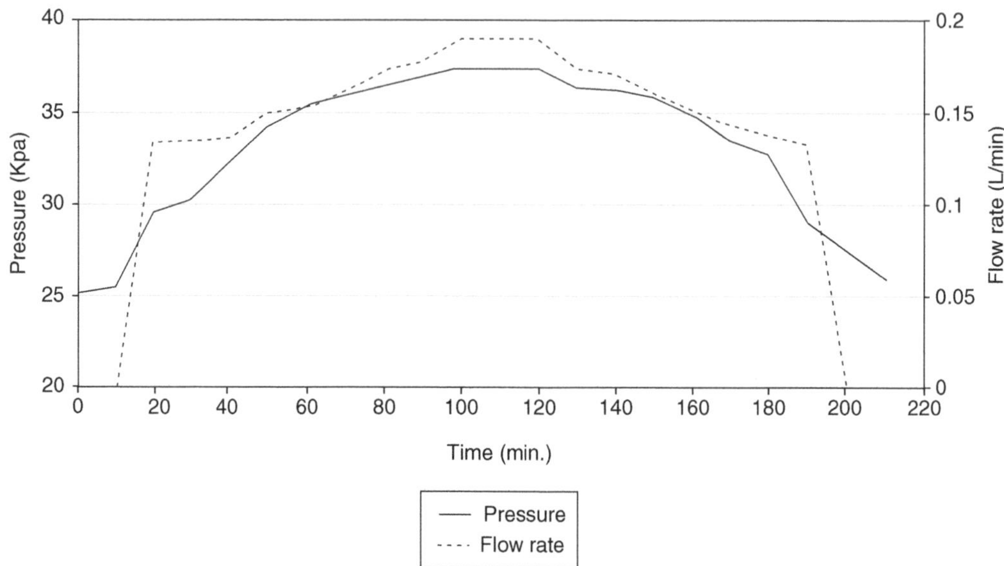

FIGURE 5.19   Variation of flow rate and pressure with time in batch reactor assembly during depolymerization of LDPE with 10 g of activated carbon as catalyst.

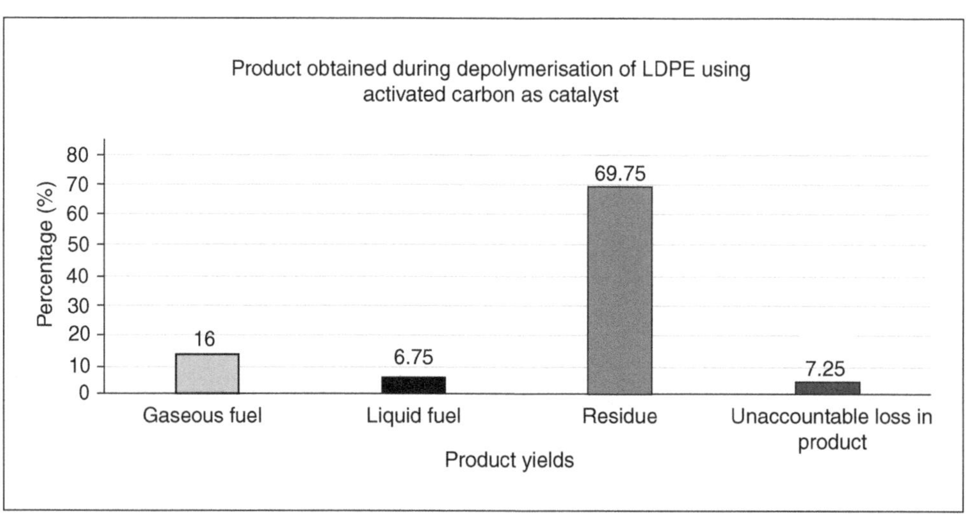

FIGURE 5.20   Products obtained during the depolymerization of LDPE with activated carbon as catalyst.

respectively. The temperature increases steadily from 30°C to 220°C till when there was no flow in the reactor. The retention time in Figure 5.20 in this energy system, when 2 kg feedstock was used is 22 min. The percentage of the gaseous, liquid, and biochar fuel is 16%, 6.75%, and 69.75%, respectively, while the unaccountable loss was found to be 7.25%.

5.3.9   DEPOLYMERIZATION OF LDPE WITH CALCIUM OXIDE AS CATALYST IN THE BATCH REACTOR

The graph in Figures 5.21 and 5.22 show that the maximum flow rate of vaporized product coming out from the system is 0.1894 L/min at a temperature of 175°C and pressure of 37.3 kPa at 110 min

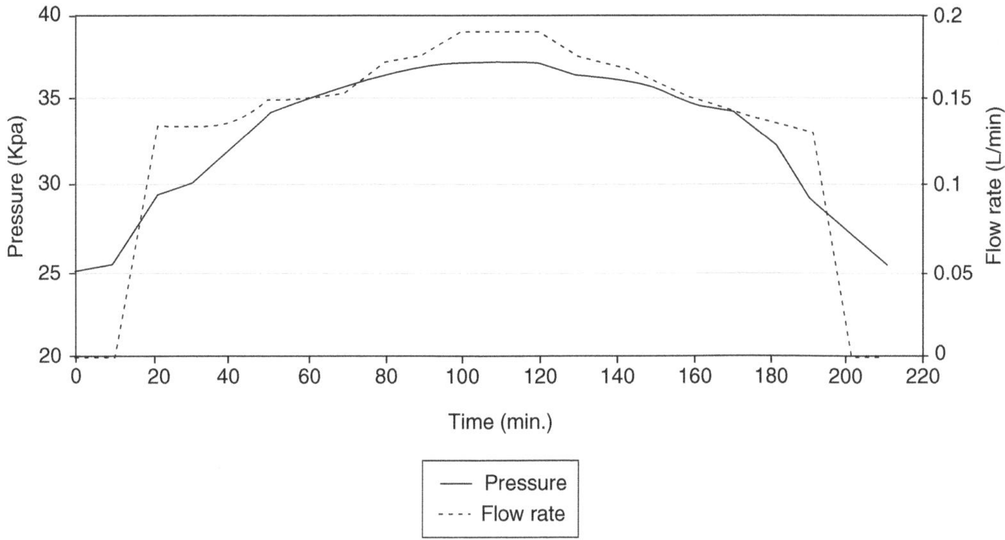

FIGURE 5.21　Variation of flow rate and pressure with time in batch reactor assembly during depolymerization of LDPE with 10 g of calcium oxide as catalyst.

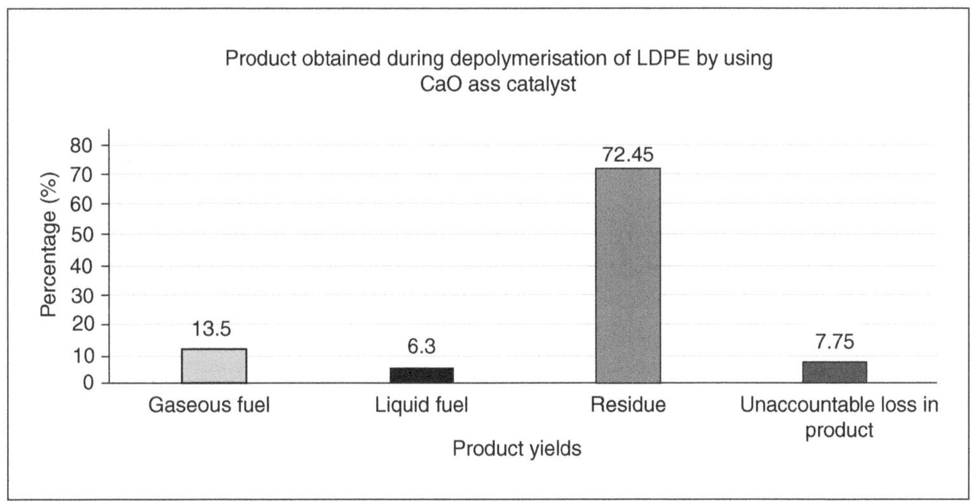

FIGURE 5.22　Product obtained during depolymerization of LDPE by using CaO as catalyst.

of operating the system. There was a pressure drop in the system as there was no flow immediately after 190 min of heating the reactor. Also, the percentages of the gaseous, liquid, and biochar (solid) fuels obtained were 13.5%, 6.3%, and 72.45%, respectively. The unaccountable loss was 7.75%, while the retention time was 31 min. It was shown in Figure 5.23 that the highest retention time was obtainable in un-catalyzed LDPE, while the least retention time of 22 min was obtained in LDPE catalyzed with activated carbon.

At 37.5 kPa, any further increase in the temperature causes a decrease in the flow rate.

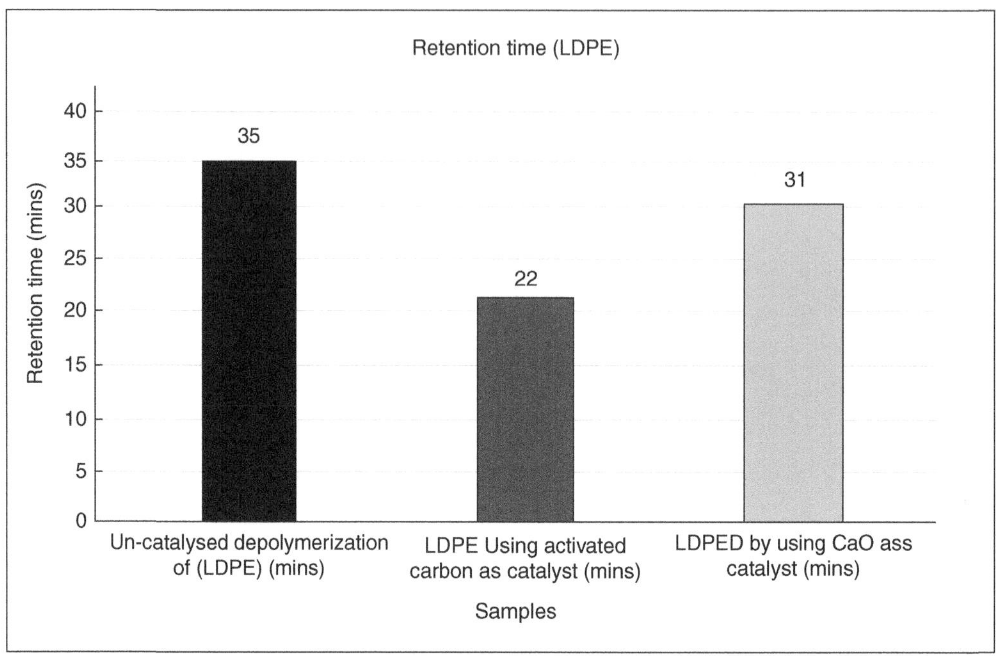

FIGURE 5.23   Retention time of HDPE during depolymerization process in batch reactor.

5.4 CONCLUSION AND RECOMMENDATIONS

The effects of the two chosen catalysts: activated carbon and calcium oxide on the catalytic depolymerization of polymeric waste comprise polyethylene therepthalate (PET), LDPE, and HDPE in the locally fabricated batch reactor whose its fuel conversion efficiency ranged from 49.2% to 98.5% at a temperature below $400°C$ which is 13.86% more than the mean efficiency obtained by the previous researchers was evaluated there was an increase in the rate of depolymerization process at a temperature immediately above the melting point of the polymeric waste samples and it proceeds faster than that of un-catalyzed depolymerization process. The temperature inside the reactor during the process increases as time increases, and the pressure also increases steadily till when the whole polymeric waste melts and causes a decrease in the pressure; consequently, the flow rate decreases as there is a fall in the pressure inside the reactor. The findings established that more gaseous fuel was produced in the reactor when activated carbon was used as a catalyst with PET polymeric waste, while more liquid fuel was also produced when HDPE was catalyzed with the same activated carbon. More residual ash was produced at the same time of firing the reactor in the un-catalyzed LDPE polymeric sample. The retention time for the un-catalyzed depolymerization process of PET polymeric waste was the highest. The very small quantity of liquid fuel produced from polyethylene therepthalate (PET) was attributed to the clogging of pipe by the yellowish jellylike matter coming out with the vaporized product during the process and often made it more difficult for it to be used as a feedstock for liquid fuel production but good for gaseous fuel. The material energy recovery is feasible through waste to energy-by using a depolymerization process.

REFERENCES

[1] Rominiyi, O.L. (2015): *Evaluation of Energy Content of Municipal Solid Waste in Ado-Ekiti Metropolis Nigeria*, M. Eng Thesis. Department of Mechanical Engineering, The Federal University of Technology, Akure, Ondo State, Nigeria.

[2] https://datatopics.worldbank.org

[3] Onwudili, J. A., Insura, N., and Williams, P. T. (2009): Composition of products from the pyrolysis of polyethylene and polystyrene in a closed batch reactor: Effects of temperature and residence time. *Journal of Analytical and Applied Pyrolysis*, 86, 293–303.

[4] Rominiyi, O. L. (2023): *Development of a Batch Reactor for Catalytic Depolymerization of Polymeric Waste for Liquid and Gaseous Fuel Production*. Ph.D Thesis Department of Mechanical Engineering, The Federal University of Technology, Akure, Ondo State, Nigeria.

[5] Rominiyi, O. L., Ikumapayi, O. M., Orumwense, E. O. (2022): Design and fabrication of a gasifier for the production of liquid fuel: A case study of *Spondias mombins*. *Advances in Mechanical Science and Engineering, Lecture Notes in Mechanical Engineering*, 131–145. https://doi.org/10.1007/978-981-19-3307-3_12

[6] Pushkaraj, R. P. (2010): Understanding the product distribution from biomass fast pyrolysis", Environmental Science, Chemistry, 1–162, https://doi.org/10.31274/etd-180810-1964

[7] Unapumnuk, K. (2006): *A Study of the Pyrolysis of Tire Derived Fuels and an Analysis of Derived Chars and Oils*. University of Cıncınati.

[8] Yin, L. J., Chen, D. Z., Wang, H., Ma, X. B., and Zhou, G. M. (2014): Simulation of an innovative reactor for waste plastics pyrolysis. *Chemical Engineering Journal*, 237, 229–235.

[9] Low, S. L., Connor, M. A., and Covey, G. H. (2001): *"Turning mixed plastic wastes into a useable liquid fuel"* 6th World Congress of Chemical Engineering, 23–27 September 2001, Melbourne, Australia: Department of Chemical Engineering, University of Melbourne.

[10] Aguado, J., Serrano, D. P., Miguel, G. S., Castro, M.C., and Madrid, S. (2007): Feedstock recycling of polyethylene in a two-step thermo-catalytic reaction system. *Journal of Analytical and Applied Pyrolysis*, 79, 415–423.

[11] Gao, F. (2010): *Pyrolysis of Waste Plastics into Fuels*. UC Research Repository, Faculty of Engineering | Te Kaupeka Pūhanga.

[12] Demirbas, A. (2004): Pyrolysis of municipal plastic wastes for recovery of gasoline-range hydrocarbons. *Journal of Analytical and Applie Pyrolysis*, 72, 97–102.

[13] Lee, K. H. (2009): Thermal and catalytic degradation of pyrolytic oil from pyrolysis of municipal plastic wastes. *Journal of Analytical and Applied Pyrolysis*, 85, 372–379.

[14] Rapsing, Jr., E.C. (2016): Design and fabrication of waste plastic oil converter. *International Journal of Interdisciplinary Research and Innovations*, ISSN 2348-1226 (online), Memorial State College of Agriculture and Technology (DEBESMSCAT), Mandaon, Masbate, Philippines.

[15] Hopewell, J., Dvorak, R., Kosior, E. (2009): Plastics recycling: Challenges and opportunities. *Philosophical Transactions of the Royal Society B*, 364, 2115–2126.

[16] Das, P., and Tiwani, P. (2018): Effect of slow pyrolysis on the conversion of packaging waste plastixc into fuel. *Science Direct*, 79, 615–624.

[17] Ceamanos, J., Mastral, J. F., Millera, A., and Aldea, M. E. (2018): Kinetics of pyrolysis of high density polyethylene. Comparison of isothermal and dynamic experiments. *Journal of Analytical and Applied Pyrolysis*, 65(2), 99–110.

[18] Kunwar, B., Cheng H. N., Chandrashekaran, S. R., Sharma, B. K. (2016): Plastics to fuel: A review. *Renewable and Sustainable Energy Reviews*, 54, 421–428.

[19] Mazloum, S., Awad, S., Allam, N., Aboumsallem, Y., Loubar, K., and Tazerout, M. (2021): Modelling plastic heating and melting in a semi – Batch pyrolysis reactor. *Applied Energy*, 283, 1–10, https://doi.org/10.1016/j.apenergy.2020.116375

[20] Sharuddin, S. D. A., Abnisa, F., Daud, W. M. A. W., and Aroua, M. K. (2016): A review on pyrolysis of plastic wastes. *Energy Conversion and Management*, 115, 308–326.

[21] Manos, G., Garforth, A., and Dwyer, J. (2000): Catalytic degradation of high density polyethylene over different zeolitic structures. *Industrial & Engineering Chemistry Research*, 39, 1198–1202.

[22] Manos, G., Yusof, Y. I., Papayannakos, N., and Gangas, H. N. (2002): Tertiary recycling of polyethylene to hydrocarbon fuel by catalytic cracking over aluminum pillared clays. *Energy Fuel*, 16, 485–489.

[23] Quesada, L., Calero, M., Martin-Lara, M. A., Perez, A., and Blazquez, G. (2019): Characterization of fuel produced by pyrolysis of plastic film obtained of municipal solid waste. *Energy*, 186, 115874, https://doi.org/10.1016/j.energy.2019.115874

[24] Kassargy, C., Awad, S., Burnens, G., Kahine, K., and Tazerout, M. (2017): Experimental study of catalytic pyrolysis of polyethylene and polypropylene over USY zeolite and separation to gasoline and diesel-like fuels. *Journal of Analytical and Applied Pyrolysis*, 127, 31–37. https://doi.org/10.1016/j.jaap.2017.09.005.

[25] Al-Salam, S. M., and Lettieri, P. (2010): Kinetics of polyethylene terephthalate (PET) and polystyrene dynamic pyrolysis. *World Academy of Science, Engineering and Technology*, 42, 1253–1261.

[26] Yan, J., Chou, S., Desideri, U., and Lee, D. (2015, December 15): Transition of clean energy systems and technologies towards a sustainable future (Part I). *Applied Energy*, 160, 619–622.

[27] Djafar, R., and Darise, F. (2018): Pengaruh Jumlah Aliran Udara terhadap Nyala Api Efektif dari Reaktor Gasifikasi Biomassa Tipe Fixed Bed Downdraft Menggunakan Bahan Bakar Tongkol Jagung. *Jurnal Technopreneur (JTech)*, 6(2), 94. https://doi.org/10.30869/jtech.v6i2.211

[28] Rudy, S., Pandri, P., Ida, B. A., and Mirmanto, M. (2017, June): Effect of air-vapor flow rate on syngas productioncompositions of updraft horse manure gasification. *Asian Journal of Applied Sciences*, 05(03), ISSN: 2321-0893.

[29] Aditya, K., Bambang, S., and Andri, P. (2020): Design of bench scale pyrolysis reactor for plastic waste conversion into liquid fuel using biomass as heating source. *Eksergi*, 17(1), 1–6

[30] Aboulkas, A., Harfi, K. E., and Bouadili, A. E. (2010) Thermal degradation behaviors of polyethylene and polypropylene. Part I: Pyrolysis kinetics and mechanisms. *Energy Conversion and Management*, 51, 1363–1369.

[31] Miskolczi, N., and Nagy, R. (2012): Hydrocarbons obtained by waste plastic pyrolysis: Comparative analysis of decomposition described by different kinetic models. *Fuel Processing Technology*, 104, 96–104.

[32] Hussain, Z., Khan, K. M., Perveen, S., Hussain, K., and Voelter, W. (2012): The conversion of waste polystyrene into useful hydrocarbons by microwave metal interaction pyrolysis. Fuel *Processing Technology*, 94, 145–150.

[33] Akinola, A. O. (2016): Evaluation of the efficiency of a thermochemical reactor for wood pyrolysis. *European Journal of Engineering and Technology*, 4(4), 17–25.

[34] Ogedengbe, T. I., Oroye, O., and Akinola, A. O. (2018): Modelling the products yield of used tyre pyrolyzed in fixed bed reactor. *Leonardo Electronic Journal of Practices and Technologies*, 32, 103–118.

[35] Ibrahim, H. A. H. (2020): Recent advances in pyrolysis. *Recent Advances in Pyrolysis*, 122. https://doi.org/10.5772/intechopen.77528. ISBN: 178984942X, 9781789849424.

[36] www.polymerscience.com

6 Analysis of a Hybrid Solar Power System as a Potential Substitute in Selected Southern Nigerian Economic Activity Areas

Zubairu Ismaila, Olugbenga A. Falode, Chukwuemeka J. Diji, Rasaq A. Kazeem, Omolayo M. Ikumapayi, Tien-Chien Jen, Opeyeolu Timothy Laseinde, and Esther T. Akinlabi

6.1 INTRODUCTION

Hybrid renewable energy system (HRES) combines and converts various energy sources into a single energy type, generally, AC or DC electricity [1–3]. Examples of HRES include biomass/wind/fuel cell systems, wind/PV/fuel cell hybrid systems, wind systems/diesel generators, wind electric/PV systems, etc. However, most hybrid systems use diesel generators as backup generators. Hybrid power systems provide clients with options and value that cannot be replicated by individual technologies, but they are not ideal for rural applications owing to their low end-use and high capital cost productivity [4,5]. The foundation of an integrated renewable energy system (IRES) is the idea that distinct energy demands require diverse energy types and quality. The model employs locally available RE resources as well as end-use technology to satisfy a variety of needs. Low-grade thermal energy, irritation water, potable and residential water supply, cold storage, cooking, communication, educational technology, community and domestic lighting, and small-scale businesses are a few of these demands [6,7]. This method requires meticulous and systematic planning to meet needs and the resources available to maximize benefits and end-user benefits. A combination of renewable energy sources, energy conservation, and energy efficiency may result in an IRES that offers overall advantages. The Integrated Renewable Energy System - Multi-MicroGrid (IRES-MMG) concept is used in this study to evaluate the cost-effectiveness of incorporating fossil fuel and renewable energy technologies in Nigeria's power sector, in addition to the customer-owned business strategy and a wide range of case reports to validate the advanced technical and economic business models [8]. Because Renewable energy resources are unstable, it is currently impractical to completely replace conventional diesel generators [8,9]. Hybrid energy systems that combine renewable and conventional sources of power may be an effective way to maximize each system's strengths [9]. However, to design a hybrid RE system that produces the desired results, several factors, including technical, institutional, and social considerations must be considered. The optimum hybrid systems, including PV/wind/diesel/battery systems, have been described in several works of literature [10–13]. In the meantime, PV systems have proven to be less expensive and more practical than wind turbines [14]. Hybrid PV/diesel systems were recommended for grid connection in numerous research [9,10,15–18].

DOI: 10.1201/9781032651958-6

Studies have investigated the optimized designing techniques of hybrid systems using various simulation software, such as HOMER, and several other optimization techniques, such as differential evolution (DE), artificial neural networks (ANN), and particle swarm optimization. These studies evaluate the appropriate sizing and operating strategies for various combinations [19–25]. In their study on a PV/diesel hybrid power system with battery backup for a remote village in Djanet, Algeria, Yahiaoui et al. [26] presented their findings by comparing optimization algorithms and observed that higher capabilities, lower costs, and quicker convergence are all features of the Grey Wolf Optimization. Using the small Saharan village of Tiberkatine, in southern Algeria's Tamanrasset province, as a case study, Fodhil et al. [27] presented a technique to maximize and assess the outcome of an autonomous mixed energy system, which is the diesel/photovoltaic/battery, and a comparative evaluation between Hybrid Optimization of Multiple Energy Resources (HOMER) and Particle Swarm Optimization (PSO) was then performed, demonstrating that the PSO-based method with more PV penetration than HOMER was significantly more cost Government gasoline subsidies have reduced the cost of fuel pumps in many underdeveloped nations, but this does not reflect what may be purchased on a competitive marketplace. Babatunde et al. [28] offered an efficient combined renewable off-grid energy system to sustain a traditional rural health facility throughout Nigeria's six geo-political regions. The diesel/photovoltaic/battery combination was the most economically feasible of all the sites, according to the results acquired. Ogunjuyigbe and Ayodele published a technology-economic evaluation of the Nigerian integrated energy system for the base transceiver plants [29]. Using the HOMER simulation platform, the integrated energy system was calculated for the entire net present cost (NPC) over a chosen 25-year project life cycle. The simulation's results revealed that the economically feasible diesel/photovoltaic/battery hybrid energy systems offered the best cost-benefit ratio. The technology-economic feasibility of a combined photovoltaic off-grid diesel/battery system, as a solution to such risks, was evaluated by Tijani et al. [30]. As a result, a case study in a remote region of the far north of Nigeria was carried out, which included a global organization with a peak demand of 90 kilowatts. The proposed technology keeps electricity costs down while generating a lot of energy and emitting fewer greenhouse gases. In a hypothetical combined diesel, off-grid PV, and battery hybrid system, Ciez and Whitacre [31] employed a device-level lifetime cost-of-use optimization approach to analyze the effects of various battery technologies. To account for device degradation during a 20-year system lifespan, three separate batteries utilized a time-step model of battery deterioration. The proposed method offers a high generation capacity, low electricity costs, and lower carbon emissions. As a result, a case study in a remote region of the far north of Nigeria was carried out, which included a global organization with a peak demand of 90 kW. The proposed technology keeps electricity costs down while generating a lot of energy and emitting fewer greenhouse gases. Das et al. [32] evaluated the potential for sustainable power in Sarawak, East Malaysia, noting that the region had a lot of solar power but inadequate wind speed. The feasibility of fuel cell, PV, and battery-based systems was therefore investigated for loading a long-housed hamlet of 50 residents in Kapit, Sarawak. The HOMER maximization and sensitivity analysis results indicated how cost-effective and environmentally safe the PV/battery system was. Thus, a system powered by diesel can be a viable option. Ghazvini and Olamaei [33] developed and studied a heuristic optimization approach to address the issue of the proper scale of the combined PV/diesel/battery network concerning the Vehicle-to-Grid (V2G) parking controlled load. According to the modeling results, the V2G parking lot would lower the overall cost of the PV/diesel/battery network optimization by 5.21%. Roth et al. [34] offered a reactive HRES grassroots metamodel design and software deployment. To determine the best size of HRES, one can use the metamodel to construct a MILP-configured optimization problem with a custom objective function, but the metamodel's flexibility depends on its ability to handle many hybrid variants in three different types of HRES applications: on-site demand, remote demand, or combination applications. Two case studies were presented: a self-contained HRES comprised of photovoltaic panels, a battery array, a diesel generator, and a tropical island facility. Akyüz et al.

[35] studied Balikesir's solar radiation data in Turkey to assess the technology-economic viability and ecological performance of a combined PV/diesel/battery device to meet the load demands of a typically isolated farmhouse. It was discovered that as PV capacity increases, the time required for a hybrid system's diesel generators to run reduces.

Most of the research published in the literature is mainly concentrated on the PV/battery/diesel hybrid systems that are currently available due to their cost-effectiveness in providing access to electricity in mostly remote areas across various nations. Also, Previous studies have considered the technical and economic performance of the hybrid shipboard power system (HSPS) in agricultural, industrial, and urban-residential economic activity area (EAA)s to improve electricity access with little consideration for quaternary EAAs such as medical [36] and academic institutions. Therefore, this research study needs to critically assess Nigeria's various sources of conventional and renewable energies. An effective hybrid renewable off-grid energy system model is described in this study that serves two distinct areas of economic operation, such as industrial and, most importantly, academic institutions. Based on grid supply, diesel power generation (DPG), and solar-photovoltaic (SPV), both technological and economic evaluations were carried out to boost access to electricity in these selected areas in Nigeria. It is worth noting that the local currency is herein presented in United States dollar (USD).

6.2 METHODOLOGY

6.2.1 Description of the Site

The economic activity region under consideration is situated in Oyo and Ogun States, Nigeria. Both states were selected for the research because of their popularity, vast geographical land area, and industrial prowess. The study locations chosen for this work represent the hub of commercial activities in Southern Nigeria. There are lots of economic activities occurring in these areas when compared to other states in Southern Nigeria, and this forms a basis for their selection for this work. In Nigeria and the African continent, Oyo state occupies a size of nearly 28,454 km^2, gradually rising from around 500 m above sea level in the southern part to about 1,200 m. On the other side, Ogun state is noteworthy for being a significant manufacturing hub and for having many industrial estates in Nigeria. The National Aeronautics and Space Administration (NASA) website has generated the ambient temperature, solar irradiation, and wind speed of the locations under study. The hypothetical EAA considered for Oyo state is the University of Ibadan located at 7.4432°N, 3.9003°E while Taju Industrial Nigeria Limited located at 6.9980°N, 3.4737°E was considered for that of Ogun state.

6.2.2 Components of the System

Microgrid devices, diesel generators, PV, batteries, and a converter have been included in the planned HRES. As specified in the HOMER program, the generator considered in the research work has the overall specifications. The designed generator's capital and replacement cost is \$0.34/kW, its cost of operation and maintenance is \$0.03/h, and its operating lifetime is 15000 hrs. The cost of fuel for this model is \$0.64/L. The specifications of the battery include the following: each battery cost is \$193.19; replacement costs and, operation and maintenance costs are set to \$0.00 at the time of purchase. The lifespan of the project is 10 years, and it is estimated that the annual real interest rate is 2%. Solar resources are activated from the NASA surface meteorology and solar energy database system when Global Positioning System (GPS) coordinates are entered on the NASA website. To have a stable supply of energy, the mean radiation from solar panels should have a steady pattern and yearly radiation over 4 kWh/m^2/day. For the study areas included in this analysis, the radiation values were calculated. In this analysis, it is assumed that the shortage capacity factor is zero, therefore there is no load unmet. It improves the efficiency of the system as this is a critical component in the sectors selected. The optimum system is rated by HOMER based on the NPC produced for the individual module. The NPC is the deducted amount over the system's operating life of the investment, replacement, and repair costs. The energy needs or utilization of an EAA were determined by performing an energy audit at the location. The energy audit provides an estimate of the energy consumed by the EAA as well as the energy required to power all the electrical devices in the EAA conveniently. The carbon emission of the generator and the grid was computed with HOMER. Because both the generators and the boiler use fuel with established qualities, HOMER models their emissions in a similar way. It has a slightly altered grid model. Before simulating the power system, HOMER determines the emissions factor (kg of pollutant emitted per unit of fuel consumed) for each pollutant. After the model, the annual emissions of that pollutant are computed by dividing the emissions factor by the total fuel use for the year. When simulating a grid-connected system, HOMER determines the net grid purchases, which are calculated as total grid purchases minus total grid sales.

6.2.3 Optimization of the Process

The optimization of an energy system is concerned with making the system maximally effective and fully perfect. Subject to other defined technical limitations, HOMER explores an energy system that effectively meets the requirements for electrical energy with a total required NPC. Input parameters can also be analyzed for sensitivity. By maintaining energy equilibrium measures for each phase in the year, the facility's operating conditions are simulated by HOMER. The energy the system can supply at every level in that stage is contrasted with the thermal and electric load during the time phase. HOMER processes the activity in the generator for each phase of the period when fuel-powered and battery-powered generators are applied to the grid to discharge or charge the batteries, use cycle charges, and load following. The system's lifespan cost (discounted amount of original capital, replacement, maintenance and operation, interest costs, and fuel) is determined until the demand is satisfied by the combination of the modules. In addition, sensitivity analysis on the non-linear input can be implemented. HOMER will be applied for this analysis. Different assessment parameters may be used that are appropriate for the intended purpose of assessing the efficiency of Hybrid Energy System (HES). These are economic (levelized cost of electricity [LCOE] and NPC), environmental (emission minimization), or technical (reliability). The system's cumulative NPC is the difference between the current values of all expenses over the lifetime of the project and the current values of all profits received over the life of the project. While this analysis presents the effects of other assessment efficiency metrics, the NPC is used in the optimal method of raking. On the other hand, LCOE calculates the amount of life divided by energy generation, as shown in Figure 6.1. LCOE computes the current value of the total construction and operation of the power station over the expected lifetime. LCOE makes it possible to compare emerging methods (e.g., wind, solar, natural gas) with uneven service life, size of the project, varying capital costs, uncertainty, revert, and efficiency.

6.2.4 System Control Inputs

The diesel generator is the only source of energy used in these HRESs, which is not dispatchable or renewable. As renewable energy output is highly erratic and cannot be managed by the consumer, it must be used when necessary to meet the load demand or to recharge the power bank. If RESs or the battery bank cannot meet the load, the diesel generator must be auto-started to meet the load demand without triggering power disruptions. In hybrid systems, using a diesel generator is essential for supplying the load in a controlled mode to boost the system's availability. However, due to various factors, regulating the output of a diesel generator is very complex. Therefore, the diesel generator operates under low load conditions, resulting in lower performance if the generator supplies

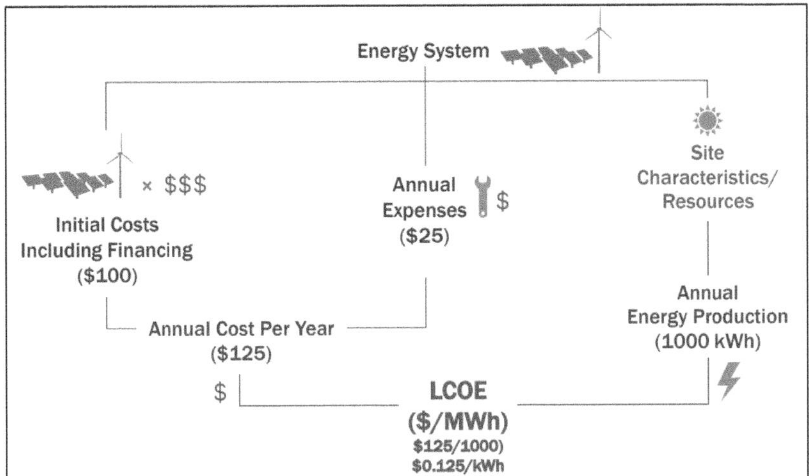

FIGURE 6.1   Simple LCOE concept [37].

the charge that renewable energy sources cannot supply. Usually, running the power generator at maximum capacity if necessary, and using the excess energy to charge the battery bank would be more advantageous. This method is suitable based on numerous factors, such as generator power. Therefore, the hybrid device should also be configured to establish a distribution plan. There are two types of dispatch plans, i.e., "Cycle-charging": When required, the diesel generator starts and works at maximum capacity, and excess energy will be sent to the battery system to recharge the battery; and "Load following": When needed, the diesel generator is turned on and generates only the required amount of energy that the renewables and the storage battery cannot generate to provide the charge.

6.2.5   Integrated Renewable Energy Mini/Microgrid Model (IREMMM)

In both areas of substantial economic activity, the IREMMM produced in this work will be employed as an empirical model: a residential building and a manufacturing company are among the buildings. IREMMM is a composite electric power system that receives electricity from 3 primary power plants: distribution firms' grid supply, self-generation from diesel generators and gas, and single photovoltaic (PV) systems with a battery bank in various combinations to meet existing energy needs and usage in the two distinct EAAs. The combination is determined by availability, technological restrictions, and costs. Table 6.1 shows the expected combined electrical power system combination.

Below are the identified areas of research illustrated.

6.2.5.1   University of Ibadan (UI)

Colleges are information and learning centers. They are the main contributors of skilled labor to any country's economy. They enhance the economy of the students and academic and non-academic personnel who live on campus, in addition to providing a teaching and research platform in Nigeria. The University of Ibadan, the research institution, was established in 1948 and has grown to serve a diverse community of students and faculty over the last 70 years. There are 600 professors and 12 educational learning centers at the University, with over 40,000 enrolled undergraduate and graduate students. There are 12 residential halls and 3 others that are under development at the school, offering housing for around 30% of the student population. To serve its employees, the school also has over 1,212 residential units; senior staff occupies 609 housing units, while junior staff occupies 603 units. The university offers its staff and students essential services such as an electricity and water system, defense, healthcare facilities, and other public infrastructure to carry out its important

TABLE 6.1
Combination of IREMMM Power Source

	Power-Supply Sources		
Combinations	Supply From Grid	Self-Generation	Solar PV
I	0.6	0.0	0.4
II	0.6	0.4	0.0
III	0.6	0.2	0.2
IV	0.4	0.3	0.3
V	0.4	0.1	0.5
VI	0.4	0.5	0.1
VII	0.2	0.4	0.4

functions of training, study, and social service. These facilities are provided with a robust 33/11 kV substation that is the university's main public power supply. Their electricity system provides robust energy. On campus, there are more than 57 transformers of different capabilities that deliver power to the school. The transformer power at the university ranges from 250 kVA to 1 MVA and 2 out of 10 MVA for the 33 kV sub-station precisely. The university built and managed a power station with around five diesel units, and operated generator sets, to provide a substitute power supply to the school when power outages from the public supply occur. This research explores the economic effects of incorporating a renewable supply of energy into the campus's existing energy system.

6.2.5.2   Taju Industries Limited (TIL)

The economic potential of incorporating renewable energy into Taju Industries' energy production activities in Agbara, Ogun was studied in the second case study. Taju Industries is a food manufacturing company that produces 24,000 tons of food each year. The industry operates two plants, Chanco and Chevron Nigeria Limited (CNL), one for the manufacture of savory seasonings and the other for the production of bread and cake-ready mix flour. In addition to the factories, it also has an administrative and logistics office block. Electrical, thermal, and mechanical sources are the major sources of power utility for the factory. It is either the national grid or the power supply itself, which is the main source of electricity. The firm has 6 generators to serve as backup generators, including four gas generators and two diesel generators. With only three of these generators generating power for the savoury season factory and tea bags facility, each one of these generators has a rating of 1,000 kW.

6.3   RESULTS AND DISCUSSION

The simulation results generated from the HOMER Pro for five study areas are being presented. Each study area contains one standalone diesel generator, hybrid PV-grid connected system, and one standalone solar PV system (SSPVS). Each study area had different electricity demands, and different energy system sizes, and as such, any comparison could only be carried out between the standalone diesel generator system (SDGS), SSPVS, and hybrid diesel generator-solar PV-grid system (HDG-SPV-GS) in that area.

6.3.1   University of Ibadan (UI)

6.3.1.1   Standalone Diesel Generator

The system consisted of a diesel generator connected to one busbar that supplied electricity to the load. The energy demand was 696 kWh/d, and the peak demand was 85.01 kW, but the dispatch

TABLE 6.2
UI Diesel Generator Annualized Cost

Name	Capital ($)	Operating ($)	Replacement ($)	Salvage ($)	Resource ($)	Total ($)
UI GEN	3.17	25,421.99	14.60	−0.43	56,521.74	81,841.43

TABLE 6.3
UI Hybrid Annualized Costs

Name	Capital ($)	Operating ($)	Replacement ($)	Salvage ($)	Resource ($)	Total ($)
Generic flat PV	21,074.17	0.00	0.00	−10,767.26	0.00	10,281.33
Grid	0.00	−18542.20	0.00	0.00	0.00	−18,542.20
System converter	4,987.21	0.00	0.00	−1420.03	0.00	3,580.56
System	26,086.96	−18542.20	0.00	−12,199.49	0.00	−4,654.73

strategy used was cycle-charging. The total electricity production for the year was 254,040 kWh/yr, and the electricity consumed was also 254,040 kWh/yr because the generator automatically sized itself. The mean electrical output was the average electrical power output of the generator over the hours it ran. There was 0% unmet load and 0% excess electricity because the system did not produce any excess electricity, as the generator was auto-sized to meet the load. The amount of fuel the system consumed in a year was 74,966 L with a specific fuel consumption of 0.350 kWh. The mean electrical output was 29 kW but the minimum and maximum electrical output showed the lowest and highest electrical power output of the year 0.610 and 85.0 kW, respectively. The quantity of CO_2 released from the stand-alone generator was 239,821 kg/yr HOMER Pro analyzed the NPC of the diesel generator at $751,230.18, and LCOE of $0.32 for the entire lifetime of the system. The LCOE was calculated on the economic input, considering the diesel price at 250/L. The total annualized system costs $35,294.12, and the highest costs arose from the resource, which was the cost of buying diesel. The annualized cost for diesel generators is shown in Table 6.2.

6.3.1.2 PV-DG-Grid Hybrid System

The daily energy demand was 696 kWh/d, and the peak load was 85.01 kW. The hybrid system consisted of two busbars connected to the PV array, the diesel generator, and the grid. Excess energy produced by the system was also sold to the grid. A converter was connected to the busbar and acted as an inverter, as well as a rectifier. The busbar was connected to the load to supply electricity to the load, while the diesel generator was meant to operate as an alternative power supply when solar power was not available or when there was a power outage from the grid. The hybrid system gave a total NPC of −$42,767.47, and the COE obtained was −$0.00. The cost summary for the entire system per year indicated that the solar PV system had the highest costs of $10,281.33, due to high initial capital costs. The system converter had annual costs of $3,580.56, giving the total system cost of −$4,654.73. The capital costs for the grid were always zero but the grid had the lowest total costs of −$18,542.20 from the operations and maintenance costs. Grid operations and maintenance costs were equal to the annual cost of buying electricity from the grid, minus any income from the sales to the grid. The negative value indicated that more grid sales were made from the excess energy produced in the hybrid system. Table 6.3 shows the cost summary for the UI PV-DG-Grid hybrid system.

The total electricity produced in a year was 1,094,659 kWh/yr, with 980,620 kWh/yr from PV, grid purchases were 114,038 kWh/yr. The excess electricity produced was 77,490 kWh/yr, and it was used to charge the battery. The unmet electrical load was loading that the PV was unable to serve, and in this case, there was no unmet electrical load. There was also no emission of pollutants because it was renewable energy. The battery was used in the absence of a grid to ensure the continuity of the electricity supply. The renewable energy penetration was 89.6%, and the grid was 10.4%. The grid provided backup power if solar energy proved inadequate to satisfy the load. Solar was employed as the primary option to provide energy to the load. When solar power was not sufficient or there was a grid power failure, the diesel generator was designed to function as a backup power source. Because of this, the hybrid system's results showed no energy coming from the generator. The grid-supplied power when there was not enough power from renewable energy sources to meet load demand, and the grid also consumed power when excess power was produced. The total energy purchased was 114,038 kWh, while 717,972 kWh of energy was sold. The grid rates are shown in Table 6.4.

6.3.1.3 Standalone Solar PV System

The daily energy demand was 696 kWh/d, and the peak load was 85.01 kW. The hybrid system consisted of two busbars connected to the PV array and the battery. A converter was connected to the busbar, which was connected to the load, to supply electricity to the load. The simulation results showed that the solar PV system gave a total NPC of $121,561.99 and COE of $0.06, although, the total annualized cost was $13,248.08. Storage systems such as batteries were expensive because of seasonal variations, but they were essential. The highest costs for the entire system arose from battery capital costs at $11,304.35, but the annualized costs of the UI SSPVS are shown in Table 6.5. The total electricity produced in a year was 403,375 kWh/yr, with excess electricity amounting to 151,219 kWh/yr. The excess energy was used to charge the battery, but the unmet electrical load was 25,652 kW, while the unmet electrical load was loading that the PV was unable to serve. This usually happens when electrical demand exceeds supply due to seasonal variations. There was no emission of pollutants because it was renewable energy.

6.3.2 Taju Industries Nigeria Limited (TINL)

6.3.2.1 Standalone Diesel Generator System

The system's architecture consisted of a diesel generator connected to one busbar that supplied electricity to the load. The energy demand was 12,000 kWh/d, and the peak demand was 916.61 kW but the dispatch strategy used was cycle-charging. The total electricity production for the year was 4,380,000 kWh/yr, and the electricity consumed was also 4,380,000 kWh/yr, as the generator automatically sized itself. However, the total hours of operation for a year were 8,760 h, though the mean electrical output was 500 kW. The minimum and maximum electrical output showed the lowest and highest electrical power output of the year 24.1 and 917 kW, respectively. The quantity of CO_2 released from the stand-alone generator was 3,349,957 kg/yr. There was 0% unmet load and 0% excess electricity because the system did not produce any excess electricity, as the generator was auto-sized to meet the load. The system consumed 1,234,136 L of fuel in a year, with a specific fuel consumption of 0.282 kWh. The standalone diesel system gave a total NPC of $9,730,421.99, and LCOE of $0.42, following the diesel price of ₦94.75/L. The annual operating costs consisted of operating, capital, salvage, and resource costs, which were $35,294.12. The total annualized system cost $1,061,381.07, and the highest costs arose from the resource, which was the cost of buying diesel. The annualized costs are shown in Table 6.6.

TABLE 6.4
UI Grid Rates

Month	Energy Purchased (kWh)	Energy Sold (kWh)	Net Energy Purchased (kWh)	Peak Demand (kW)	Energy Charge ($)	Demand Charge ($)
January	9,298	69,998	−60,700	84.8	−1,862.91	0.00
February	8,158	64,923	−56,765	79.1	−1,742.16	0.00
March	9,466	69,236	−59,770	78.2	−1,834.37	0.00
April	9,297	65,153	−55,855	85.0	−1,714.24	0.00
May	9,410	59,309	−49,899	79.9	−1,531.43	0.00
June	9,436	47,885	−38,449	80.6	−1,180.03	0.00
July	9,293	47,378	−38,085	80.1	−1,168.86	0.00
August	9,727	47,756	−38,029	76.4	−1,167.14	0.00
September	9,790	49,463	−39,673	74.3	−1,217.59	0.00
October	9,942	61,156	−51,214	72.1	−1,571.78	0.00
November	10,104	64,312	−54,208	84.4	−1,663.66	0.00
December	10,117	71,403	−61,286	78.7	−1,880.91	0.00
Annual	**114,038**	**717,972**	**−603,934**	**85.0**	**−18,535.06**	**0.00**

TABLE 6.5
UI Solar PV System Annualized Costs

Name	Capital ($)	Operating ($)	Replacement ($)	Salvage ($)	Resource ($)	Total ($)
Generic 1 kW ion	11,304.35	0.00	0.00	−2,787.72	0.00	8,516.62
Flat plate PV	8,670.08	0.00	0.00	−4,424.55	0.00	4,245.52
System converter	708.54	0.00	0.00	−201.29	0.00	507.25
System	20,664.96	0.00	0.00	−7,416.88	0.00	13,248.08

TABLE 6.6
TINL Diesel Generator Annualized Cost

Name	Capital	Operating	Replacement	Salvage	Resource	Total
TAJU Gen. ($)	33.88	271,099.74	156.17	0.00	790,281.33	1,061,381.07

TABLE 6.7
TINL Hybrid Annualized Costs

Name	Capital ($)	Operating ($)	Replacement ($)	Salvage ($)	Resource ($)	Total ($)
PV	322,250.64	0.00	0.00	−164,194.37	0.00	157,033.25
Grid	0.00	−205,115.09	0.00	0.00	0.00	−205,115.09
Converter	53,708.44	0.00	0.00	−15,242.97	0.00	38,363.17
System	375,959.08	−205,115.09	0.00	−179,539.64	0.00	−9,641.94

6.3.2.2 PV-Generator-Grid Hybrid System

The hybrid system consisted of two busbars connected to the PV array, the diesel generator, and the grid. The daily energy demand was 12,000 kWh/d, and the peak load of 916.61 kW. Excess energy produced by the system was also sold to the grid, as a converter was connected to the busbar, which acted as an inverter, as well as a rectifier. The busbar was connected to the load to supply electricity to the load, but the diesel generator was meant to operate as an alternative power supply when solar power was not available or when there was a power outage from the grid. The cost summary for the entire system indicated that the solar PV system had the highest costs of -$9,641.94, due to high initial capital costs, while the system converter had annual costs of $38,363.17. The capital costs for the grid were always zero but the total costs for the grid were −$205,115.09 from the operations and maintenance. Grid operations and maintenance costs were equal to the annual cost of buying electricity from the grid minus any income from the sales to the grid. The negative value indicated that more grid sales were made from the excess energy produced in the hybrid system. The total system costs were thus −$9,641.94. The annualized costs for the hybrid system are shown in Table 6.7. From the simulation results, the system gave a total NPC of $88,494.76.

The total electricity produced in a year was 17,164,992 kWh/yr, with 14,857,298 kWh/yr from PV, and grid purchases were 2,307,694 kWh/yr. The renewable energy penetration was 86.6%, while the grid produced only 13.4% of the total energy. The excess electricity produced was 3,214,981 kWh/yr, as there was no unmet electrical load; though, a battery was used in the absence of a grid

TABLE 6.8
TINL Grid Rates

Month	Energy Purchased (kWh)	Energy Sold (kWh)	Peak Demand (kW)
January	193,542	834,021	871
February	167,515	783,385	858
March	194,652	838,782	917
April	187,618	816,789	854
May	189,336	771,949	803
June	192,306	643,348	810
July	190,999	634,049	805
August	200,092	629,370	799
September	194,660	626,014	772
October	200,675	780,357	762
November	196,172	775,933	848
December	200,127	853,898	825
Annual	**2,307,694**	**8,987,895**	**917**

to ensure continuity of energy supply. The renewable energy penetration was 95.3%, and the grid was 4.75%. The grid provided support if solar energy proved inadequate to satisfy the load as the primary alternative for supplying electricity to the load. When a PV system was not available or there was a grid power outage, the diesel generator was designed to function as a backup power source. Because of this, the hybrid system's results showed no energy coming from the generator. The grid-supplied power when there was not enough power from renewable energy sources to meet load demand, and the grid also consumed power when excess power was produced. The total energy purchased was 14,303 kWh, while 203,700 kWh of energy was sold. Table 6.8 shows the grid rates for annual energy produced and supplied.

6.3.2.3 Standalone Solar PV System

The daily energy demand was 12,000 kWh/d and the peak load was 916.61 kW. The hybrid system consisted of two busbars connected to the PV array and the battery 1 kW LI. A converter was connected to the busbar, while the busbar was connected to the load, to supply electricity to the load. From the simulation results, the Standalone PV system gave a total NPC of $2,250,991.05 and Cost Of Energy (COE) of $0.06. The total annualized costs were $245,524.30 but storage systems such as batteries were expensive, and they were essential, due to seasonal variations. The highest costs for the entire system arose from battery capital costs at $169,820.97, while TINL Solar PV System annualized costs are shown in Table 6.9. The total electricity produced in a year was 6,520,580 kWh/ yr with excess electricity used to charge the battery being 2,140,222 kWh/yr. The unmet electrical load was 7,456 kW, but the unmet electrical load was loading that the PV was unable to serve. This usually happens when electrical demand exceeds supply, due to seasonal variations, although, there was no emission of pollutants because it was renewable energy.

6.3.3 Environment Impacts

The impacts on the environment from the pollutants were also obtained, as the diesel generator caused significant emissions of pollutants, and HOMER Pro calculated emissions of six gasses: nitrogen oxide, sulphur dioxide, unburned hydrocarbons, carbon monoxide, and carbon dioxide. Carbon dioxide had the largest percentage of emissions, and quantities of emissions from other gasses were almost negligible. The summary of the number of emissions is shown in Table 6.10.

TABLE 6.9
Taju Solar PV System Annualized Costs

Name	Capital ($)	Operating ($)	Replacement ($)	Salvage ($)	Resource ($)	Total ($)
Generic 1 kWh Li-Ion	228,388.75	0.00	0.00	−58,567.77	0.00	169,820.97
Generic flat plate PV	140,920.72	0.00	0.00	−72,122.76	0.00	68,797.95
System converter	9,565.22	0.00	0.00	−2,711.00	0.00	6,854.22
System	378,516.62	0.00	0.00	−133,248.08	0.00	245,524.30

TABLE 6.10
Quantity of Emissions

Quantity (kg/yr)	Carbon-Dioxide	Carbon-Monoxide	Unburned Hydrocarbon	Particulate Matter	Sulphur-Dioxide	Nitrogen-Oxides
UI Gen	239,821	4,458	63.6	8.84	587	1,369
Taju Gen	3,349,957	20,363	889	123	8200	19,129

6.3.4 SUMMARY OF RESULTS

For each study area, the total hours of operation for a year was 8760 h. Simulated data were used to determine the key operating parameters, including annual electrical loads, annual electrical energy output, renewable energy portion, excess electricity, capacity shortfall, and unmet loads. The diesel generators used in this work were automatically sized, to meet the loads; therefore, they did not produce excess energy or had unmet loads. HOMER pro calculated the economy of the project over the entire lifetime, including capital costs, operating costs, resource costs, salvage costs, and replacement costs. The annualized expenses at the fuel price of 250/L and the net current costs were also obtained. The system with the lowest NPC was the most cost-effective because the net present cost (NPC; also known as life-cycle cost) of a component was the present value of all the costs of installing and operating the component over the project lifetime, minus the present value of all the revenues it earned over the project lifetime. The Levelized cost of energy, as defined by HOMER, is the average price per kWh of usable electrical energy generated by the system. A system with the lowest NPC and LCOE was usually the best. However, the hybrid systems in both study areas had the highest initial capital costs, compared to standalone diesel generators and solar PV systems. Even, with the high capital costs, the hybrid Generator-Solar PV-Grid Systems still had the lowest NPC and LCOE. Hybrid systems also produce the highest amount of energy. Therefore, it is safe to say that hybrid systems are more cost-efficient than standalone diesel generators or solar PV systems.

6.4 CONCLUSION

Renewable electricity supply improved the standard of living and reduced greenhouse gas emissions. Power supply from the grid and Standalone Solar Photovoltaic systems in Nigeria is presently inconsistent; thus, several options for improving electricity have been proposed, one of which is the incorporation of alternative energy sources into the country's overall energy and electricity mix. Hybrid systems generally supply more reliable power; therefore, this assessment attempted to analyze the financial and technical issues involved with the use of fossil fuels and promote the integration of renewable energy sources in Nigeria's power sector. This analysis also tried to show the difficulties involved in integrating different energy systems and tried to resolve the issues. HOMER Pro was used to assess the systems, as it had been proven to be a very useful tool for analyzing energy systems. The NPC and the Levelized cost of energy were selected as the main economic indicators. An assessment of energy systems was carried out in two study areas in Nigeria, and the assessment considerations resulted in three power-generating scenarios (SDG, SSPVS, and HDG-SPV-GS) to be compared among each other in terms of performance and costs. After a thorough analysis, it was determined that the hybrid setup appeared to be the most cost-effective and reliable alternative for producing electricity. The findings indicated that employing HRES over standalone systems had significant advantages over standalone systems, even with high starting expenditures, due to hybrid systems' lower operating and maintenance expenses. Compared to standalone diesel generators, hybrid systems generated power at the lowest cost and were also viewed as the most environmentally friendly because they emit less carbon dioxide.

REFERENCES

[1] Khare, V., Nema, S., & Baredar, P. (2016). Solar–wind hybrid renewable energy system: A review. *Renewable and Sustainable Energy Reviews*, *58*, 23–33.

[2] Bajpai, P., & Dash, V. (2012). Hybrid renewable energy systems for power generation in stand-alone applications: A review. *Renewable and Sustainable Energy Reviews*, *16*(5), 2926–2939.

[3] Sorrenti, I., Rasmussen, T. B. H., You, S., & Wu, Q. (2022). The role of power-to-X in hybrid renewable energy systems: A comprehensive review. *Renewable and Sustainable Energy Reviews*. DOI: 10.1016/j.rser.2022.112380

[4] Vanek, F. M., Albright, L. D., & Angenent, L. T. (2016). *Energy systems engineering: evaluation and implementation*. McGraw-Hill Education.

[5] Allan, G., Eromenko, I., Gilmartin, M., Kockar, I., & McGregor, P. (2015). The economics of distributed energy generation: A literature review. *Renewable and Sustainable Energy Reviews*, *42*, 543–556.

[6] Hamilton, J. A. (2017). *Investigation into discontinuous low temperature waste heat utilisation from a renewable power plant in rural India for absorption refrigeration* (Doctoral dissertation, University of Nottingham).

[7] Roberts, T. (2012). *Benefits and disadvantages of using a holistic development approach vs streamlined rural electrification programme in using renewable energy to support social development objectives in the Himalayan context* (Doctoral dissertation, Murdoch University).

[8] Phuangpornpitak, N., & Kumar, S. (2007). PV hybrid systems for rural electrification in Thailand. *Renewable and Sustainable Energy Reviews*, *11*(7), 1530–1543.

[9] Olatomiwa, L., Mekhilef, S., Huda, A. N., & Sanusi, K. (2015). Techno-economic analysis of hybrid PV–diesel–battery and PV–wind–diesel–battery power systems for mobile BTS: The way forward for rural development. *Energy Science & Engineering*, 3(4), 271–285. https://doi.org/10.1002/ese3.71

[10] Khatib, T., Mohamed, A., Sopian, K., & Mahmoud, M. (2011). Optimal sizing of building integrated hybrid PV/diesel generator system for zero load rejection for Malaysia. *Energy and Buildings*, *43*(12), 3430–3435.

[11] Olatomiwa, L., Mekhilef, S., Huda, A. S. N., & Ohunakin, O. S. (2015). Economic evaluation of hybrid energy systems for rural electrification in six geo-political zones of Nigeria. *Renewable Energy*, *83*, 435–446.

[12] Olatomiwa, L. J., Mekhilef, S., & Huda, A. N. (2014). Optimal sizing of hybrid energy system for a remote telecom tower: A case study in Nigeria. In *2014 IEEE Conference on Energy Conversion (CENCON)* (pp. 243–247). IEEE.

[13] Olatomiwa, L. (2016). Optimal configuration assessments of hybrid renewable power supply for rural healthcare facilities. *Energy Reports*, *2*, 141–146.

[14] Ismaila, Z., Falode, O. A., Diji, C. J., Ikumapayi, O. M., Awonusi, A. A., Afolalu, S. A., & Akinlabi, E. T. (2022). A global overview of renewable energy strategies. *AIMS Energy*, *10*(4), 718–775.

[15] Wong, S. Y., & Chai, A. (2012). An off-grid solar system for rural village in Malaysia. In *2012 Asia-Pacific Power and Energy Engineering Conference* (pp. 1–4). IEEE.

[16] Ajan, C. W., Ahmed, S. S., Ahmad, H. B., Taha, F., & Zin, A. A. B. M. (2003). On the policy of photo-voltaic and diesel generation mix for an off-grid site: East Malaysian perspectives. *Solar Energy, 74*(6), 453–467.

[17] Lena, G. (2013). Rural electrification with PV hybrid systems. *International Energy Agency*, Report IEA-PVPS T9-13, 7–10.

[18] Neves, D., Silva, C. A., & Connors, S. (2014). Design and implementation of hybrid renewable energy systems on micro-communities: A review on case studies. *Renewable and Sustainable Energy Reviews, 31*, 935–946.

[19] Olatomiwa, L., Mekhilef, S., & Ohunakin, O. S. (2016). Hybrid renewable power supply for rural health clinics (RHC) in six geo-political zones of Nigeria. *Sustainable Energy Technologies and Assessments, 13*, 1–12.

[20] Banos, R., Manzano-Agugliaro, F., Montoya, F. G., Gil, C., Alcayde, A., & Gómez, J. (2011). Optimization methods applied to renewable and sustainable energy: A review. *Renewable and Sustainable Energy Reviews, 15*(4), 1753–1766.

[21] Belfkira, R., Zhang, L., & Barakat, G. (2011). Optimal sizing study of hybrid wind/PV/diesel power generation unit. *Solar Energy, 85*(1), 100–110.

[22] Erdinc, O., & Uzunoglu, M. (2012). Optimum design of hybrid renewable energy systems: Overview of different approaches. *Renewable and Sustainable Energy Reviews, 16*(3), 1412–1425.

[23] Haidar, A. M., John, P. N., & Shawal, M. (2011). Optimal configuration assessment of renewable energy in Malaysia. *Renewable Energy, 36*(2), 881–888.

[24] Zahboune, H., Zouggar, S., Krajacic, G., Varbanov, P. S., Elhafyani, M., & Ziani, E. (2016). Optimal hybrid renewable energy design in autonomous system using Modified Electric System Cascade Analysis and Homer software. *Energy Conversion and Management, 126*, 909–922.

[25] Olatomiwa, L., Mekhilef, S., Ismail, M. S., & Moghavvemi, M. (2016). Energy management strategies in hybrid renewable energy systems: A review. *Renewable and Sustainable Energy Reviews, 62*, 821–835.

[26] Yahiaoui, A., Fodhil, F., Benmansour, K., Tadjine, M., & Cheggaga, N. (2017). Grey wolf optimizer for optimal design of hybrid renewable energy system PV-diesel generator-battery: Application to the case of Djanet city of Algeria. *Solar Energy, 158*, 941–951.

[27] Fodhil, F., Hamidat, A., & Nadjemi, O. (2019). Potential, optimization and sensitivity analysis of photovoltaic-diesel-battery hybrid energy system for rural electrification in Algeria. *Energy, 169*, 613–624.

[28] Babatunde, O. M., Adedoja, O. S., Babatunde, D. E., & Denwigwe, I. H. (2019). Off-grid hybrid renewable energy system for rural healthcare centers: A case study in Nigeria. *Energy Science & Engineering, 7*(3), 676–693.

[29] Ogunjuyigbe, A. S. O., & Ayodele, T. R. (2016). Techno-economic analysis of stand-alone hybrid energy system for Nigerian telecom industry. *International Journal of Renewable Energy Technology, 7*(2), 148–162.

[30] Tijani, H. O., Wei Tan, C., & Bashir, N. (2014). Techno-economic analysis of hybrid photovoltaic/diesel/battery off-grid system in northern Nigeria. *Journal of Renewable and Sustainable Energy, 6*(3), 033103.

[31] Ciez, R. E., & Whitacre, J. F. (2016). Comparative techno-economic analysis of hybrid micro-grid systems utilizing different battery types. *Energy Conversion and Management, 112*, 435–444.

[32] Das, B. K., & Zaman, F. (2019). Performance analysis of a PV/diesel hybrid system for a remote area in Bangladesh: Effects of dispatch strategies, batteries, and generator selection. *Energy, 169*, 263–276.

[33] Ghazvini, A. M., & Olamaei, J. (2019). Optimal sizing of autonomous hybrid PV system with considerations for V2G parking lot as controllable load based on a heuristic optimization algorithm. *Solar Energy, 184*, 30–39.

[34] Roth, A., Boix, M., Gerbaud, V., Montastruc, L., & Etur, P. (2019). A flexible metamodel architecture for optimal design of hybrid renewable energy systems (HRES) – Case study of a stand-alone HRES for a factory in tropical island. *Journal of Cleaner Production, 223*, 214–225.

[35] Akyüz, E., Oktay, Z., & Dincer, I. (2009). The techno-economic and environmental aspects of a hybrid PV-diesel-battery power system for remote farmhouses. *International Journal of Global Warming. 1*, 392–404.

[36] Ismaila, Z., Falode, O. A., Diji, C. J., Kazeem, R. A., Ikumapayi, O. M., Petinrin, M. O., Awonusi, A. A., Adejuwon, S. O., Jen, T-C., Akinlabi, S. A., & Akinlabi, E.T. (2023). Evaluation of a hybrid solar power system as a potential replacement for urban residential and medical economic activity areas in southern Nigeria. *AIMS Energy*, *11*(2), 319–336.

[37] Office of Indian Energy. (2015). Levelized cost of energy (LCOE). *U.S. Department of Energy.* www.energy.gov/sites/default/files/2015/08/f25/LCOE.pdf. 1–9.

7 From Domestic Sewage Sludge to Clean Energy, Useful Resources, Zero Waste and Circular Economy Approach

J. K. Bwapwa

7.1 INTRODUCTION

The majority of the residual material removed during the wastewater treatment process is referred to as sewage sludge. It is the solid, semi-solid, or liquid residue generated during the treatment of domestic sewage, and it is actually an amalgamation of all liquid wastes from society after aerobic biological treatment; it is not tightly controlled, so it may contain highly undesirable polluting trace contaminants. Across the world, sewage sludge is being generated in large quantities as a result of municipal wastewater treatment. The rising volume of sewage sludge from wastewater treatment plants in many countries is becoming a major concern around the world. Because of the high organic and toxic content, including heavy metals among its constituents, sewage sludge management is particularly challenging and may constitute a serious environmental threat. When large amounts of sewage sludge are not disposed of properly, they cause major problems such as increased odors, migration of pathogenic microorganisms in the soil and groundwater, air pollution during windy weather, and pollution of soil and groundwater by traces of heavy metals. Sustainable management of sewage sludge is a major concern in wastewater treatment plants due to the increasing urban population. Sewage sludge is disposed of worldwide via landfilling, agricultural utilization, landscaping, thermal treatment, and many other relevant approaches. In some developing countries without proper regulations on sewage disposal, sewage sludge can sometimes be disposed via improper dumping in the sea, rivers, or anywhere else. It is important to stress that sewage sludge management is not enough to deal with the environmental impact of sewage sludge. In the circular economy era, it is important that sewage should be considered as a resource rather than waste. The main options for conversion of sewage include anaerobic digestion, co-digestion, incineration with energy recovery, co-incineration in coal-fired power plants, co-incineration with organic waste focused on energy recovery, use as an energy source in the production of cement or building materials, pyrolysis, gasification, thermoliquefaction, supercritical (wet) oxidation, high-temperature hydrolysis, production of hydrogen, acetone, butanol, or ethanol, and direct generation. Sewage sludge can be considered a renewable energy source, and its incineration generates substantially lower greenhouse gas emissions than energy generation from fossil fuels.

To avoid excessive cumulation of biosolids, sewage sludge must be systematically removed from a treatment plant on a regular basis. Sewage sludge can be disposed of in the following ways: application to land as a soil conditioner or fertilizer, agricultural use, disposal to ocean (in the sea) (not

DOI: 10.1201/9781032651958-7

permitted in the EU), disposal on land by placing it in a surface disposal site (not permitted in the EU, to be phased out), disposal in a municipal solid waste landfill unit (not permitted in the EU, to be phased out) (no longer permitted in the EU) Production of "bio-soils" for market sale, composting, land reclamation, and so on. Incineration (some disposal in the air as a result of incineration), and sludge to energy (newly applied novel technologies). However, the disposal options mentioned here should take into consideration the quality and the content of the sludge prior to the disposal process. Sewage sludge may have a fluctuating composition, and there are existing and emerging issues that may be of concern when it comes to the reuse of biodegradable sewage sludge. For instance, the use of treated sewage sludge as a low-cost fertilizer is an issue of public perception (Apedaile, 2001). Concerns have been raised over potential health, safety, quality of life, and environmental impacts that the land spreading of sludge may have (Robinson et al., 2012). This perception could be, in part, due to the fact that treated sewage sludge is heavily regulated or that animal manure is more commonly seen and used. There are also the nutrient and metal losses; in this case, phosphorus and reactive N losses to a surface waterbody originate from either the soil (chronic) or in a runoff where episodic rainfall events follow the land application of fertilizer (incidental sources) (Brennan et al., 2012). Such losses to a surface waterbody occur via primary drainage systems (end-of-pipe discharges, open drain networks (Ibrahim et al., 2013), runoff, and/or groundwater discharges). The presence of pathogens is also part of the concerning issues. As the survival of many microorganisms and viruses in wastewater is linked to the solid fraction of the waste, the number of pathogens present in sludge may be much higher than the water component (Straub et al., 1992). Although treatment of municipal sewage sludge using lime, AD, or temperature may substantially reduce pathogens, complete sterilization is difficult to achieve (Sidhu & Toze, 2009), and some pathogens, particularly enteric viruses, may persist. Pharmaceuticals are likely to be found in any body of water influenced by raw or treated wastewater, including rivers, lakes, streams, and groundwater, many of which are used as a drinking water source (Yang et al., 2011).

This study aims to show that sewage sludge is an important ingredient that should be considered as a resource; it is an important source of energy that can be explored to supply energy to WWTPs or communities that surround them. Also, the study aims to show the possibilities of product recovery from sewage and energetic clean resources. The focus is also on how to achieve a zero waste strategy from WWTPs by using sewage sludge, which is the main waste and aligns with the circular economy principle. Figure 7.1 presents a simplified approach from sewage to energy and resourceful products.

7.2 ZERO WASTE STRATEGY AND CIRCULAR ECONOMY APPROACH TO BE USED ON WASTEWATER TREATMENT PLANT

Water and wastewater management is one of the biggest challenges for the circular economy as many kinds of industries depend on water, and limited access to clean water resources can limit both production capacity and profits (Mauchauffee et al., 2012). Moreover, the disposal of wastewater, which can cause environmental damage, is an inherent element of water management as more than half of the global freshwater (2212 km^3 per year) is released into the environment as wastewater in the form of municipal and industrial effluent and agricultural drainage water. Kowalewski et al. (2018). The remaining 44% of global freshwater (1716 km^3 per year) is mainly consumed by agriculture through evaporation in irrigated cropland (UN Report, 2017). Feedstocks, such as food, agriculture, beverage, and municipal solid waste act as promising resources to produce renewable energy. Similarly, the concept of microbial fuel cells employing biowaste has clearly gained research focus in the past few decades. The major principle of circular economy is to reuse and recycle the products. Developing countries use most of the environmental resources and produce goods with environmental and economic benefits (Teigiserova et al., 2020). It is therefore important to integrate the zero-waste strategy combined with the circular economy principles on wastewater treatment plants to reduce operating costs and energy consumption.

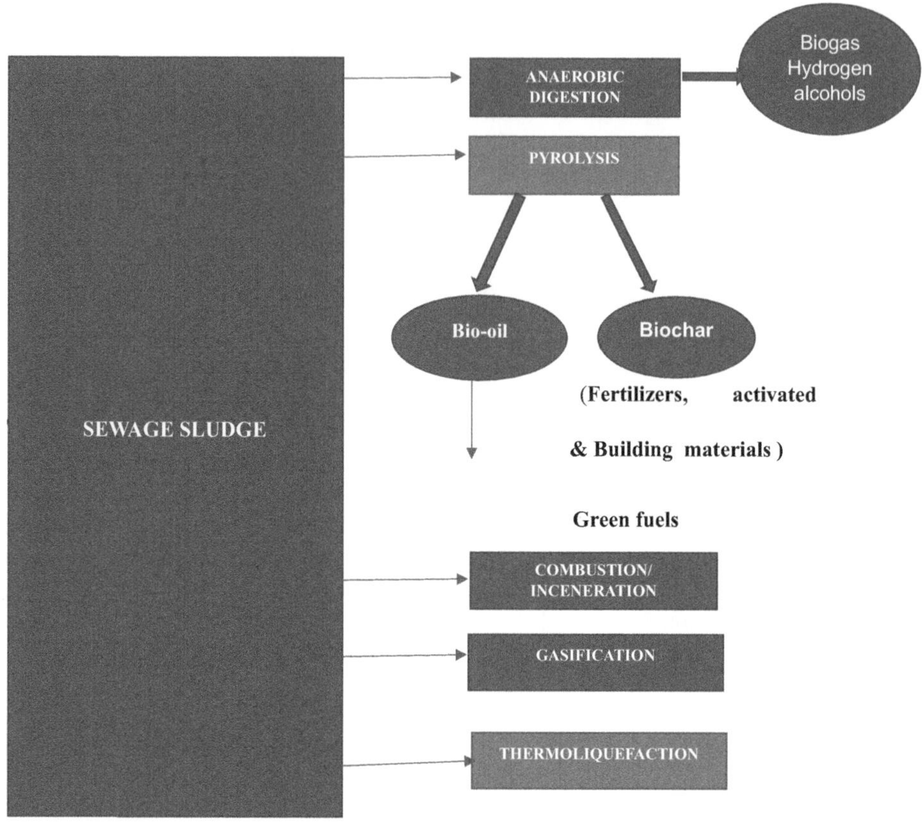

FIGURE 7.1   Options from sewage to energy and resourceful products.

7.3   PROCESSES FROM SLUDGE TO CLEAN ENERGY/TO PRODUCT RECOVERY

7.3.1   SEWAGE SLUDGE AS A RESOURCE

The two components in sewage sludge that are technically and economically feasible to recycle are nutrients (primarily nitrogen (N) and phosphorus (P)) and energy (carbon) (Tyagi & Lo, 2013). As sewage sludge contains organic matter, energy can be recovered while treating it. There is a considerable amount of nutrients within sewage sludge, especially P and N. However, due to depleting sources, P is fast becoming the most significant nutrient. Emerging technologies have been developed to extract this valuable resource, including Kemwater Recycling Process (KREPO), Aqua-Reci, Kemicond, BioCon, Sequential Precipitation of Phosphorus (SEPHOS), and Sustainable and Safe Re-use of Municipal Sewage Sludge for Nutrient Recovery (SUSAN), and are based on physical-chemical and thermal treatment to dissolve the P, with final recovery by precipitation (Cordell et al., 2011; Tyagi & Lo, 2013). Other resources include the reuse of sludge for construction materials, heavy metals, polyhydroxyalkanoates (PHA), proteins, enzymes, and volatile fatty acid (VFA). Advances in technology have revealed innovative emerging products from treated sewage sludge, including VFA, polymers, and proteins in the form of worms, larvae, and fungi.

7.3.2   FERMENTATIVE HYDROGEN PRODUCTION FROM SLUDGE

One of the promising biological options is the creation of hydrogen from wastewater. As a result, utilizing wastewater as a substrate for hydrogen synthesis while concurrently treating wastewater is an appealing and practical method of obtaining clean energy from renewable resources in a sustainable manner. Significant interest is being shown in this direction in various biological routes of hydrogen production using bio-photolysis, photo fermentation, heterotrophic dark fermentation, or a combination of these processes. As a result, the use of industrial wastewater as the principal substrate for the dark fermentation process may be assessed, as can the various parametric factors related to this sustainable strategy for improved energy generation. Industrial wastewater, such as rice slurry, food and domestic, citric acid, and paper mill wastewaters, could be used to generate biohydrogen. Microorganisms such as bacteria break down organic matter to generate hydrogen in fermentation-based systems. Organic matter can be refined sugars, raw biomass like maize stover, or even wastewater. These methods are commonly referred to as "dark fermentation" procedures since no light is needed for them. The hydrogen is produced by the bacteria themselves in direct hydrogen fermentation. These microorganisms can degrade complex compounds via a variety of routes, and the by-products of some of these pathways can be combined by enzymes to generate hydrogen. Researchers are investigating how to make fermentation systems create more hydrogen from the same quantity of organic matter (increase the yield) while producing it faster (raising the rate). Microbial electrolysis cells (MECs) are devices that use the energy and protons produced by microbes breaking down organic matter, along with a tiny electric current, to generate hydrogen. This technology is still in its early stages, and researchers are attempting to improve several parts of the system, from discovering lower-cost materials to determining the most effective bacteria to use (Hawkes et al., 2002; Yang and Wang, 2017).

7.3.4 BIOGAS FROM ANAEROBIC DIGESTION OF SLUDGE

The most often used sludge-stabilizing process is anaerobic digestion (AD). In the absence of air, AD organically stabilizes the sludge. It decreases the amount of volatile substances by converting them to biogas (methane CH_4, carbon dioxide CO_2, and water H_2O). The biogas then requires additional processing to extract and utilize its methane content. The stabilized solids product, like the solids product formed by lime dosage, can be used for soil conditioning. AD can be completed at mesophilic (30–39°C) or thermophilic temperatures (49–57°C). Psychrophilic operation is defined as operation at temperatures below 20°C. The low-temperature operation may be more energy efficient. However, because biochemical reaction rates and biogas production rates decrease at lower temperatures, the required tank size increases. Because anaerobic digestion is a biological process, high concentrations of toxic or potentially toxic materials, such as sulfates and heavy metals, can have a negative impact. Foaming and over-acidification are two well-known problems in anaerobic digestion (AD) operations (Abdel-Shafy et al., 2014; Dohányos et al., 2004). These two elements are frequently intertwined because they both pertain to the regulation of microbiology through the balancing of organic and nutritional loads. They also cause a drop in methane output, which can be detected by monitoring biogas methane and carbon dioxide concentrations.

7.3.5 THERMOLIQUEFACTION OF SEWAGE SLUDGE OR MUNICIPAL SOLID WASTE

The technology is based on hydrothermal liquefaction, a thermochemical process in an aqueous solution that converts the initial biomass into a low sulfur bio-oil with high calorific value, with a yield of up to 16% (depending on the composition of the original feedstock), with properties like those of low American Petroleum Institute (API) degree fossil fuels, as described in the dedicated paragraph. This can be blended straight into low-sulfur fuel for shipping or converted into advanced biofuels. In more detail, hydrothermal liquefaction converts biomass or sewage sludge to liquefied products by heating in the presence of a solvent, often the component water of the raw biomass. This method is specifically developed for wet materials, lowering energy expenses associated with drying feedstock prior to transformation. Because of this, the liquefaction method is suitable for valorizing wet waste biomass such as OFMSW (Organic Fraction of Municipal Solid Waste). The process is optimized to concentrate the beginning carbon content of OFMSW into bio-oil while producing additional by-products. There is a water phase, a residue phase, and a gas phase. Indeed, the process generates a biogenic CO_2-rich gas (the gas phase that can be captured and reused using a Bioenergy with Carbon Capture and Utilization (BECCU scheme) and an aqueous stream with dissolved organic content that, once purified, can generate biomethane. The entire process takes place at lower temperatures than its counterparts: 250–310°C rather than 400–500°C for pyrolysis and 800–1000°C for gasification. The yield of energy is high: 80% for hydrothermal liquefaction, 50% for biogas, and 10% to 30% for incineration (D'Aquila et al., 2021; Knapczyk et al., 2020; Rawat et al., 2016)

7.3.6 Sewage Sludge Used as Construction Materials

Due to land scarcity and increasingly rigorous environmental control rules, sludge disposal by landfilling may no longer be appropriate in densely developed cities. As a result, future sludge management tendencies will be toward the reduction and reutilization of usable resources. One viable method for reducing bulky sludge to essentially inert, odorless, and sterile ash is incineration. This much-reduced volume of sludge ash can thus be simply handled on a much smaller scale. If sludge or sludge ash can be used in large-scale economic applications, disposal difficulties will be greatly minimized. Several thermal solidification techniques have been created. Lightweight aggregates, brick, interlocking tile, char, and slag are examples. A full-scale plant of these has been running successfully for over ten years. The resulting goods are of higher quality than traditional ones. They are all replacements for existing ones. Overall, it appears that the majority of methods are technically feasible but not commercially viable. Their production costs are always higher than the market price. Furthermore, they consume a lot of energy. However, if they are identified for a sludge disposal process, they are all worth considering for a large city, landfilling sludge may no longer be an option. The average payout per 1,000 kg of ash or sludge cake is US$100. At the same price, the Portland cement manufacturer accepts either cake or ash. It accounts for approximately 50 to 30% of the energy cost of thermal solidification. The topic is whether dewatered cake or incinerated ash is preferable for Portland cement application scale plant of them has been successfully (Haustein et al., 2022; Lynn et al., 2018; Monzó et al., 1996; Rusănescu et al., 2022; Świerczek et al., 2021)

7.3.7 Sewage Sludge to Biochar: Potential for Wastewater Treatment

Biochar is a carbonaceous porous product formed from biomass, the amount and chemical composition of which depends on the source material. Agricultural leftovers, wood waste, cow dung, and sewage sludge have all been used to make biochar (Mohan et al., 2014; Wang et al., 2020; Xiang et al., 2020). Biochar has also been widely employed as an adsorbent in the removal of harmful metals, organic contaminants, and nutrients from wastewater. Manufactured biochar has a greater surface area, stronger adsorption capability, or more numerous surface functional groups (SFG) than pure biochar, representing a novel type of carbon material with tremendous application possibilities in various wastewater treatments. The production of biochar from sewage sludge (SS) is consistent with the goal of sustainable resource recovery and encourages a circular economy based on wastewater. Thermochemical conversion of SS to biochar addresses two main challenges at the same time: lowering disposal costs and acting as a resource to remove hazardous pollutants from water and wastewater. Biochar is an economically viable material for wastewater treatment because of its reusability and is readily available throughout the year. The production of biochar from sewage sludge (SS) is consistent with the use of SS-derived biochar as an adsorbent for contaminants present in wastewaters, the potential use of biochar as a catalyst and support material in advanced oxidation processes, and the use of biochars as an electrode material. The significance of SS-generated biochar in the adsorption and activation of hydrogen peroxide and persulfate in advanced oxidation processes requires more studies, with a particular emphasis on the mechanistic aspects objective of sustainable resource recovery. Sewage sludge ash (SSA) can be used as a replacement for the main binder or as an ingredient in mortar. The primary goal of partial binder replacement is to reduce its demand. After mortar, concrete is the most often utilized building material. The impact of SSA on concretes is being studied using the following methods: major binder substitution, direct addition of SSA without removing the fine aggregate, and partial replacement of the fine aggregate (Enaime et al.,2020; Wang and Wang, 2019; Wang et al., 2020; Xiang et al., 2020; Yang et al., 2020)

7.3.8 Nutrient Recovery from Sewage Sludge

Because it contains organic matter and inorganic elements, treated sewage sludge can be used as an agricultural fertilizer (Girovich, 1996). Because there are only 50–100 years of phosphorus reserves depending on future demand, recycling treated sewage sludge to agriculture as a source of the fundamental nutrients and metals required for plant growth will be critical for future sustainable development (Cordell et al., 2009). Spreading treated sewage sludge on arable or grassland, providing it is treated to acceptable standards, may provide an excellent source of nutrients and metals essential for plant and agricultural growth (Jeng et al., 2006). Treated sewage sludge can also help to improve the physical and chemical properties of soil (Mondini et al., 2008). It improves water absorption and tilth and may minimize soil erosion (Meyer et al., 2001).Depending on future demand, phosphorus (Cordell et al., 2009). Spreading treated sewage sludge on arable or grassland, providing it is treated to acceptable standards, may provide an excellent source of nutrients and metals essential for plant and agricultural growth (Jeng et al., 2006). Treated sewage sludge can also help improve soil physical and chemical properties (Mondini et al., 2008). It improves water absorption and tilth and may minimize soil erosion (Meyer et al., 2001). One of the main concerns about using treated sewage sludge as an organic fertilizer on grassland is the loss of nutrients, metals, and pathogens to a water body via direct discharges, surface and near-surface pathways, and/or groundwater discharge (Wall et al., 2011). Recently, so-called "emerging pollutants," which may include antibiotics, medicines, and other xenobiotics, have been studied because they pose health hazards. As a result, nutrient recovery from treated sewage sludge must be balanced against the potential negative consequences of its use.

7.3.9 Volatile Fatty Acids

Short-chained fatty acids with six or fewer carbon atoms that can be distilled at atmospheric pressure are known as volatile fatty acids (Lee et al., 2014). Proteins and carbohydrates in sewage sludge can be transformed into VFA, increasing the generation of methane, hydrogen, and polyhydroxyalkanoates (Yang et al., 2012). Su et al. (2009) describe an anaerobic process involving hydrolysis and acidogenesis (or dark fermentation) to produce VFA from biosolids. Complex polymers in garbage are hydrolyzed into comparable organic monomers by enzymes secreted by hydrolytic bacteria. Acidogenesis then converts these monomers into mostly VFA, such as acetic, propionic, and butyric acids. Bacteriocides, Clostridia, Bifidobacteria, Streptococci, and Enterobacteriaceae are among the obligatory and facultative anaerobes involved in both processes (Lee et al., 2014).

7.3.10 Polymers

Extracellular polymeric substances (EPS) are the primary organic matter elements in sewage sludge floc, along with polysaccharides, proteins, nucleic acids, lipids, and humic acids (Jiang et al., 2011). They are found in the intercellular space of microbial aggregation, specifically at or near the cell surface (Neyens et al., 2004), and can be extracted using physical (such as centrifugation, ultrasonication, and heating) or chemical methods (such as ethylenediamine tetraacetic acid), although formaldehyde plus NaOH has been shown to be effective in extracting EPA from most types of sludge (Liu & Fang, 2002). Extracellular polymeric molecules have a significant impact on the physical characteristics of microbial aggregates (Seviour et al., 2009). EPS has numerous biotechnological applications, including the manufacture of food, paints, and oil drilling "muds"; their moisturizing characteristics are also employed in cosmetics and pharmaceuticals. Furthermore, EPS may have potential applications as biosurfactants, such as tertiary oil production and biological glue. Extracellular polymeric compounds are an intriguing component of all biofilm systems and continue to have significant biotechnological promise (Flemming & Wingender, 2001). Aerobic granular sludge (AGS) technology is a relatively new approach for treating sewage sludge (Morgenroth et al., 1997). AGS has a high concentration of alginate-like exopolysaccharides (ALE), which have distinct properties than converted activated sludge. Aerobic granular sludge technology yields a substance with properties similar to alginate, a polymer derived from brown seaweed. Exopolysaccharides that resemble alginate can be harvested and employed as a gelling agent in textile printing, food preparation, and the paper industry (Hogendoorn, 2013). Lin et al. (2010) demonstrated a possible production of 160 4 mg/g (VSS ratio) volatile suspended solid (VSS) of extractable alginate-likeexopolysaccharides. They were also discovered to be one of the most abundant exopolysaccharides in aerobic granular sludge.

7.3.11 PROTEINS

Vermicomposting (earthworm sludge reduction) is a very popular technology, particularly in poor nations with small-scale settings. This procedure produces vermicompost, which is composed of earthworm feces and can be utilized as a fertilizer due to its high N content, high microbial activity, and low heavy metal concentration (Ndegwa & Thompson, 2001). Vermicomposting is the process by which waste streams are bioconverted into two useful products: earthworm biomass and vermicompost. Elissen et al. (2010) discovered that aquatic worms reared on treated municipal sewage sludge had excellent protein levels with a variety of amino acids. These proteins can be used as non-food animal feed for aquarium fish or other ornamental aquatic species. Other applications for the protein could include coatings, glues, and emulsifiers. The research also discovered that dead worm biomass can be used as an energy source in anaerobic digestion. Experiments have revealed that worm biogas output is three times that of sewage sludge. Fat and fatty acid extraction are two other applications. The use of earthworms to treat sewage sludge is extensively known; however, research investigations on protein extraction from earthworms grown on sewage sludge are quite sparse. For many years, researchers have been studying the bioconversion of biosolids using fly larvae. Organic waste has a high nutritional and energetic potential and can be used as a larval diet substrate. Grown larvae are a good protein source in animal feed, in addition to significantly reducing organic waste. The insect protein could be utilized to substitute fishmeal in animal feed (Lalander et al., 2013). The larvae of the Black Soldier fly (*Hermetia illucens* L.) are one of the most studied species. This nonpest fly's larvae feed on and decompose organic material of various origins (Diener et al., 2011a). The prepupa, the sixth instar, migrates from the sludge to pupate and can thus be easily retrieved. Because prepupae contains 44% crude protein and 33% fat on average, it is a suitable alternative to fishmeal in animal feed (St-Hilaire et al., 2007). There have been suggestions for other applications for the pupae than animal feed. The pupae's other components (protein, fat, and chitin) might be fractionated and sold separately. Chitin is of commercial importance because of its high percentage of N (6.9%) compared to synthetically substituted cellulose (1.25%) (Diener et al., 2011b). There has been extensive research on *H. illucens* and its role in drastically lowering organic wastes; nonetheless, there are various knowledge gaps regarding the pupae's potential use in terms of protein, fat, and chitin. Filamentous fungi are frequently grown in the food industry as a source of valuable goods such as protein and a variety of biochemicals, using relatively expensive substrates such as starch or molasses (More et al., 2010). The biomass produced during fungal wastewater treatment has a far larger potential value in terms of useful fungal by-products such as amylase, chitin, chitosan, glucosamine, antimicrobials, and lactic acids than that produced during the bacterial activated sludge process (van Leeuwen et al., 2012).

7.3.12 IMPACT OF NUTRIENT RECOVERY, ENERGY/PRODUCT GENERATION ON ENERGY AND COST SAVINGS IN A SEWAGE TREATMENT PLANT

It is widely acknowledged that the potential energy available in raw wastewater influent exceeds the electrical requirements of the treatment operations. The energy captured in organics entering the plant can be linked to the chemical oxygen demand load of the influent flow. Calorific measurements can be used to calculate a capita-specific energy intake of 1760 KJ per population equivalent (PE) in terms of the 120 g chemical oxygen requirement of organic matter (Wett et al., 2007). This one-of-a-kind organic load is broken down aerobically and anaerobically, releasing some of the captured energy. Traditional wastewater treatment plants (WWTP) utilize a lot of energy and have problems with sewage sludge and chemical residue disposal. Water treatment is predicted to account for 3–5% of total electrical energy consumption in several developed and emerging countries (Chen & Chen, 2013). Kapshe et al. (2013) demonstrated how methane recovery by anaerobic digestion might create 1.5–2.5 million kWh of captive-use electricity per year in four Indian WWTPs. Another benefit is a reduction in CO_2 emissions of 80,000 tonnes per year. Dewatered sludge (15–35% D.S.) cannot currently be employed in energy recovery or incineration due to its extremely low Lower Heating Value (LHV). Dried sludge (about 70–75% D.S.). However, if mixed with fuels (e.g., natural gas) and/or other waste with a high calorific value (e.g., Residue Derived Fuels, RDF), it may be a valuable energy source, as its LHV may reach up to 16 MJ/kg, allowing its use as a secondary fuel in, for example, the cement industry. 344 WWTPs in North Rhine Westphalia (NRW) in Germany have undergone energy analysis (Wett et al., 2007), which consists of two stages: a first stage in which operational data is collected and energy consumption rates and biogas yields are targeted; and a second stage in which optimization measures are implemented. Energy costs can be lowered by using this protocol. Through the reuse of energy produced during wastewater treatment, the long-term sustainability of the WWTPs is enhanced, while also contributing to offset installation and ongoing operational costs. In Southern European countries, including the Mediterranean area, cultural, social, and economic reasons mean that the management of sewage sludge is not necessarily the same as in other EU countries. Here, recycling to agriculture is the main route for final disposal. For example, in Portugal and Spain about 50% of the sewage sludge is recycled in agriculture (Milieu et al., 2013a, 2013b, 2013c). Therefore, sewage sludge management in these countries should be governed by the following objectives: (1) provision of solutions that are technically and economically adapted to the economic realities of these countries (lower investment and operating costs); (2) full legal compliance, including the ability to adapt to future restrictions, which may be placed on the disposal of treated sludge in agriculture; (3) diversification of the final disposal of sludge with new sludge treatment systems; (4) reduction in the quantity of sewage sludge to be disposed of; (5) optimization of the utilization of weather conditions for sludge treatment, which makes solar drying an appealing solution

7.4 CONCLUSION

Although capturing energy from waste and sludge is a relatively new phenomenon in many countries, the successful implementation of conversion processes is dependent on sludge composition, operating costs, and an understanding of various environmental factors that have a direct impact on the microorganisms responsible for waste degradation. Sustainable and renewable energy sources are recognized as the best energy and conventional fuel substitutes. The only option to achieve environmental sustainability is to switch from non-renewable to renewable energy sources. Biogas and hydrogen are two of the options being explored to address the supply-demand gap for renewable energy sources. They could be a significant source of energy derived from household sewage sludge. The production of biogas from household sewage sludge is a suitable approach to generat energy in a sustainable manner from a biodegradable raw material available from wastewater treatment plants. Sustainable and renewable energy sources are being explored at the moment. As wastewater treatment plants generate more sewage sludge on a daily basis, its availability is increasing. Lifestyle, industrialization, and rapid population increase are all factors that influence sewage sludge output in various countries across the world. Despite significant limitations, anaerobic digestion, gasification, combustion, pyrolysis, and dark fermentation have all been shown to be effective methods for producing biogas and hydrogen. There is also a tremendous opportunity to obtain new products from sludge that can be utilized as fertilizers, proteins, volatile fatty acids, polymers, and many more items that can be employed in the construction sector as building materials. The zero-waste approach and the circular economy concept can be effective techniques for reducing the environmental impact of trash. This will also aid in the development of new technology to decarbonize or minimize the environmental carbon footprint. Domestic sludge should be regarded as a valuable resource rather than a waste.

REFERENCES

Abdel-Shafy, H. I. and Mansour, M. S. (2014). Biogas production as affected by heavy metals in the anaerobic digestion of sludge. *Egyptian Journal of Petroleum*, 23(4), 409–417.

Apedaile, E. (2001). A perspective on biosolids management. *Canadian Journal of Infectious Diseases*, 12(4), 202–204.

Brennan, R. B., Healy, M. G., Grant, J., Ibrahim, T. G., and Fenton O. (2012). Incidental phosphorus and nitrogen loss from grassland plots receiving chemically amended dairy cattle slurry. *Science of the Total Environment*, 441, 132–140.

Chen S., and Chen B. (2013). Net energy production and emissions mitigation of domestic wastewater treatment system: a comparison of different biogas–sludge use alternatives. *Bioresource Technology*, 144, 296–303

Cordell, D., Rosemarin, A., Schröder J. J., and Smit A. L. (2011). Towards global phosphorus security: A systems framework for phosphorus recovery and reuse options. *Chemosphere*, 84(6), 747–758.

Cordell, D., Schmid-Neset, T., White, S., and Drangert, J. O. (2009). Preferred future phosphorus scenarios: a framework for meeting long-term phosphorus needs for global food demand. International Conference on Nutrient Recovery From Wastewater Streams, IWA Publishing, pp. 23–43.

D'Aquila, G., Botta, M., Bosetti, A., and Stefanucci, M. (2021, September). Waste to fuel process a sustainable and circular solution for organic urban waste. In *OMC Med Energy Conference and Exhibition*. OnePetro.

Diener, S., Studt, Solano N., Roa Gutiérrez, F., Zurbrügg, C., and Tockner K. (2011a). Biological treatment of municipal organic waste using black soldier fly larvae. *Waste Biomass Valorization*, 2(4), 357–363.

Diener, S., Zurbrügg, C., Gutiérrez, F. R., Nguyen, D. H., Morel, A., and Koottatep, T. (2011b). Black soldier fly larvae for organic waste treatment – prospects and constraints. In: *Proceedings of the WasteSafe – 2nd International Conference on Solid Waste Management in the Developing Countries*, Khulna, Bangladesh, M. Alamgir, Q. H. Bari, I. M. Rafizul, S. M. T. Islam, G. Sarkar and M. K. Howlader (eds), pp. 1–8.

Doháyos, M., Zabranska, J., Kutil, J., and Jeníček, P. (2004). Improvement of anaerobic digestion of sludge. *Water Science and Technology*, 49(10), 89–96.

Elissen, H. J. H., Mulder, W. J., Hendrickx, T. L. G., Elbersen, H. W., Beelen, B., Temmink, H., and Buisman, C. J. N. (2010). Aquatic worms grown on biosolids: Biomass composition and potential applications. *Bioresource Technology*, 101(2), 804–811.

Enaime, G., Baçaoui, A., Yaacoubi, A., and Lübken, M. (2020). Biochar for wastewater treatment—Conversion technologies and applications. *Applied Sciences*, 10(10), 3492.

Flemming H. C., and Wingender, J. (2001). Relevance of microbial extracellular polymeric substances (EPSs) – Part II: Technical aspcets. *Water Science and Technology*, 43, 9–16.

Girovich M. J. (1996). *Biosolids Treatment and Management: Processes for Beneficial Use*. CRC Press, New York.

Haustein, E., Kuryłowicz-Cudowska, A., Łuczkiewicz, A., Fudala-Książek, S., and Cieślik, B. M., (2022). Influence of cement replacement with sewage sludge ash (SSA) on the heat of hydration of cement mortar. *Materials*, 15(4), 1547.

Hawkes, F. R., Dinsdale, R., Hawkes, D. L., and Hussy, I. (2002). Sustainable fermentative hydrogen production: Challenges for process optimisation. *International Journal of Hydrogen Energy*, 27(11–12), 1339–1347.

Hogendoorn, A. (2013). *Enhanced Digestion and Alginate-Like-Exopolysaccharides Extraction from Nereda Sludge*. Masters in Science Civil Engineering, University of Technology.

Ibrahim, T. G., Fenton, O., Richards, K. G., Fealy, R. M., and Healy, M. G. (2013). Loads and forms of nitrogen and phosphorus in overland flow and subsurface drainage on a marginal land site in south east Ireland. *Biology and Environment: Proceedings of the Royal Irish Academy*, 113B(2), 169–186.

Jiang, J., Zhao, Q., Wei, L., Wang, K., and Lee D.-J. (2011). Degradation and characteristic changes of organic matter in sewage sludge using microbial fuel cell with ultrasound pretreatment. *Bioresource Technology*, 102(1), 272–277.

Jeng, A. S., Haraldsen, T. K., Grønlund, A., and Pedersen, P. A. (2006). Meat and bone meal as nitrogen and phosphorus fertilizer to cereals and rye grass. *Nutrient Cycle in Agroecosystems*, 76, 183–191.

Kapshe, M., Kuriakose, P. N., Srivastava, G., and Surjan, A. (2013). Analysing the co-benefits: Case of municipal sewage management at Surat, India. *Journal of Cleaner Production*, 58, 51–60.

Knapczyk, A., Francik, S., Jewiarz, M., Zawislak, A., and Francik, R., (2020). Thermal Treatment of biomass: A bibliometric analysis—The Torrefaction Case. *Energies*, 14, 162.

Kowalewski, Z., Neverova-Dziopak, E., and Preisner, M. (2018). An attempt at development of a regression model for estimating BOD5 values of municipal wastewater. *Ochrona Srodowiska*, 40(1), 21–27.

Lalander, C., Diener, S., Magri, M. E., Zurbrügg, C., Lindström, A., and Vinnerås B. (2013). Faecal sludge management with the larvae of the black soldier fly (*Hermetia illucens*)—From a hygiene aspect. *Science of the Total Environment*, 458–460, 312–318.

Lee, W. S., Chua, A. S. M., Yeoh, H. K., and Ngoh G. C. (2014). A review of the production and applications of waste-derived volatile fatty acids. *Chemical Engineering Journal*, 235, 83–99.

Lin, Y., de Kreuk, M., van Loosdrecht, M. C. M., and Adin A. (2010). Characterization of alginate-like exopolysaccharides isolated from aerobic granular sludge in pilot-plant. *Water Research*, 44(11), 3355–3364.

Liu, H., and Fang, H. H. P. (2002). Extraction of extracellular polymeric substances (EPS) of sludges. *Journal of Biotechnology*, 95(3), 249–256.

Lynn, C.J., Dhir, R.K., and Ghataora, G.S. (2018). Environmental impacts of sewage sludge ash in construction: Leaching assessment. *Resources, Conservation and Recycling*, 136, 306–314.

Mauchauffee, S., Denieul, M. P., and Coste, M. (2012). Industrial wastewater re-use: Closure of water cycle in the main water consuming industries—The example of paper mills. *Environmental Technology*, 33(19), 2257–2262.

Meyer V. F., Redente E. F., Barbarick K. A. and Brobst R. (2001). Biosolids applications affect runoff water quality following forest fire. *Journal of Environmental Quality*, 30, 1528–1532

Milieu, WRC, RPA (2013a). *Environmental, Economic and Social Impacts of the Use of Sewage Sludge on Land. Final Report – Part I: Overview Report*. Service contract No 070307/2008/517358/ETU/G4.

Milieu, WRC, RPA (2013b). *Environmental, Economic and Social Impacts of the Use of Sewage Sludge on Land. Final Report – Part II: Report on Options and Impacts*. Service contract No 070307/2008/517358/ETU/G4.

Milieu, WRC, RPA (2013c). *Environmental, Economic and Social Impacts of the Use of Sewage Sludge on Land. Final Report – Part III: Project Interim Reports*. Service contract No 070307/2008/517358/ETU/G4.

Mohan, D., Sarswat, A., Ok, Y.S., and Pittman Jr, C.U., 2014. Organic and inorganic contaminants removal from water with biochar, a renewable, low cost and sustainable adsorbent – A critical review. *Bioresource Technology*, 160, 191–202.

Mondini C., Cayuela M. L., Sinicco T., Sánchez-Monedero M. A., Bertolone E., and Bardi L. (2008). Soil application of meat and bone meal. Short-term effects on mineralization dynamics and soil biochemical and microbiological properties. *Soil Biology and Biochemistry*, 40, 462–474.

Monzó, J., Paya, J., Borrachero, M. V., and Córcoles, A., 1996. Use of sewage sludge ash (SSA)-cement admixtures in mortars. *Cement and Concrete Research*, 26(9), 1389–1398.

More, T. T., Yan, S., Tyagi, R. D., and Surampalli, R. Y. (2010). Potential use of filamentous fungi for wastewater sludge treatment. *Bioresource Technology*, 101(20), 7691–7700.

Morgenroth, E., Sherden, T., van Loosdrecht, M. C. M., Heiinen, J. J., and Wilderer, P. A. (1997). Aerobic granular sludge in sequencing batch reactors. *Water Research*, 31(12), 3191–3194.

Ndegwa, P. M., and Thompson, S. A. (2001). Integrating composting and vermicomposting in the treatment and bioconversion of biosolids. *Bioresource Technology*, 76(2), 107–112.

Neyens, E., Baeyens, J., and Dewil, R. (2004). Advanced sludge treatment affects extracellular polymeric substances to improve activated sludge dewatering. *Journal of Hazardous Materials*, 106(2–3), 83–92.

Rawat, J., Kaalva, S., Rathore, V., Gokak, D. T., and Bhargava, S. (2016). Environmental friendly ways to generate renewable energy from municipal solid waste. *Procedia Environmental Sciences*, 35, 483–490.

Robinson K. G., Robinson C. H., Raup L. A., and Markum T. R. (2012). Public attidudes and risk perception toward land application of biosolids within the south-eastern United States. *Journal of Environmental Management*, 98, 29–36.

Rusănescu, C. O., Voicu, G., Paraschiv, G., Begea, M., Purdea, L., Petre, I. C., and Stoian, E. V. (2022). Recovery of sewage sludge in the cement industry. *Energies*, 15(7), 2664.

Seviour, T., Pijuan, M., Nicholson, T., Keller, J., and Yuan, Z. (2009). Gel-forming exopolysaccharides explain basic differences between structures of aerobic sludge granules and floccular sludges. *Water Research*, 43(18), 4469–4478.

Sidhu, J. P. S., and Toze, S. G. (2009). Human pathogens and their indicators in biosolids: A literature review. *Environment International*, 35, 187–120.

St-Hilaire, S., Sheppard, C., Tomberlin, J. K., Irving, S., Newton, L., McGuire, M. A., Mosley, E. E., Hardy, R. W., and Sealey, W. (2007). Fly prepupae as a feedstuff for rainbow trout, oncorhynchus mykiss. *Journal of the World Aquaculture Society*, 38(1), 59–67.

Straub T. M., Pepper I. L., and Gerba C. P. (1992). Persistence of viruses in desert soils amended with anaerobically digested sewage sludge. *Applied Environmental Microbiology*, 58, 636–641.

Su, H., Cheng, J., Zhou, J., Song, W., and Cen, K. (2009). Improving hydrogen production from cassava starch by combination of dark and photo fermentation. *International Journal of Hydrogen Energy*, 34(4), 1780–1786.

Świerczek, L., Cieślik, B. M., and Konieczka, P. (2021). Challenges and opportunities related to the use of sewage sludge ash in cement-based building materials – A review. *Journal of Cleaner Production*, 287, 125054.

Teigiserova, D. A., Hamelin, L., and Thomsen, M. (2020). Towards transparent valorization of food surplus, waste and loss: Clarifying definitions, food waste hierarchy, and role in the circular economy. *Science of the Total Environment*, 706, 136033.

The United Nations World Water Development Report. (2017). https://unesdoc.unesco.org/images/0024/002 475/247553e.pdf. (UNWWDR, 2017). Accessed March 25, 2023.

Tyagi, V. K., and Lo, S.-L. (2013). Sludge: A waste or renewable source for energy and resources recovery? *Renewable and Sustainable Energy Reviews*, 25, 708–728.

Van Leeuwen, J. H., Rasmussen, M. L., Sankaran, S., Koza, C. R., Erickson, D. T., Mitra, D., and Jin, B. (2012). Fungal treatment of crop processing wastewaters with value-added co-products. In: *Sustainable Bioenergy and Bioproducts*, K. Gopalakrishnan, J. (Hans) van Leeuwen and Robert C. Brown (Eds.), Springer, Germany, pp. 13–44.

Wall, D., Jordan, P., Melland, A. R., Mellander, P.-E., Buckley, C., Reaney, S. M., and Shortle, G. (2011). Using the nutrient transfer continuum concept to evaluate the European Union Nitrates Directive National Action Programme. *Environmental Science and Policy*, 14(6), 664–674.

Wang, X., Guo, Z., Hu, Z., and Zhang, J. (2020). Recent advances in biochar application for water and wastewater treatment: A review. *PeerJ*, 8, e9164.

Wang, J., and Wang, S. (2019). Preparation, modification and environmental application of biochar: A review. *Journal of Cleaner Production*, 227, 1002–1022.

Wett, B., Buchauer, K., and Fimml, C. (2007). Energy self-sufficiency as a feasible concept for wastewater treatment systems. In: *IWA Leading Edge Technology Conference*, Singapore.

Yang, G., and Wang, J. (2017). Fermentative hydrogen production from sewage sludge. *Critical Reviews in Environmental Science and Technology*, 47(14), 1219–1281.

Yang, X., Du, M., Lee, D.-J., Wan, C., Zheng, L., and Wan, F. (2012). Improved volatile fatty acids production from proteins of sewage sludge with anthraquinone-2,6-disulfonate (AQDS) under anaerobic condition. *Bioresource Technology*, 103(1), 494–497.

Yang, H., Ye, S., Zeng, Z., Zeng, G., Tan, X., Xiao, R., Wang, J., Song, B., Du, L., Qin, M., and Yang, Y. (2020). Utilization of biochar for resource recovery from water: A review. *Chemical Engineering Journal*, 397, 125502.

Yang, X., Flowers, R. C., Weinberg, H. S., and Singer P. C. (2011). Occurrence and removal of pharmaceuticals and personal care products (PPCPs) in an advanced wastewater reclamation plant. *Water Research*, 45(16), 5218–5228.

Xiang, W., Zhang, X., Chen, J., Zou, W., He, F., Hu, X., Tsang, D.C., Ok, Y.S., and Gao, B. (2020). Biochar technology in wastewater treatment: A critical review. *Chemosphere*, 252, 126539.

8 Nanofluid Flow Through Non-Circular Cross-Section for Thermo-Hydraulic Enhancement of Energy Systems

Daniel R. E. Ewim, Temiloluwa O. Scott,
Andrew C. Eloka-Eboka, Sogo M. Abolarin, and
Adekunle O. Adelaja

8.1 INTRODUCTION

Heat transfer performance improvement is of critical importance in many engineering applications because the performance enhancement of thermal systems leads to increased interest in improvement efforts and overall energy efficiency of the process value chain [1–14]. Several systems, including electronic devices, heat exchangers, boilers, turbines, automobile engines, and others, are involved in transferring heat. As a result, it is essential to utilise suitable heat transfer enhancement techniques capable of optimising these energy devices. The heat transfer in a thermal system is generally improved using two approaches. The first is the active technique, which includes mechanical mixing, vibration of surfaces, electrostatic field, jet impingement, etc. While these methods contribute significantly to heat transfer enhancement, they require an external power source, which eventually leads to increased energy consumption [15,16]. The second technique is referred to as the passive, where the heat transfer surface area is increased by the insertion of turbulators of different configurations in order to partition the flow passage, increase the flow velocity of the working fluid and or circulate thermal enhancement nanofluid in the flow passage [17–19]. The passive technique does not require external power, which contributes to its advantages. However, the technique sometimes leads to higher pressure drop, and the effectiveness needs to be compared to its heat transfer counterpart. Nonetheless, the introduction of solid nanoparticles into heat transfer fluids represents a novel approach [13, 20] when compared with traditional working fluid, such as water. Nanofluids are engineered fluids that are formed by suspending nanoscale particles in a base fluid. Nanofluids are produced using either a one-step or a two-step method [21]. Literature showed nanofluids exhibit unique thermal and rheological properties compared to their base fluids like water, oil, or ethylene glycol, making them a popular choice in many engineering applications [13, 22–25]. Water is the most widely used base fluid due to its low cost, high heat capacity, and thermal stability. Oil-based nanofluids are applied during lubrication and metal cutting, where they can improve both the cooling and lubrication properties of the fluid [26, 27]. Ethylene glycol is mostly deployed as a base fluid in automotive and industrial cooling systems due to its low freezing point and corrosion resistance [28]. The enhancement of heat transfer with the circulation of nanofluids is affected by particle material, size, shape, concentration, base fluid type, the temperature of the bulk fluid, and so on [14, 29].

DOI: 10.1201/9781032651958-8

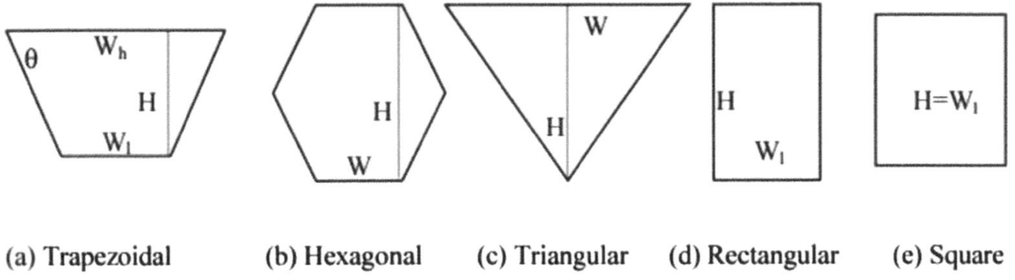

FIGURE 8.1 Non-circular shapes.

According to the literature, several researchers have investigated the flow of nanofluids in circular pipes and channels. For instance, an experimental study was conducted by Liu and Yu [30] on forced convection heat transfer single-phase flow by circulating Al_2O_3-water nanofluid in a mini-circular cross-section. They measured the friction factors and coefficients of heat transfer for various volume concentrations of nanofluids from 0.01 to 0.05 and compared them with the base fluid. Numerically investigated the laminar flow regime forced convection flow of an alumina-water nanofluid in a tube on which a uniform wall heat flux boundary condition was applied [31]. The authors studied the effects of Reynolds number, tube diameter, and volume fraction of the alumina nanoparticle. Mei et al. [32] investigated the thermo-hydraulic enhancement of Fe_3O_4-Water nanofluids circulated in a tube, investigating the influence of mass fractions of the nanoparticle as well as Reynolds numbers. Sahin et al. [33] also conducted experiments with an Al_2O_3 aqueous nanofluid, utilising volume concentrations of 0.005, 0.01, 0.02, and 0.04 in a tube. However, the flow of nanofluids through non-circular shapes such as elliptical, rectangular, and triangular shapes is more complicated due to the formation of secondary flow patterns.

Recently, the number of studies conducted on the flow of nanofluids has increased, particularly in non-circular shapes (Figure 8.1), which are commonly found in many industrial processes, and the majority of these studies have focused on laminar or turbulent flow regimes [34, 35]. By studying the flow of nanofluids in non-circular channels, researchers can better understand the fundamental mechanisms of heat transfer and fluid flow in these systems and optimise the design and operation of various industrial processes, which will eventually contribute to thermal performance improvement. In particular, numerical simulations and experimental studies provide valuable insights into the contributions of particle size and concentration, channel geometry, and flow velocity on the heat transfer and flow characteristics of nanofluids in non-circular channels. This study aims to provide an overview of the investigations that have been conducted numerically and experimentally on the flow of nanofluids in non-circular shapes.

8.2 NANOFLUIDS

Nanofluids are a type of fluid consisting of a base fluid such as oil or water that is dispersed with nanoparticles, typically with sizes between 1 and 100 nanometres. Adding nanoparticles to a base fluid have been reported to significantly contribute to the base fluid's thermal and dynamic properties, making nanofluids promising candidates for various industrial applications, including heat transfer, energy conversion, and cooling systems [13,14,36,37].

8.2.1 PREPARATION OF NANOFLUIDS

The procedure followed in preparing nanofluids is critical and is the first step towards analysing them experimentally. The degree of particle dispersion uniformity is primarily dependent on the preparation process employed and is likely to significantly influence the nanofluid's thermophysical properties. The primary objective of nanofluid preparation is to obtain a smooth and stable suspension with no particle agglomeration in either the particles or fluid. The two primary methods that are generally used to prepare nanofluids are one-step method and two-step method [38, 39].

8.2.1.1   One-Step Technique

The one-step approach, which is also referred to as the single-step method, involves the simultaneous creation and direct nanoparticle dispersion into the base fluid. Literature shows that researchers prefer this method because it reduces the challenge of producing stable nanofluids [37]. The single-step technique is also known to be the simplest process for producing nanofluids since it eliminates complicated stages like drying, transporting, storing, and dispersing; as a result, it reduces the agglomeration of nanoparticles and ensures that the nanofluid is stable [40]. On the contrary, one of the drawbacks of one-step nanofluid preparation is the high deposition of residual reactants. In addition, this method has not been scaled up for big-volume production [41].

8.2.1.2   Two-Step Technique

The two-step approach is another method used in the preparation of nanofluids. It is also known as the double-step method. This method is the most used to prepare nanofluids. In this approach, the nanoparticles are first formed as dry fine particles through a chemical or physical process. This step is then followed by the dispersion of nano-sized powders in the base fluid by deploying various techniques such as ultrasonic agitation and high-shear mixing [42]. This two-step approach is the most economical and simple compared to the one-step approach. However, the primary drawback of the two-step approach is the tendency of nanoparticles to agglomerate because they are small in size and possess high surface area, leading to reduced stability in the fluid. The most crucial approach to improve the stability of nanoparticles in the fluid is by utilising surfactants [43]. Table 8.1 presents some of the outcomes of the production of nanofluids for conducting experimental studies on non-circular ducts in recent literature.

8.2.2   Application of Nanofluids for Thermo-hydraulic Enhancement in Channels

Nanofluids have been widely used in different applications across various industries because of their exceptional properties compared to conventional fluids, such as enhanced thermal conductivity, higher surface area, and improved heat transfer characteristics. According to literature, some of its applications are not limited to heat exchangers, solar energy, microfluidics, oil and gas industry, etc. The use of non-circular channels can further enhance heat transfer efficiency by increasing the contact area between the fluid and the channel walls.

8.2.2.1   Coolant in Automobiles and Electronics

Nanofluid is regarded as an advanced coolant for automotive vehicles because it possesses the characteristics needed to improve the thermal performance of engines, fuel economy, maintenance, and reduce pollution [3, 8]. Additionally, nanofluid can be used as a refrigerant to improve the refrigeration system's heat transfer properties and reduce domestic refrigerators' energy consumption [5, 37]. They are also an important application in the electronics industry, where their use is driven by the high heat flux and limited space for cooling in electronic devices [37]. This advanced working fluid can effectively cool down electronic components [50]; this is because conventional cooling methods do not meet the high cooling requirements of modern electronic devices [51]. For instance, in an application of elliptical channel with aspect ratio 1–3 for electronic cooling purposes investigated in the work of Olakoyejo et al. [52] through which Al_2O_3 nanofluid concentration 0–4% was circulated over the Reynolds number 100–500, the authors found that the heat transfer coefficient increased by 78% with the use of the nanofluid, while the increase in the aspect ratio only

TABLE 8.1

Some Outcomes of the Production of Nanofluids for Conducting Experimental Studies on Non-Circular Ducts

Nanoparticles	NP Size (nm)	Base Fluid	Volume Concentration (%)	Mixing Procedure	Preparation Technique	References
TiO_2	13	Water	0.2 and 0.5	Ultrasonic (pulses 400 W at 24 kHz for 5 h)	Two-step	[44]
CuO	30–50	Water	0.1, 0.2, 0.5, 0.8, 1.0 and 1.5	Ultrasonic	Two-step	[45]
SiO_2	<20	Water	0.05, 0.07 and 0.2	Ultrasonic	One-step	[46]
TiO_2	13	Water	0.2 and 0.5	Ultrasonic (pulses 400 W at 24 kHz for 5 h)	Two-step	[47]
Al_2O_3	30	Distilled water	0.20, 0.50, 1.0, 1.5, 2.0 and 2.5	Ultrasonic	Two-step	[45]
CuO	40	Water	0.25, 0.50 and 1.0	Ultrasonic	Two-step	[48]
Al_2O_3	–	Water	0.5 to 2.0	Ultrasonic (pulses 750 W at 20 kHz for 1 h)	Two-step	[49]

caused the pressure drop to increase by 2%. This is very important as the heat transfer enhancement was greater than the pressure drop. Some of the nanofluids being used for cooling applications contain carbon nanotubes as well as graphene [53]. Non-circular channels have been used in solar thermal systems to increase the thermal efficiency of the system. Solar thermal systems energy takes advantage of the energy from the sun to generate heat, which is then used to generate electricity or provide heating and cooling. Nanofluids have been shown to improve the efficiency of solar thermal systems by enhancing heat transfer rates and reducing the amount of energy lost during the conversion process [54]. Nanofluids have also been found to improve the solar thermal collectors performance and increase the overall system efficiency [55].

8.2.2.2 Thermal Storage Systems

Nanofluids can be deployed in thermal energy storage (TES) systems, which can enhance the efficiency and dependability of energy systems. TES systems are utilised to store thermal energy for later use, and nanofluids have unique heat storage capabilities and can function as phase change materials (PCMs) in these systems. Studies have shown that TES systems using nanofluid-based PCMs have higher energy storage capacity, quicker charging and discharging rates, and better thermal conductivity compared to conventional TES systems [56, 57]. Nanofluids are not only being investigated for use in solar thermal and TES systems but have been deployed in nuclear reactors, heat exchangers, and fuel cells. For instance, they have demonstrated the ability to enhance the effectiveness of heat exchangers by increasing heat transfer rates and minimising fouling [58]. Research studies have examined the potential use of nanofluids in fuel cells, which can help increase the fuel cell's efficiency by enhancing heat transfer and minimising the concentration polarisation at the electrode interface [59, 60].

8.2.2.3 Secondary Flow Characteristics

Furthermore, non-circular channels have been shown to be characterised with improved heat transfer efficiency compared with circular channels [61]. This is due to the generation of secondary flow patterns in non-circular channels, which enhance mixing. Mixing enhancement is crucial in many industrial processes, particularly those involving nanofluids. Non-circular channels can be used to improve mixing in such processes. The shape of the channel significantly contributes to the mixing of fluids. Non-circular channels can induce turbulence and chaotic flow, leading to better mixing of the fluids [62, 63]. The secondary flow patterns in non-circular channels also promote convective heat transfer. This enhanced heat transfer can be particularly important in high-performance microfluidic systems, such as those used for thermal management in electronics, energy production, or medical devices.

8.2.2.4 Flow Control

Non-circular channels may be used to control nanofluids' flow rate and direction, which is useful in microfluidic devices and nanofluidic transport. Flow control is the manipulation of fluid motion in order to achieve desired outcomes, such as reducing drag or increasing lift. Various methods of flow control have been developed for different applications. These methods can be broadly classified as passive and active flow control techniques.

The passive flow control methods deal with geometric modifications to an object's surface or shape to alter fluid flow [64, 65]. Passive flow control techniques include vortex generators, riblets, and dimples. Vortex generators are tiny structures placed on an object's surface that generate vortices in the flow. These vortices can reduce drag. Riblets are small grooves or ridges on an object's surface that can reduce turbulence and drag. Dimples are tiny depressions on an object's surface that can create an air boundary layer, thereby reducing drag [64].

On the other hand, active flow control techniques involve using external energy sources, such as actuators, to manipulate the flow of fluid [66]. It comprises a range of strategies, including the implementation of jets, suction, and oscillatory blowing. In the case of jets, fluid streams are introduced into the flow to change its velocity or direction. Alternatively, suction may be employed to remove fluid from the flow, thereby altering its direction or speed. Oscillatory blowing relies on the periodic emission of fluid pulses via a device, which can induce changes in the flow. Flow control techniques offer significant potential for diverse applications in multiple fields, including aerospace, automotive, and biomedical engineering. For instance, flow control can help enhance the aerodynamic performance of an aircraft wing, resulting in reduced drag and improved fuel efficiency. In the automotive industry, flow control can improve fuel efficiency and decrease emissions. Additionally, flow control can be beneficial in the biomedical field, specifically in enhancing the performance of medical devices like blood pumps and artificial hearts.

8.2.2.5 Oil and Gas Pipelines

Non-circular channels are also commonly used in oil and gas pipelines due to their ability to transport fluids with high viscosity. The circulation of nanofluids in these channels can augment the flow properties of the fluids, reducing friction and increasing efficiency. Several reports show that the flow of nanofluids led to a significant improvement in heat transfer performance, with an increase in the heat transfer coefficient compared with the use of pure water as the working fluid [13, 14, 20, 36, 38, 40, 41, 43, 67, 68]. Although the choice of base fluid could play a significant role in impacting the properties and performance of the resulting nanofluid, it is important to carefully consider the properties and requirements of the application when selecting a base fluid [69].

The combination of nanofluids and non-circular channels has been shown to have the potential for various applications. The enhanced thermophysical properties of nanofluids combined with the improved heat transfer efficiency of non-circular channels could lead to increased efficiency and performance in various systems. However, further research is needed to fully understand the behaviour of nanofluids in non-circular channels and optimise their performance for specific applications.

8.2.3 Energy Systems Derivable from Nanofluids

The production and consumption of energy have been critical to the progress of society, but it has also resulted in harm to the environment and climate change. As a result, there is a requirement for alternative energy sources that are sustainable, efficient, and environmentally friendly [70]. Nanofluids contain nano-size particles suspended in a base fluid and have arisen as a potential medium for energy systems owing to their distinctive thermal properties.

8.2.3.1 Nanofluids for Energy Systems

Nanofluids have exhibited significant potential in various energy systems like solar thermal, geothermal, and nuclear energy. Nanofluids could enhance the efficiency of solar thermal energy systems by serving as heat transfer fluid. The addition of nanoparticles like copper oxide or alumina to the base fluid has been demonstrated to increase the heat transfer coefficient and decrease the required fluid flow rate. These enhancements have resulted in about 30% increase in the efficiency of solar thermal collectors compared with conventional fluids, as per research studies [71–73]. A recent study was done by Hu et al. [74]. The authors reviewed the recent advances in the use of hybrid nanofluids for solar thermal harvesting.

Geothermal energy systems generate power by extracting heat from the earth's subsurface. The efficiency of these systems can be enhanced by taking advantage of the high thermal conductivity of nanofluids to transfer the heat from geothermal systems using channels. The addition of nanoparticles like carbon nanotubes or silica to the base fluid, improved thermal conductivity, and reduced the pumps power requirements . Studies indicate that using nanofluids can lead to a 25% increase in the efficiency of geothermal systems as compared to conventional fluids [75–77].

Nuclear energy is another energy system where nanofluid has exhibited significant potential. One application of nuclear energy is to generate electricity from the energy produced by nuclear reactions. Using nanofluids as coolants in nuclear energy systems could potentially increase both their safety and overall efficiency. Adding nanoparticles such as diamond or boron nitride to the base fluid could enhance thermal conductivity, and the risk of nuclear fuel melting could be reduced. Research has demonstrated that nanofluids can contribute to the safety and efficiency of nuclear systems by a significant amount compared to traditional coolants [55, 78, 79].

Despite the benefits of circulating nanofluids through energy systems, some challenges remain that need to be addressed. A major challenge is related to the stability of nanofluids, which could affect their heat transfer properties. If nanoparticles aggregate or settle, it could reduce thermal conductivity and increase pumping power requirements leading to high energy consumption [80]. The thermal conductivity of nanoparticles aggregate could be increased when the conventional compact spherical structure of nanoparticles is changed to a chain-like structure [81]. Hence, future research should concentrate on developing techniques to enhance the stability of nanofluids and minimise the cost of nanoparticles [72, 73, 76] with their use in nuclear systems.

8.2.4 PROPERTIES AND CHARACTERISTICS OF NANOFLUIDS

A detailed understanding of the properties and characteristics of nanofluids is essential for their efficient application in various fields. This review provides an overview of recent research on nanofluids' properties and characteristics, such as thermal conductivity, viscosity, stability, and flow behaviour, which is crucial for developing effective strategies for their practical implementation.

An important thermophysical property that determines the effectiveness of nanofluids in heat transfer applications is thermal conductivity [5, 82]. Studies have shown that the thermal conductivity of nanofluids improves with increasing particle volume fraction and decreasing particle size [83]. The viscosity of a nanofluid could be influenced by adding nanoparticles to the base fluid, and could result in shear-thickening and shear-thinning behaviour, depending on the particle size and volume fraction. The viscosity of nanofluids is dependent on particle size, volume fraction, and temperature [14, 84, 85]. The specific heat capacity of nanofluids is also an important property that can affect their heat transfer characteristics. Several studies have investigated the effect of various parameters, such as particle size, shape, and volume fraction, on the specific heat capacity of nanofluids [86, 87]. Studies have shown that particle volume fraction increases with the specific heat capacity of nanofluids.

The stability of nanofluids is another important characteristic that can affect the use of nanofluids in various applications. The stability of nanofluids is a function of particle size, shape, surface charge, and dispersion method [88, 89]. It has been found that the addition of surfactants or dispersants can improve the stability of nanofluids, and alter the flow behaviour of the resulting nanofluid. Several studies have investigated the effect of various parameters, such as particle size, shape, and volume fraction, on the flow behaviour of nanofluids [90, 91]. It has been found that the addition of nanoparticles can result in a decrease in pressure drop and an increase in the heat transfer coefficient in certain flow conditions [2].

In summary, nanofluids exhibit unique properties and characteristics that can significantly alter the thermophysical properties and behaviour of the base fluid. The properties and characteristics of nanofluids are dependent on various parameters such as particle size, shape, volume fraction, and temperature. Understanding these properties and characteristics is crucial for the effective utilisation of nanofluids in various applications.

8.3 NUMERICAL STUDIES

Computational fluid dynamics (CFD) is a commonly used numerical method for simulating fluid flow and heat transfer in nanofluid flows. CFD simulations generally provide a detailed understanding of the flow and temperature fields in complex geometries, such as non-circular channels, and can be used to optimise the design and operation of various industrial processes. In CFD simulations, the fluid and particle motion are described by the Navier-Stokes, continuity, momentum and energy equations, which are solved numerically with finite difference, finite volume, or finite element methods. Adding nanoparticles to the base fluid introduces additional transport equations, such as those for the particle concentration and temperature, which must be solved simultaneously with the Navier-Stokes equations.

One common approach for modelling nanofluid flows in CFD simulations is the single-phase approach, where the nanofluid is treated as a homogeneous mixture with effective thermophysical properties [7, 11, 12]. In this approach, the particle effects on the fluid properties, such as viscosity and thermal conductivity, are accounted for using empirical or theoretical models. Ahmed, et al. [92] used numerical simulations to study the effects of various parameters, including nanofluid concentration, Reynolds number, and channel geometry, on the heat transfer performance of the corrugated channel. The study was carried out using a single-phase flow simulation. The results showed that the use of nanofluids in the corrugated channel led to significant improvements in the heat transfer rate and overall thermal performance. Buongiorno [93] investigated the convective heat transfer of nanofluids in a single-phase flow. The researcher explored the mechanisms of heat transfer enhancement in nanofluids and developed a model for predicting the effective thermal conductivity of nanofluids. The findings of the study suggested that nanofluids have the potential to significantly enhance convective heat transfer and that the effective thermal conductivity of nanofluids can be predicted based on the thermal conductivity and volume fraction of the nanoparticles. Ahmed et al. [94] investigated the use of nanofluids to enhance heat transfer in a wavy channel through numerical simulations. The authors used a three-dimensional model to simulate the flow and heat transfer of a water-based nanofluid flowing through a wavy channel and compared the results to those obtained with pure water. The effects of various parameters, such as the nanoparticle volume fraction, Reynolds number, and wave amplitude, on the heat transfer enhancement were investigated. The study found that circulating nanofluids in the wavy channel, significantly enhanced the heat transfer, with the magnitude of the enhancement depending on the nanoparticle volume fraction and Reynolds number.

Another approach is the two-phase analysis, where the nanofluid is considered as a two-phase system with separate equations for the fluid and particle phases. In this approach, the particle motion and concentration are determined by additional transport equations, and the particle effects on the fluid properties are accounted for using volume fraction weighting functions. Manavi et al. [95] numerically investigated the turbulent forced convection of nanofluid in a wavy channel using a two-phase model. The study investigated the effect of nanofluid concentration and Reynolds number on the fluid flow and heat transfer characteristics in the wavy channel. The results revealed that the Nusselt number increased with the nanofluid concentration, and the highest heat transfer enhancement was achieved at a concentration of 2%. Furthermore, the pressure drop increased with the Reynolds number, and the nanofluid concentration had a negligible effect on the pressure drop. More importantly, it was noted that the presence of nanoparticles enhanced the heat transfer rate in the wavy channel, and the heat transfer enhancement was more significant at higher Reynolds numbers.

Rashidi et al. [96] compared the accuracy of single-phase and two-phase models for predicting the heat transfer characteristics of nanofluids in a wavy channel. The study considered Al_2O_3-water, Cu-water, and TiO_2-water as the nanofluids. The results showed that the two-phase model predicts heat transfer characteristics of nanofluids more accurately compared to the single-phase model. The study also found that the heat transfer coefficient of nanofluids is higher than that of the base fluid due to the presence of nanoparticles, and it indicated that the heat transfer enhancement of nanofluids is dependent on the particle size and volume fraction. Toosi and Siavashi [97] investigated the natural convection of nanofluid flow in a cavity that is partially filled with porous media to augment heat transfer using a two-phase mixture numerical simulation. The researchers examined the effects of different parameters such as Rayleigh number, Darcy number, and volume fraction of nanoparticles on the fluid flow and heat transfer characteristics of the system. The results showed that increasing the Rayleigh number and volume fraction of nanoparticles enhances the natural convection heat transfer coefficient, whereas the Darcy number decreases the heat transfer rate. The authors also found that adding nanoparticles to the fluid increased the thermal conductivity of the mixture, which in turn enhanced the heat transfer rate.

In addition to CFD simulations, other numerical methods such as molecular dynamics (MD) simulations and lattice Boltzmann method (LBM) simulations have also been used to study nanofluid flows in thermal systems. MD simulations provide a more detailed understanding of the nanoscale fluid, particle interactions as well as the influence of nanoparticle aggregation on thermal conductivity [80, 81, 98], heat transfer enhancement of Copper surface in water through surface charge application [99], while LBM simulations can handle complex geometries and fluid-particle interactions. Overall, numerical simulations provide a valuable tool for studying nanofluid flows in thermal systems with non-circular cross section and can be used to optimise the design and operation of various industrial processes. However, validating the numerical results with experimental data is important to ensure their accuracy and reliability.

8.4 EXPERIMENTAL STUDIES

Experimental investigations of nanofluid flow in non-circular channels have been relatively limited in comparison to numerical investigations. The few studies that have been conducted involved the investigation of the effects of nanoparticle concentration, particle size, and channel geometry on fluid flow and heat transfer characteristics. Mehrjou et al. [100] investigated the heat transfer performance of CuO/water nanofluid in a square cross-section duct under turbulent flow conditions. The experiment was conducted by measuring the temperature difference between the inlet and outlet of the test section, and the effects of the Reynolds number, nanoparticle concentration, and heat flux on the heat transfer performance of the nanofluid were examined. The results showed that the heat transfer coefficient and Nusselt number of the nanofluid increased with increasing Reynolds number and nanoparticle concentration. Moreover, the enhancement in heat transfer coefficient was higher at lower heat fluxes. The study concluded that CuO/water nanofluid could augment the turbulent heat transfer performance in the square cross-section duct.

Attalla [101] conducted an experimental study on heat transfer and pressure drop characteristics of SiO_2/water nanofluid flow through conduits with different cross-sectional shapes, including circular, square, rectangular, triangular, hexagonal, and octagonal. The results showed that the heat transfer coefficient and Nusselt number of the nanofluid increased with the increase in the mass concentration and Reynolds number. It was noted that heat transfer enhancement and pressure drop characteristics of the nanofluid flow were affected by the cross-sectional shape of the conduit. The circular conduit showed the best performance in terms of heat transfer enhancement, while the hexagonal conduit showed the worst performance. Furthermore, the friction factor and pressure drop of the nanofluid flow increased with the increase in Reynolds number, while the effect of the cross-sectional shape on the friction factor was negligible.

Nassan et al. [45] investigated the heat transfer characteristics of two different types of nanofluids, Al_2O_3/water and CuO/water, in a square cross-section duct. The researchers studied the nanofluids' heat transfer coefficient and pressure drop with a range of particle concentrations, Reynolds numbers, and heat fluxes. The results showed that the heat transfer coefficient of both nanofluids increases with increasing particle concentration, Reynolds number, and heat flux. The study also noted that the heat transfer coefficient of the CuO/water nanofluid is higher than that of the Al_2O_3/water nanofluid at the same particle concentration, Reynolds number, and heat flux. The pressure drop of the nanofluids also increases with increasing particle concentration and Reynolds number, with the CuO/water nanofluid exhibiting a higher pressure drop than the Al_2O_3/water nanofluid. It was concluded that CuO/water nanofluid might be a more effective coolant than Al_2O_3/water nanofluid in certain applications.

Bhadouriya et al. [102] conducted both experimental and numerical investigations to study the fluid flow and heat transfer characteristics in a twisted square duct. The findings showed that a twisted square duct exhibits a more complex flow and heat transfer behaviour than a straight duct. The twist in the duct creates a swirling flow, which enhances the heat transfer, particularly on the twisted side of the duct. The experimental results show that the Nusselt number, which is a measure of heat transfer, is higher on the twisted side of the duct compared to the straight side. This suggests that the twist in the duct can be used to enhance heat transfer in certain applications. Heris et al. [49] conducted an experimental investigation of the heat transfer characteristics of Al_2O_3/water nanofluid flowing through an equilateral triangular duct with constant wall heat flux in laminar flow conditions. The experimental results showed that the heat transfer coefficient of the nanofluid is higher than that of the base fluid for all tested conditions. It was observed that the enhancement in heat transfer coefficient increases with increasing nanoparticle volume fraction and Reynolds number. The maximum enhancement in heat transfer coefficient is found to be 21.2% for a nanoparticle volume fraction of 2% and a Reynolds number of 140. The study also investigated the flow regime's effect on the nanofluid's heat transfer characteristics. The authors find that the heat transfer coefficient increases with increasing Reynolds number, which is consistent with the expected behaviour of the laminar flow.

It is worth noting that a significant amount of research has been carried out on the flow of nanofluids, particularly in non-circular shapes, and has focused on laminar or turbulent flow regimes. The transitional flow regime, which occurs in many industrial processes, has not been well investigated. The transitional flow regime is a critical regime for nanofluid flow through non-circular shapes, as it involves the transition from laminar to turbulent flow, which affects the heat transfer and pressure drop characteristics [35, 103]. The existing literature suggests that the flow behaviour of nanofluids in non-circular shapes in the transitional flow regime is highly complex and depends on various factors such as particle size, shape, concentration, and the geometry of the non-circular shape.

8.5 TRANSITIONAL FLOW REGIME

Previous studies on the circulation of nanofluids have generally been conducted in the laminar and turbulent flow regimes. However, Meyer and Abolarin [18], in their research investigation found that the transitional flow regime provides a compromise between the laminar and turbulent flow regimes. This is because the heat transfer coefficient in the transitional flow regime is higher than that of the laminar, and the pressure drop is lower than that of the turbulent [104,105]. A search on Scopus with the quoted words "nanofluid heat transfer enhancement in channels in the transitional flow regime" reveals the research gaps that exist in the transitional flow regime with respect to heat transfer enhancement investigation of nanofluid circulation in channels. From the search, only three articles were reported to have been conducted in this area. These transitional flow regime studies are the experiments of Osman et al. [106] published at a conference for a flow of dilute alumina-water with the volume fraction of 0.1% circulated in a rectangular channel over the Reynolds number 200 to 5000. The detailed version is published as a journal article in Osman et al. [107]. Osman et al. [107] conducted using alumina-water nanofluid of volume fraction 0.3, 0.5 and 1% circulated over Reynolds number 200 to 7000 in a rectangular channel. The study found that the heat transfer coefficient was enhanced in the transitional flow regime by 54% when the nanofluid with a volume concentration of 1.0% was circulated in the rectangular channel. In the turbulent flow regime, the maximum heat transfer enhancement was 11%. The third article found on Scopus is the numerical work of Xie et al. [108], where Al_2O_3 was circulated in a rectangular channel for two configurations of dimples as well as protrusions over the Reynolds number 1000 to 10000. This Reynolds number covered the laminar, transitional and turbulent flow regimes. The study focused on heat transfer enhancement and entropy generation in the deployment of the nanofluid and the channel configurations. The dimple configuration resulted in a decrease in entropy generation when compared with the smooth channel. The protrusion configuration resulted in a higher pressure drop. These limited studies in the transitional flow regime show that more studies focusing on the different channel configurations with nanofluid with different volume fractions are needed in order to advance knowledge in the thermal science field.

8.6　FUTURE RESEARCH DIRECTIONS

Nanofluid flow in non-circular channels has attracted increasing interest in recent years due to its potential applications in various fields, such as energy production, electronics cooling, and biomedical devices. However, several challenges and future directions need to be addressed to further enhance the understanding of nanofluid flow in non-circular channels. Below are some suggested recommendations for future researchers analysing numerically and experimentally the use of nanofluids in non-circular ducts:

- Although several experimental studies have been conducted in recent years, there is still a need for more extensive and systematic experimental investigations to validate numerical models and obtain a better understanding of nanofluid flow in non-circular channels. In addition, selecting suitable nanofluid properties for numerical models is still a challenge, as the properties of nanofluids can vary significantly depending on the particle size, concentration, and shape.
- Non-circular channels can exhibit highly complex and unpredictable flow patterns, which can significantly affect the heat transfer performance of nanofluids. Therefore, more sophisticated numerical models need to be developed to accurately predict the fluid flow and heat transfer characteristics in non-circular channels.
- The effect of nanoparticles on the friction factor and pressure drop in non-circular channels is still not fully understood. Some studies have reported an increase in the pressure drop due to the addition of nanoparticles in non-circular channels, while others have reported a decrease. Thus, more research is needed to better understand the effect of nanoparticles on the friction factor and pressure drop in non-circular channels, which can have significant implications for the design and optimisation of heat transfer systems.
- To investigate the potential of novel nanofluid-based heat transfer systems for various applications. For instance, recent studies have shown that nanofluids can be used to enhance the performance of microfluidic systems, which can have significant implications for micro-electronic cooling and lab-on-a-chip devices. In addition, nanofluid-based heat transfer systems can be used in renewable energy applications, such as solar thermal and geothermal energy systems, to increase efficiency and reduce the cost of energy production.

8.7　CONCLUSION

The study of nanofluid flow in non-circular channels is a rapidly growing field with great potential for various applications. Although there are several challenges that need to be addressed, there are also promising future directions for further research and development. By addressing these challenges and pursuing these future directions, researchers, designers and operators of energy systems can gain a better understanding of nanofluid flow in non-circular channels and develop more efficient and sustainable heat transfer systems for various applications. Moreso, nanofluids have shown promising results in enhancing the fourth industrial revolution, sustainability and efficiency of different energy systems by improving the thermal properties of the fluids, leading to a reduction in energy consumption and enhanced heat transfer. Although certain challenges remain, more research can result in the development of stable and cost-effective nanofluids that can transform the energy industry.

REFERENCES

[1] A. O. Adelaja, D. R. Ewim, J. Dirker, and J. P. Meyer, "Heat transfer, void fraction and pressure drop during condensation inside inclined smooth and microfin tubes," *Experimental Thermal and Fluid Science*, vol. 109, p. 109905, 2019.

[2] S. A. Adio, T. A. Alo, R. O. Olagoke, A. E. Olalere, V. R. Veeredhi, and D. R. Ewim, "Thermohydraulic and entropy characteristics of Al2O3-water nanofluid in a ribbed interrupted microchannel heat exchanger," *Heat Transfer*, vol. 50, no. 3, pp. 1951–1984, 2021.

[3] S. A. Adio et al., "Thermal and entropy analysis of a manifold microchannel heat sink operating on CuO–water nanofluid," *Journal of the Brazilian Society of Mechanical Sciences and Engineering*, vol. 43, pp. 1–15, 2021.

[4] S. Bhattacharyya, M. Pathak, M. Sharifpur, S. Chamoli, and D. R. Ewim, "Heat transfer and exergy analysis of solar air heater tube with helical corrugation and perforated circular disc inserts," *Journal of Thermal Analysis and Calorimetry*, vol. 145, pp. 1019–1034, 2021.

[5] D. Ewim, A. S. Shote, E. J. Onyiriuka, S. A. Adio, S. Bhattacharyya, and A. Kaood, "Thermal performance of nano refrigerants: A short review," *Journal of Mechanical Engineering Research*, vol. 44, pp. 89–115, 2021.

[6] D. R. E. Ewim, J. P. Meyer, and S. N. R. Abadi, "Condensation heat transfer coefficients in an inclined smooth tube at low mass fluxes," *International Journal of Heat and Mass Transfer*, vol. 123, pp. 455–467, 2018.

[7] O. G. Fadodun, D. R. E. Ewim, and S. M. Abolarin, "Investigation of turbulent entropy production rate with SWCNT/H$_2$O nanofluid flowing in various inwardly corrugated pipes," *Heat Transfer*, vol. 51, no. 8, pp. 7862–7889, 2022, https://doi.org/10.1002/htj.22671

[8] I. A. Fetuga, O. T. Olakoyejo, D. E. Ewim, J. K. Gbegudu, A. O. Adelaja, and O. O. Adewumi, "Computational investigation of thermal behaviors of the automotive radiator operated with water/ anti-freezing agent nanofluid based coolant," *The Journal of Engineering and Exact Sciences*, vol. 8, no. 2, pp. 13977-01e, 2022.

[9] J. P. Meyer and D. R. E. Ewim, "Heat transfer coefficients during the condensation of low mass fluxes in smooth horizontal tubes," *International Journal of Multiphase Flow*, vol. 99, pp. 485–499, 2018.

[10] G. K. Muteba, D. R. Ewim, J. Dirker, and J. P. Meyer, "Heat transfer and pressure drop investigation for prescribed heat fluxes on both the inner and outer wall of an annular duct," *Experimental Thermal and Fluid Science*, p. 110907, 2023.

[11] E. Onyiriuka, O. Ighodaro, A. Adelaja, D. Ewim, and S. Bhattacharyya, "A numerical investigation of the heat transfer characteristics of water-based mango bark nanofluid flowing in a double-pipe heat exchanger," *Heliyon*, vol. 5, no. 9, p. e02416, 2019.

[12] E. Onyiriuka, A. Obanor, M. Mahdavi, and D. R. E. Ewim, "Evaluation of single-phase, discrete, mixture and combined model of discrete and mixture phases in predicting nanofluid heat transfer characteristics for laminar and turbulent flow regimes," *Advanced Powder Technology*, vol. 29, no. 11, pp. 2644–2657, 2018.

[13] T. O. Scott, D. R. Ewim, and A. C. Eloka-Eboka, "Experimental investigation of natural convection Al2O3-MWCNT/water hybrid nanofluids inside a square cavity," *Experimental Heat Transfer*, pp. 1–19, 2022.

[14] T. O. Scott, D. R. Ewim, and A. C. Eloka-Eboka, "Hybrid nanofluids flow and heat transfer in cavities: A technological review," *International Journal of Low-Carbon Technologies*, vol. 17, pp. 1104–1123, 2022.

[15] M. Jadhav, R. Awari, D. Bibe, A. Bramhane, and M. Mokashi, "Review on enhancement of heat transfer by active method," *International Journal of Current Engineering and Technology*, no. 6, pp. 221–225, 2016. [Online]. Available: https://inpressco.com/wp-content/uploads/2016/10/Paper49 221-225.pdf

[16] S. Thapa, S. Samir, K. Kumar, and S. Singh, "A review study on the active methods of heat transfer enhancement in heat exchangers using electroactive and magnetic materials," *Materials Today: Proceedings*, vol. 45, pp. 4942–4947, 2021, https://doi.org/10.1016/j.matpr.2021.01.382

[17] P.-N. Martín, C. M.-G. Jorge, and C. Jorge Luis García, "Use of heat transfer enhancement techniques in the design of heat exchangers," in *Advances in Heat Exchangers*, G. Laura Castro and F. Víctor Manuel Velázquez (eds.) Rijeka: IntechOpen, 2018, ch. 3.

[18] J. P. Meyer and S. M. Abolarin, "Heat transfer and pressure drop in the transitional flow regime for a smooth circular tube with twisted tape inserts and a square-edged inlet," *International Journal of Heat and Mass Transfer*, vol. 117, pp. 11–29, 2018, https://doi.org/10.1016/j.ijheatmasstransfer.2017.09.103

[19] K. Orhan and O. Veysel, "A review of heat transfer enhancement methods using coiled wire and twisted tape inserts," in *Heat Transfer*, V. Konstantin (ed.) Rijeka: IntechOpen, 2018, ch. 10.

[20] K. V. Wong and M. J. Castillo, "Heat transfer mechanisms and clustering in nanofluids," *Advances in Mechanical Engineering*, vol. 2, p. 795478, 2010.

[21] L. M. Amoo and R. Layi Fagbenle, "12 – Advanced fluids – A review of nanofluid transport and its applications," in *Applications of Heat, Mass and Fluid Boundary Layers*, R. O. Fagbenle, O. M. Amoo, S. Aliu, and A. Falana (eds.) Woodhead Publishing, 2020, pp. 281–382.

[22] J. Chevalier, O. Tillement, and F. Ayela, "Rheological properties of nanofluids flowing through microchannels," *Applied Physics Letters*, vol. 91, no. 23, p. 233103, 2007, https://doi.org/10.1063/1.2821117

[23] E. A. Chavez Panduro, F. Finotti, G. Largiller, and K. Y. Lervåg, "A review of the use of nanofluids as heat-transfer fluids in parabolic-trough collectors," *Applied Thermal Engineering*, vol. 211, p. 118346, 2022, https://doi.org/10.1016/j.applthermaleng.2022.118346

[24] T. Huq, H. C. Ong, B. T. Chew, K. Y. Leong, and S. N. Kazi, "Review on aqueous graphene nanoplatelet Nanofluids: Preparation, Stability, thermophysical Properties, and applications in heat exchangers and solar thermal collectors," *Applied Thermal Engineering*, vol. 210, p. 118342, 2022, https://doi.org/10.1016/j.applthermaleng.2022.118342

[25] P. K. Das and Z. Ahmed, "Experimental and molecular dynamics approach to evaluate the thermo-rheological properties of CuO nanofluids for heat transfer application," *Particulate Science and Technology*, pp. 1–15, 2022, https://doi.org/10.1080/02726351.2022.2124472

[26] T. P. Jeevan and S. R. Jayaram, "Tribological properties and machining performance of vegetable oil based metal working fluids—A Review," *Modern Mechanical Engineering*, vol. 8, pp. 42–65, 2018, https://doi.org/10.4236/mme.2018.81004

[27] W. Xu et al., "Electrostatic atomization minimum quantity lubrication machining: from mechanism to application," *International Journal of Extreme Manufacturing*, vol. 4, no. 4, p. 042003, 2022, https://doi.org/10.1088/2631-7990/ac9652

[28] W. N. Mutuku, "Ethylene glycol (EG)-based nanofluids as a coolant for automotive radiator," *Asia Pacific Journal on Computational Engineering*, vol. 3, no. 1, p. 1, 2016, https://doi.org/10.1186/s40540-016-0017-3

[29] X. Li et al., "Heat transfer enhancement of nanofluids with non-spherical nanoparticles: A review," *Applied Sciences*, vol. 12, no. 9, p. 4767, 2022. [Online]. Available: www.mdpi.com/2076-3417/12/9/4767

[30] D. Liu and L. Yu, "Single-phase thermal transport of nanofluids in a minichannel," *Journal of Heat Transfer*, vol. 133, no. 3, 2011.

[31] S. Tahir and M. Mital, "Numerical investigation of laminar nanofluid developing flow and heat transfer in a circular channel," *Applied Thermal Engineering*, vol. 39, pp. 8–14, 2012/06/01, https://doi.org/10.1016/j.applthermaleng.2012.01.035

[32] S. Mei, C. Qi, M. Liu, F. Fan, and L. Liang, "Effects of paralleled magnetic field on thermo-hydraulic performances of Fe3O4-water nanofluids in a circular tube," *International Journal of Heat and Mass Transfer*, vol. 134, pp. 707–721, 2019/05/01, https://doi.org/10.1016/j.ijheatmasstransfer.2019.01.088

[33] B. Sahin, G. G. Gültekin, E. Manay, and S. Karagoz, "Experimental investigation of heat transfer and pressure drop characteristics of Al$_2$O$_3$–water nanofluid," *Experimental Thermal and Fluid Science*, vol. 50, pp. 21–28, 2013.

[34] P. Keblinski, S. Phillpot, S. Choi, and J. Eastman, "Mechanisms of heat flow in suspensions of nano-sized particles (nanofluids)," *International journal of heat and mass transfer*, vol. 45, no. 4, pp. 855–863, 2002.

[35] W. Williams, J. Buongiorno, and L.-W. Hu, "Experimental investigation of turbulent convective heat transfer and pressure loss of alumina/water and zirconia/water nanoparticle colloids (nanofluids) in horizontal tubes," *Journal of Heat Transfer,* vol. 130, no. 4, 2008.

[36] B. Mehta, D. Subhedar, H. Panchal, and Z. Said, "Synthesis, stability, thermophysical properties and heat transfer applications of nanofluid–a review," *Journal of Molecular Liquids,* vol. 364, p. 120034, 2022.

[37] M. E. Haque, M. S. Hossain, and H. M. Ali, "Laminar forced convection heat transfer of nanofluids inside non-circular ducts: A review," *Powder Technology,* vol. 378, pp. 808–830, 2021/01/22/, https://doi.org/10.1016/j.powtec.2020.10.042

[38] S. Mukherjee and S. Paria, "Preparation and stability of nanofluids-a review," *IOSR Journal of Mechanical and Civil Engineering,* vol. 9, no. 2, pp. 63–69, 2013.

[39] L. S. Sundar, K. V. Sharma, M. K. Singh, and A. Sousa, "Hybrid nanofluids preparation, thermal properties, heat transfer and friction factor–a review," *Renewable and Sustainable Energy Reviews,* vol. 68, pp. 185–198, 2017.

[40] M. Gupta, V. Singh, R. Kumar, and Z. Said, "A review on thermophysical properties of nanofluids and heat transfer applications," *Renewable and Sustainable Energy Reviews,* vol. 74, pp. 638–670, 2017.

[41] W. Yu and H. Xie, "A review on nanofluids: preparation, stability mechanisms, and applications," *Journal of nanomaterials,* vol. 2012, pp. 1–17, 2012.

[42] S. Suresh, K. Venkitaraj, P. Selvakumar, and M. Chandrasekar, "Synthesis of Al2O3–Cu/water hybrid nanofluids using two step method and its thermo physical properties," *Colloids and Surfaces A: Physicochemical and Engineering Aspects,* vol. 388, no. 1–3, pp. 41–48, 2011.

[43] Y. Li, S. Tung, E. Schneider, and S. Xi, "A review on development of nanofluid preparation and characterization," *Powder technology,* vol. 196, no. 2, pp. 89–101, 2009.

[44] M. R. Salimpour and A. Dehshiri-Parizi, "Convective heat transfer of nanofluid flow through conduits with different cross-sectional shapes," *Journal of Mechanical Science and Technology,* vol. 29, no. 2, p. 707, 2015.

[45] T. H. Nassan, S. Z. Heris, and S. Noie, "A comparison of experimental heat transfer characteristics for Al2O3/water and CuO/water nanofluids in square cross-section duct," *International Communications in Heat and Mass Transfer,* vol. 37, no. 7, pp. 924–928, 2010.

[46] F. Pourfayaz, N. Sanjarian, A. Kasaeian, F. Razi Astaraei, M. Sameti, and S. Nasirivatan, "An experimental comparison of SiO 2/water nanofluid heat transfer in square and circular cross-sectional channels," *Journal of Thermal Analysis and Calorimetry,* vol. 131, pp. 1577–1586, 2018.

[47] M. Salimpour and A. Dehshiri-Parizi, "Effect of duct cross sectional shape on the nanofluid flow heat transfer," *International Journal of Mechanical, Aerospace, Industrial, Mechatronic and Manufacturing Engineering,* pp. 1943–1947, 2014.

[48] A. Javadpour, M. Najafi, and K. Javaherdeh, "Effect of magnetic field on forced convection heat transfer of a non-Newtonian nanofluid through an annulus: An experimental study," *Heat and Mass Transfer,* vol. 54, pp. 3307–3316, 2018.

[49] S. Z. Heris, Z. Edalati, S. H. Noie, and O. Mahian, "Experimental investigation of Al2O3/water nanofluid through equilateral triangular duct with constant wall heat flux in laminar flow," *Heat Transfer Engineering,* vol. 35, no. 13, pp. 1173–1182, 2014.

[50] A. Moita, A. Moreira, and J. Pereira, "Nanofluids for the next generation thermal management of electronics: A review," *Symmetry,* vol. 13, no. 8, p. 1362, 2021. [Online]. Available: www.mdpi.com/2073-8994/13/8/1362

[51] D. S. Saidina, M. Z. Abdullah, and M. Hussin, "Metal oxide nanofluids in electronic cooling: A review," *Journal of Materials Science: Materials in Electronics,* vol. 31, no. 6, pp. 4381–4398, 2020, https://doi.org/10.1007/s10854-020-03020-7

[52] O. T. Olakoyejo *et al.,* "Numerical forced convection heat transfer, fluid flow and entropy generation analyses of Al2O3-water nanofluid in elliptical channels," *Nigerian Journal of Technological Development,* vol. 19, no. 4, pp. 361–372, 2022, http://dx.doi.org/10.4314/njtd.v19i4.9

[53] M. Bahiraei and S. Heshmatian, "Electronics cooling with nanofluids: A critical review," *Energy Conversion and Management,* vol. 172, pp. 438–456, 2018, https://doi.org/10.1016/j.enconman.2018.07.047

[54] H. Xie, H. Lee, W. Youn, and M. Choi, "Nanofluids containing multiwalled carbon nanotubes and their enhanced thermal conductivities," *Journal of Applied Physics*, vol. 94, no. 8, pp. 4967–4971, 2003.

[55] A. Kasaeian, A. T. Eshghi, and M. Sameti, "A review on the applications of nanofluids in solar energy systems," *Renewable and Sustainable Energy Reviews*, vol. 43, pp. 584–598, 2015.

[56] V. P. Kalbande, P. V. Walke, and C. Kriplani, "Advancements in thermal energy storage system by applications of Nanofluid based solar collector: A review," *Environmental and Climate Technologies*, vol. 24, no. 1, pp. 310–340, 2020.

[57] J. M. Munyalo and X. Zhang, "Particle size effect on thermophysical properties of nanofluid and nanofluid based phase change materials: A review," *Journal of Molecular Liquids*, vol. 265, pp. 77–87, 2018.

[58] K. Anoop, J. Cox, and R. Sadr, "Thermal evaluation of nanofluids in heat exchangers," *International Communications in Heat and Mass Transfer*, vol. 49, pp. 5–9, 2013.

[59] R. Islam, B. Shabani, J. Andrews, and G. Rosengarten, "Experimental investigation of using ZnO nanofluids as coolants in a PEM fuel cell," *International Journal of Hydrogen Energy*, vol. 42, no. 30, pp. 19272–19286, 2017.

[60] M. R. Islam, B. Shabani, and G. Rosengarten, "Nanofluids to improve the performance of PEM fuel cell cooling systems: A theoretical approach," *Applied Energy*, vol. 178, pp. 660–671, 2016.

[61] R. Gautam, A. K. Sharma, and K. D. Gupta, "Performance analysis of non-circular microchannels flooded with CuO-water nanofluid," 2011.

[62] C.-W. Hsu, P.-T. Shih, and J. M. Chen, "Enhancement of Fluid Mixing with U-Shaped Channels on a Rotating Disc," *Micromachines*, vol. 11, no. 12, p. 1110, 2020.

[63] A. M. Abed *et al.*, "Enhance heat transfer in the channel with V-shaped wavy lower plate using liquid nanofluids," *Case Studies in Thermal Engineering*, vol. 5, pp. 13–23, 2015.

[64] B. Yagiz, O. Kandil, and Y. V. Pehlivanoglu, "Drag minimization using active and passive flow control techniques," *Aerospace Science and Technology*, vol. 17, no. 1, pp. 21–31, 2012.

[65] M. Zaresharif, F. Ravelet, D. J. Kinahan, and Y. M. Delaure, "Cavitation control using passive flow control techniques," *Physics of Fluids*, vol. 33, no. 12, p. 121301, 2021.

[66] S. N. Joshi and Y. S. Gujarathi, "A review on active and passive flow control techniques," *International Journal on Recent Technologies in Mechanical and Electrical Engineering*, vol. 3, no. 4, pp. 1–6, 2016.

[67] Y. Xuan and Q. Li, "Heat transfer enhancement of nanofluids," *International Journal of heat and fluid flow*, vol. 21, no. 1, pp. 58–64, 2000.

[68] Z. Haddad, E. Abu-Nada, H. F. Oztop, and A. Mataoui, "Natural convection in nanofluids: are the thermophoresis and Brownian motion effects significant in nanofluid heat transfer enhancement?," *International Journal of Thermal Sciences*, vol. 57, pp. 152–162, 2012.

[69] D. Wen and Y. Ding, "Experimental investigation into convective heat transfer of nanofluids at the entrance region under laminar flow conditions," *International journal of heat and mass transfer*, vol. 47, no. 24, pp. 5181–5188, 2004.

[70] D. R. E. Ewim, S. M. Abolarin, T. O. Scott, and C. S. Anyanwu, "A Survey on the Understanding and Viewpoints of Renewable Energy among South African School Students," *The Journal of Engineering and Exact Sciences*, vol. 9, no. 2, pp. 15375–01e, 2023.

[71] S. P. Tembhare, D. P. Barai, and B. A. Bhanvase, "Performance evaluation of nanofluids in solar thermal and solar photovoltaic systems: A comprehensive review," *Renewable and Sustainable Energy Reviews*, vol. 153, p. 111738, 2022.

[72] S. K. Gupta and S. Gupta, "The role of nanofluids in solar thermal energy: A review of recent advances," *Materials Today: Proceedings*, vol. 44, pp. 401–412, 2021.

[73] K. Khanafer and K. Vafai, "A review on the applications of nanofluids in solar energy field," *Renewable energy*, vol. 123, pp. 398–406, 2018.

[74] G. Hu *et al.*, "Potential evaluation of hybrid nanofluids for solar thermal energy harvesting: A review of recent advances," *Sustainable Energy Technologies and Assessments*, vol. 48, p. 101651, 2021.

[75] M. Soltani, F. M. Kashkooli, M. A. Fini, D. Gharapetian, J. Nathwani, and M. B. Dusseault, "A review of nanotechnology fluid applications in geothermal energy systems," *Renewable and Sustainable Energy Reviews*, vol. 167, p. 112729, 2022.

[76] A. Elsheikh, S. Sharshir, M. E. Mostafa, F. Essa, and M. K. A. Ali, "Applications of nanofluids in solar energy: a review of recent advances," *Renewable and Sustainable Energy Reviews*, vol. 82, pp. 3483–3502, 2018.

[77] D. Sui, V. H. Langåker, and Z. Yu, "Investigation of thermophysical properties of nanofluids for application in geothermal energy," *Energy Procedia,* vol. 105, pp. 5055–5060, 2017.

[78] I. C. Bang and J. H. Kim, "Thermal-fluid characterizations of ZnO and SiC nanofluids for advanced nuclear power plants," *Nuclear Technology,* vol. 170, no. 1, pp. 16–27, 2010.

[79] J. Buongiorno, L.-W. Hu, S. J. Kim, R. Hannink, B. Truong, and E. Forrest, "Nanofluids for enhanced economics and safety of nuclear reactors: an evaluation of the potential features, issues, and research gaps," *Nuclear Technology,* vol. 162, no. 1, pp. 80–91, 2008.

[80] R. Wang, S. Qian, and Z. Zhang, "Investigation of the aggregation morphology of nanoparticle on the thermal conductivity of nanofluid by molecular dynamics simulations," *International Journal of Heat and Mass Transfer,* vol. 127, pp. 1138–1146, 2018, doi: https://doi.org/10.1016/j.ijheatmasstransfer.2018.08.117

[81] L. Zhou, J. Zhu, Y. Zhao, and H. Ma, "A molecular dynamics study on thermal conductivity enhancement mechanism of nanofluids – Effect of nanoparticle aggregation," *International Journal of Heat and Mass Transfer,* vol. 183, p. 122124, 2022, doi: https://doi.org/10.1016/j.ijheatmasstransfer.2021.122124

[82] A. A. Hussien, M. Z. Abdullah, and M. d. A. Al-Nimr, "Single-phase heat transfer enhancement in micro/minichannels using nanofluids: Theory and applications," *Applied Energy,* vol. 164, pp. 733–755, 2016/02/15/ 2016, doi: https://doi.org/10.1016/j.apenergy.2015.11.099

[83] W. Duangthongsuk and S. Wongwises, "Measurement of temperature-dependent thermal conductivity and viscosity of TiO2-water nanofluids," *Experimental Thermal and Fluid Science,* vol. 33, no. 4, pp. 706–714, 2009/04/01/ 2009, doi: https://doi.org/10.1016/j.expthermflusci.2009.01.005

[84] R. P. Singh, H. Xu, S. Kaushik, D. Rakshit, and A. Romagnoli, "Effective utilization of natural convection via novel fin design & influence of enhanced viscosity due to carbon nano-particles in a solar cooling thermal storage system," *Solar Energy,* vol. 183, pp. 105–119, 2019.

[85] Y. Singh, N. K. Singh, A. Sharma, A. Singla, D. Singh, and E. Abd Rahim, "Effect of ZnO nanoparticles concentration as additives to the epoxidized Euphorbia Lathyris oil and their tribological characterization," *Fuel,* vol. 285, p. 119148, 2021.

[86] C. Raju, S. Mamatha, P. Rajadurai, and I. Khan, "Nonlinear mixed thermal convective flow over a rotating disk in suspension of magnesium oxide nanoparticles with water and EG," *The European Physical Journal Plus,* vol. 134, no. 5, p. 196, 2019.

[87] R. Sharma *et al.,* "Characterization of ZnO/nanofluid for improving heat transfer in thermal systems," *Materials Today: Proceedings,* vol. 62, pp. 1904–1908, 2022.

[88] R. Kumar, G. Bharadwaj, M. Kharub, V. Goel, and A. Kumar, "Performance analysis of nanofluid based direct absorption solar collector of different configurations: A two-phase CFD modeling," *Energy Sources, Part A: Recovery, Utilization, and Environmental Effects,* pp. 1–16, 2022.

[89] Y. Zhai, L. Li, J. Wang, and Z. Li, "Evaluation of surfactant on stability and thermal performance of Al2O3-ethylene glycol (EG) nanofluids," *Powder Technology,* vol. 343, pp. 215–224, 2019.

[90] V. Kumar and J. Sarkar, "Numerical and experimental investigations on heat transfer and pressure drop characteristics of Al2O3-TiO2 hybrid nanofluid in minichannel heat sink with different mixture ratio," *Powder technology,* vol. 345, pp. 717–727, 2019.

[91] D. Ebrahimi *et al.,* "Mixed convection heat transfer of a nanofluid in a closed elbow-shaped cavity (CESC)," *Journal of Thermal Analysis and Calorimetry,* vol. 144, pp. 2295–2316, 2021.

[92] M. Ahmed, N. Shuaib, M. Z. Yusoff, and A. Al-Falahi, "Numerical investigations of flow and heat transfer enhancement in a corrugated channel using nanofluid," *International Communications in Heat and Mass Transfer,* vol. 38, no. 10, pp. 1368–1375, 2011.

[93] J. Buongiorno, "Convective transport in nanofluids," *Journal of Heat Transfer,* vol. 128, no. 3, pp. 240–250, 2006.

[94] M. Ahmed, N. Shuaib, and M. Z. Yusoff, "Numerical investigations on the heat transfer enhancement in a wavy channel using nanofluid," *International Journal of Heat and Mass Transfer,* vol. 55, no. 21-22, pp. 5891–5898, 2012.

[95] S. A. Manavi, A. Ramiar, and A. A. Ranjbar, "Turbulent forced convection of nanofluid in a wavy channel using two phase model," *Heat and Mass Transfer,* vol. 50, pp. 661–671, 2014.

[96] M. Rashidi, A. Hosseini, I. Pop, S. Kumar, and N. Freidoonimehr, "Comparative numerical study of single and two-phase models of nanofluid heat transfer in wavy channel," *Applied Mathematics and Mechanics,* vol. 35, pp. 831–848, 2014.

[97] M. H. Toosi and M. Siavashi, "Two-phase mixture numerical simulation of natural convection of nanofluid flow in a cavity partially filled with porous media to enhance heat transfer," *Journal of Molecular Liquids,* vol. 238, pp. 553–569, 2017.

[98] W. Cui, Z. Shen, J. Yang, and S. Wu, "Molecular dynamics simulation on the microstructure of absorption layer at the liquid–solid interface in nanofluids," *International Communications in Heat and Mass Transfer,* vol. 71, pp. 75–85, 2016, https://doi.org/10.1016/j.icheatmasstransfer.2015.12.023

[99] B.-B. Wang, Z.-M. Xu, X.-D. Wang, and W.-M. Yan, "Molecular dynamics investigation on enhancement of heat transfer between electrified solid surface and liquid water," *International Journal of Heat and Mass Transfer,* vol. 125, pp. 756–760, 2018, https://doi.org/10.1016/j.ijheatmasstransfer.2018.04.139

[100] B. Mehrjou, S. Z. Heris, and K. Mohamadifard, "Experimental study of CuO/water nanofluid turbulent convective heat transfer in square cross-section duct," *Experimental Heat Transfer,* vol. 28, no. 3, pp. 282–297, 2015.

[101] M. Attalla, "Experimental investigation of heat transfer and pressure drop of SiO2/water nanofluid through conduits with altered cross-sectional shapes," *Heat and Mass Transfer,* vol. 55, no. 12, pp. 3427–3442, 2019.

[102] R. Bhadouriya, A. Agrawal, and S. Prabhu, "Experimental and numerical study of fluid flow and heat transfer in a twisted square duct," *International Journal of Heat and Mass Transfer,* vol. 82, pp. 143–158, 2015.

[103] M. Chandrasekar, S. Suresh, and T. Senthilkumar, "Mechanisms proposed through experimental investigations on thermophysical properties and forced convective heat transfer characteristics of various nanofluids – A review," *Renewable and Sustainable Energy Reviews,* vol. 16, no. 6, pp. 3917–3938, 2012/08/01/, https://doi.org/10.1016/j.rser.2012.03.013

[104] S. M. Abolarin, M. Everts, and J. P. Meyer, "The influence of peripheral u-cut twisted tapes and ring inserts on the heat transfer and pressure drop characteristics in the transitional flow regime," *International Journal of Heat and Mass Transfer,* vol. 132, pp. 970–984, 2019, https://doi.org/10.1016/j.ijheatmasstransfer.2018.12.051

[105] S. M. Abolarin, M. Everts, and J. P. Meyer, "Heat transfer and pressure drop characteristics of alternating clockwise and counter clockwise twisted tape inserts in the transitional flow regime," *International Journal of Heat and Mass Transfer,* vol. 133, pp. 203–217, 2019, https://doi.org/10.1016/j.ijheatmasstransfer.2018.12.107

[106] S. Osman, M. Sharifpur, and J. P. Meyer, "Experimental examination of transitional flow of dilute alumina – water nanofluid," in *International Heat Transfer Conference 16,* Beijing, China, August, 10-15, 2018 2018, pp. 3095–3102, https://doi.org/10.1615/IHTC16.cov.023235

[107] S. Osman, M. Sharifpur, and J. P. Meyer, "Experimental investigation of convection heat transfer in the transition flow regime of aluminium oxide-water nanofluids in a rectangular channel," *International Journal of Heat and Mass Transfer,* vol. 133, pp. 895–902, 2019, https://doi.org/10.1016/j.ijheatmasstransfer.2018.12.169

[108] Y. Xie, L. Zheng, D. Zhang, and G. Xie, "Entropy generation and heat transfer performances of Al2O3-Water nanofluid transitional flow in rectangular channels with dimples and protrusions," *Entropy,* vol. 18, no. 4, p. 148, 2016. [Online]. Available: www.mdpi.com/1099-4300/18/4/148

9 In-Situ Based Observation and Reanalysis-Derived Wind Data for Offshore Wind Energy Potential Assessment in the Gulf of Guinea

Olayinka S. Ohunakin, Olaniran J. Matthew,
Windmanagda Sawadogo, Emerald U. Henry,
Victor U. Ezekiel, and Damola S. Adelekan

9.1 INTRODUCTION

The rapid increase in the exploitation of non-renewable energy resources such as petroleum has significantly increased carbon emissions, which pushes up global warming trends (Abiodun et al. 2012). Recently, this has become a topical matter, and calls for serious concern, especially in Africa, following the reliance on oil and gas (fossil fuels) for power generation (Aboobacker et al. 2021). Africa's electric energy matrix is mainly a composition of energy from hydroelectric power plants at approximately 60.8% and others (Adeniyi 2017). This large presence of hydroelectricity in the overall energy mix makes the system prone to vulnerabilities such as drought and other environmental disasters.

The literature has shown that using clean and renewable energy resources is paramount for sustainable development in any part of the world (e.g., Abiodun et al. 2012; Aboobacker et al. 2021; Adeniyi 2019; ANRC-AfDBG 2021). Wind energy presents potential among other renewable energy resources in Africa but appears not to be maximally exploited in the region. This is partly due to the high installation and maintenance costs, and the fact that little is known about wind energy potential and the wind characteristics of the region.

The basic starting point for harvesting offshore wind energy in any region is the abundance of wind resources. A major question is how ocean energy resources can contribute to renewable energy in Africa, making opportunities available to millions of Africans and keeping carbon emissions to the barest minimum. It has been revealed that continental Africa has vast offshore renewable energy resources endowed with onshore renewable energies, which are more feasible to exploit today (Ayala et al. 2020). More specifically, a few western African offshore locations have moderate (about 8 m s^{-1}, such as Senegal and Benin) to high (> = 10 m s^{-1}, such as Cape Verde) annual wind speed at 100 m height, compared to other parts of the sub-region (Bennington et al. 2014). Hence, countries in West Africa with existing offshore industries (oil and gas exploration industries), may be faced with the opportunity to develop a strong offshore capacity that will enable them to upscale offshore renewables, particularly wind power development (Ayala et al. 2020). Therefore, future advances in wind power technology could provide abundant energy for West African countries through the vast offshore resources, based on the experiences of oil and gas drilling companies.

Consequently, future investment in offshore wind power development could also have the potential to reduce costs associated with offshore wind energy generation, especially in West Africa. At the moment, the growth of offshore technology is only noticeable in countries in Europe, Asia, and America. It is worth noting that the global efforts and commitments to cut down greenhouse gas emissions will further spell a bright future for this technology in many offshore areas of the world (Chen et al. 2017; Chou et al. 2019; Elsner 2019; Emeksiz and Demirci 2019). However, the abundance of offshore wind energy on the West African coast has not been investigated. This is majorly due to a lack of long-term offshore wind observation data, insufficient funds, inadequate modern technology, and a shortfall of analytical expertise for the evaluation of the offshore wind energy potential; hence, offshore wind resources have not been harnessed. A few studies previously conducted at the inland coastline of Lagos, Nigeria, have shown that offshore wind projects could be viable at this location (Bennington et al.2014; Esteban et al. 2011). It is thus important to conduct a detailed assessment of the vast offshore wind energy potential on the coast of West Africa.

Therefore, this study aims to assess the spatial and temporal patterns of wind power resources of offshore sites in Nigeria, located within Guinea Gulf of Guinea (GoG). The specific objectives are (a) to utilize buoy-station observations of wind speed to validate the use of high-resolution RegCM4 Regional Climate Model CORDEX-CORE simulations driven by ERA-Interim (ERA) for the assessment of wind energy potentials of some selected synoptic offshore locations in the GoG, and (b) to adopt a series of standardized criteria as fully described in Section 9.2.2 to determine the suitable locations for offshore wind energy exploitations in the study area. This present study will provide a useful platform from which more targeted policies and project plans on offshore wind energy generation and utilization can be well developed in West Africa.

9.2 METHODOLOGY

9.2.1 STUDY AREA

The study area considers five offshore synoptic stations in the GoG (Figure 9.1). The offshore locations include Forcados (5.17° N, 5.18° E; on the coast), Bonny (4.40° N, 7.14° E; on the coast), Sea Eagle (4.80° N, 5.31° E; offshore), Bonga (4.56° N, 4.62° E; offshore), and Agbami (3.46° N, 5.56° E; offshore). The GoG could be described as a maritime area located within the West and Central African coastlines and the surrounding territorial waters of the Atlantic Ocean to the east of the Greenwich meridian line. The climate of the GoG, like any other region in West Africa, is predominantly influenced by the West African Monsoon (WAM), which is characterized by fluctuations of tropical maritime (mT) and continental (cT) air masses. The region is highly humid with an annual rainfall of more than 2,000 mm year^{-1} (Gallagher et al. 2016; Giorgi et al. 2012). Rainfall commences in the inland coastline of the GoG around February/March and spreads northwards from the continental coastline before it eventually reaches the extreme northern part of the inland region in May/June (Giorgi et al. 2012). The areas within this region have low-temperature variability, typically from 21°C to 30°C with a two-peaked rainy season in June and September (Ayala et al. 2020). The selected stations are scattered over the shallow and deep sea along the coastline of Nigeria, West Africa.

9.2.2 METHODS

Daily buoy-station measurements of wind speed (at 10 m) for five selected offshore synoptic stations during the 1979–2015 special observation period (SOP) were obtained and analysed. A detailed description of the measurements at the synoptic stations can be found in (Grist and Nicholson 2001). We also analysed daily sea surface wind speed of high-resolution RegCM4 Regional Climate Model CORDEX-CORE simulations driven by ERA-Interim (short-named ERA in this paper) over the continent of Africa, with a spatial resolution of 0.22° × 0.22° (*approx.* 25 km × 25 km)

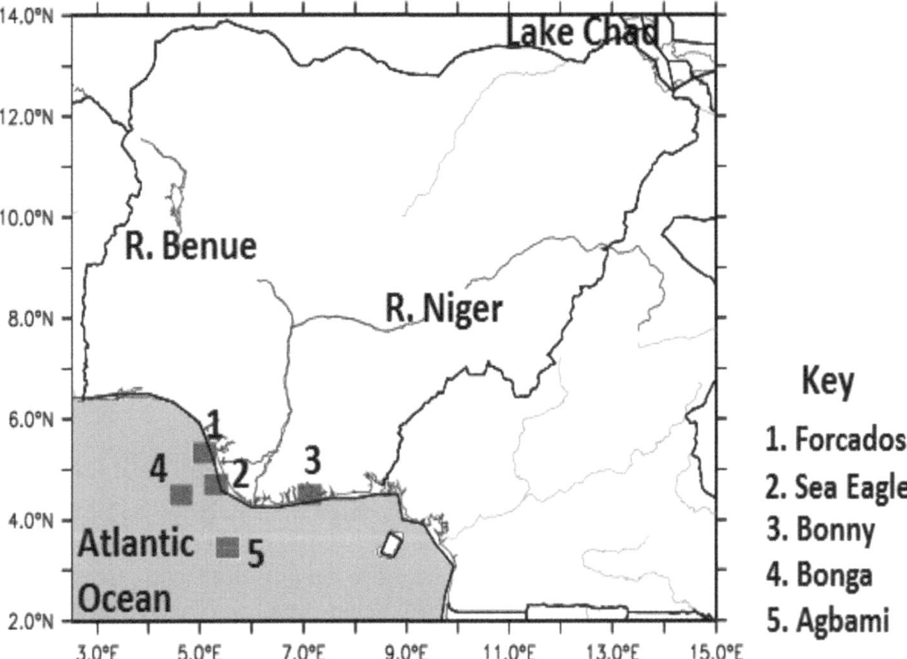

FIGURE 9.1 Geographical locations of the offshore sea-surface wind observation stations in the Gulf of Guinea used in this study.

for the SOP. In the CORDEX-CORE simulations, RegCM version 4.7 from ICTP (Abdus Salam International Centre for Theoretical Physics, Italy) was used, to downscale the ERA re-analysis (Hostetler et al. 1993; Jurasz et al. 2021). The model has 25 km grid-spacing and 41 vertical sigma levels (Kamranzad and Lin 2020). The shallow convection scheme and non-hydrostatic core of MM5 General Circulation Model (GCM) without a large-scale option were chosen as the initial boundary conditions (Kamranzad and Lin 2020; Karnauska et al. 2018; Kronenberg et al. 2013). The Lake Coupling scheme was fully documented in Hostetler et al. (1993), with modifications after Bennington et al. (2014) were employed. The sea surface wind speed from ERA was thereafter validated with the synoptic observations. Performance evaluation of the ERA model in replicating the observed wind speed at the selected sites was carried out using appropriate statistical indices. These statistical parameters include mean bias (MB), percentage mean bias (PMB), root mean square error (RMSE), correlation coefficients (r), and Nash-Sutcliff Efficiency (NSE) as expressed in Equations (9.1)–(9.5), respectively (Matthew 2022; Moriasi et al. 2007; Nie and Li 2018; Ohunakin et al. 2011).

$$\text{MB} = \frac{1}{N}\sum_{t=1}^{N}\left(SIM_t - OBS_t\right) \tag{9.1}$$

$$\text{PMB} = \frac{\frac{1}{N}\sum_{t=1}^{N}\left(SIM_t - OBS_t\right)}{\frac{1}{N}\sum_{t=1}^{N}\left(OBS_t\right)}\times 100\% \tag{9.2}$$

$$\text{RMSE} = \sqrt{\frac{\sum_{i=1}^{n}\left(\left(SIM_t - OBS_t\right)\right)^2}{N}} \tag{9.3}$$

$$r = \frac{\sum_{i=1}^{N}\left(SIM_t - \overline{SIM_t}\right)\left(OBS_t - \overline{H}_{obs_i}\right)}{\sqrt{\sum_{i=1}^{N}\left(SIM_t - \overline{SIM_t}\right)^2 \sum_{i=1}^{N}\left(OBS_t - \overline{OBS_t}\right)^2}} \tag{9.4}$$

$$\text{NSE} = 1 - \frac{\sum_{i=1}^{N}\left(OBS_t - SIM_t\right)^2}{\sum_{i=1}^{N}\left(OBS_t - \overline{OBS_t}\right)^2} \tag{9.5}$$

The time scale is represented by t, synoptic observation by OBS, ERA model simulation by SIM while N stands for the number of grid cells. The significance or otherwise of the RMSE, MB, and PMB were evaluated based on t-test statistics at 95% ($p \leq 0.05$) significant level.

We extrapolated the sea surface wind speeds (at 10 m height) of all grid points to a turbine hub height of 100 m by adopting the Power law method based on the Weibull probability density function (Ohunakin et al. 2023; Ohunakin 2013; Ohunakin and Akinnawonu, 2012).

From the Weibull probability density function, a time series of wind power density (WPD), P_D, (in W m^{-2}) was computed as:

$$P_D = \frac{1}{2}\rho c^3 \Gamma\left(1 + \frac{3}{k}\right) \tag{9.6}$$

where ρ is the density of the atmospheric air.

We also carried out a correction of wind speed for air density and humidity, as done by Karnauskas et al. (2018). A series of standardized criteria (such as WPD, coefficient of variation, monthly variability index (MVI), accessibility, extreme wind speed, distance of the wind farm to the coast, and water depth were adopted, as done by Wen et al. (2021), to identify suitable locations for offshore wind energy exploitations at shallow and deep sea sites, in the GoG. The coefficient of variation (CV) is a negative indicator defined as the ratio of the standard deviation to the average WPD. It describes wind resource volatility on a monthly time scale. A low value of CV is an indication of the high stability of wind energy. On the other hand, the MVI is the ratio of the monthly range (difference between the highest and the lowest) of WPD to its annual average value. In the literature, it is commonly used to study the intra-annual variation of wind energy resources. Another important consideration for offshore wind energy potential is device viability or survivability under extreme weather or climate conditions (Wen et al. (2021). It has been established that increasing frequency and intensity of extreme climate events could have significant impacts on offshore wind farms (Peng et al. 2019). Hence, good accessibility to offshore wind farm facilities (the percentage of time the sea or ocean parameters such as wind speed, wave height, and peak period equal to or less than certain limits) is very crucial for the operation and maintenance of the farm (Gallagher et al. 2016). Therefore, in this present study, the thresholds of 16 m s^{-1} (for wind speed), 2 m (wave height), and 13 s (peak period) were adopted after Wen et al. (2021) and Gallagher et al. (2016).

It has also been established in previous studies that shallow water depth could reduce the transportation costs of offshore power generation to the land, and as such, offshore locations less than

50 km to the coast as recommended in the literature (e.g.,Wen et al. 2021; Sant'Anna de Sousa Gomes et al. 2019; Sawadogo et al. 2020), were considered. Furthermore, the Mann–Kendal test was employed, in this study, to evaluate the statistical significance (or otherwise) of trends and inter-annual variability of offshore wind energy.

9.3 RESULTS AND DISCUSSION

9.3.1 COMPARISONS BETWEEN BUOY-STATION OBSERVATIONS AND ERA SIMULATIONS OF WIND SPEED

Figure 9.2 depicts the observed (OBS) and simulated (ERA) monthly means of sea surface wind speed at the five selected offshore synoptic stations along the coastline of Nigeria, located within the GoG. The annual cycle of ERA simulations of the wind speed shows a similar pattern to the OBS at the five offshore locations. Although ERA simulations were found to be slightly lower than the OBS in most of the months except at Bonga, the results indicated that the model adequately replicates the annual cycle of surface wind speed (Figure 9.2). For both OBS and ERA, the annual cycles of the wind speed in the GoG show bimodal peaks; the first/secondary peak was recorded in February and the second/primary peak in August. The secondary peak ranged between 3 ± 0.5 and 11 ± 0.6 m s^{-1}, while the primary ranged between 4.07 ± 1.0 m s^{-1} and 14.5 ± 1.0 m s^{-1} across the stations (Figure 9.2). In addition, the wind speeds were generally higher in the wet season (June to September) than in other periods of the year. The time of maximum wind speed coincided with the peak of the rainy season and WAM (August–September) in the study region. This could be attributed to the fact that mesoscale convective complexes have been shown to be more dominant and wave amplitudes found to be the largest at the peak of the rainy season in West Africa (Gallagher et al. 2016; Soukissian et al. 2021; Teichmann et al. 2020; Tian et al. 2021). Also, the highest and the lowest offshore wind speeds were recorded at the farthest offshore location (i.e., Agbami) and the nearest (i.e., Bonny) to the coast, respectively. This finding further corroborates the assertion in a previous study that sea surface wind speed increases with increasing distance to the coast (Wen et al. 2021).

Furthermore, the performance evaluation of the sea surface wind speed of ERA compared with buoy-station observations is presented in Table 9.1. The findings revealed that the temporal and spatial patterns of downscaled wind speed are highly consistent with the observation data. There were close agreements and fits between the ERA simulations and observations of the annual cycle of wind speeds at the five offshore locations. The model slightly underestimates both the daily and monthly observations of wind speed, except at Bonga station (Table 9.1). For example, the estimated mean biases (MB) in ERA daily wind speed were −0.07 (Forcados), −0.39 (Sea Eagle), −0.09 (Bonny), 0.91 (Bonga), and −0.28 m s^{-1} (Agbami), while the percentage mean biases (PMBs) were −2.12, −7.06, −3.13, 17.45, and −2.76% at Forcados, Sea Eagle, Bonny, Bonga, and Agbami, respectively. The RMSE ranged between 0.89 and 3.71 m s^{-1}. As expected, the MB, PMB, and RMSE obtained with monthly datasets were lower than those obtained daily. Also, the correlation coefficients ($0.52 \leq r \leq 0.96$) obtained for the monthly averages were positive and significant at three out of the five stations (i.e., 60% of the stations) while the values of NSE ranged between ±0.62 and ±0.91, except at Bonga where NSE was found to be −0.39.

In general, the results indicated that the performance of ERA in simulating sea surface wind speed was satisfactory except at Bonga, where the discrepancies in the model simulations were significant at $p = 0.05$. Substantial discrepancies in the model simulations at Bonga could be attributed to potential noise in the model grid that falls within the station. In addition, the probable reason for the poor estimate might be attributed to the fact that a few stations fall within a relatively low wind speed region, and simulation of sub-daily wind speed had been discussed to be significantly different from the observations with the poorest performance, for periods and regions with wind

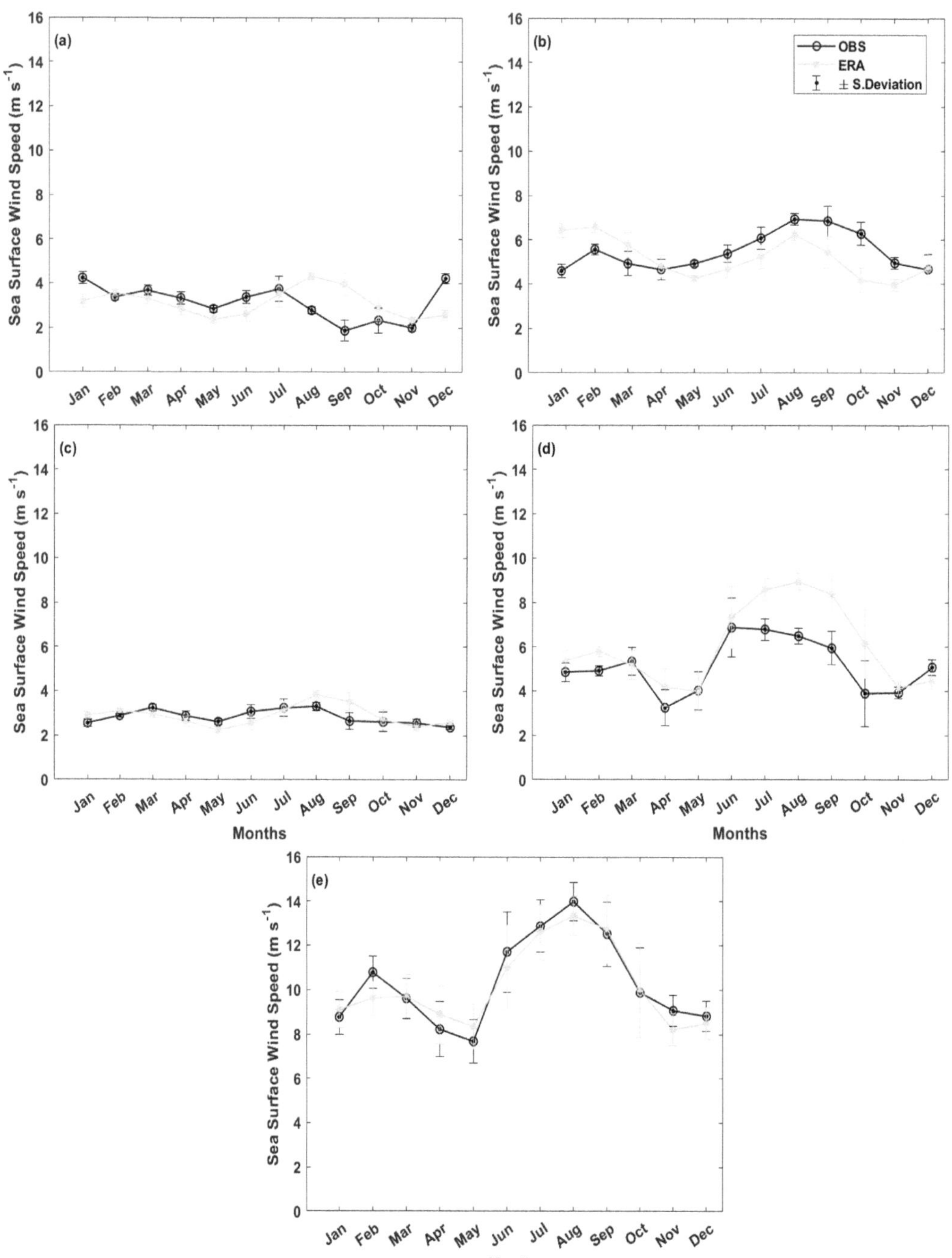

FIGURE 9.2 Ground observations and estimated monthly means of sea surface wind speed at the selected stations in the Gulf of Guinea during the study period for (a) Forcados, (b) Sea Eagle, (c) Bonny, (d) Bonga, and (e) Agbami.

TABLE 9.1

Performance Evaluation of Sea Surface Wind Speed (in ms^{-1}) of ERA-Driven Regional Climate Model with Ground Observations (OBS) at the Selected Stations

	Daily Datasets					Monthly Datasets				
Indices	**Forcados**	**Sea Eagle**	**Bonny**	**Bonga**	**Agbami**	**Forcados**	**Sea Eagle**	**Bonny**	**Bonga**	**Agbami**
Mean of OBS	3.13	5.51	2.93	5.20	10.32	3.15	5.49	2.83	5.12	10.33
Mean of ERA	3.07	5.12	2.83	5.91	10.03	3.13	5.20	2.78	6.01	10.17
RMSE	1.51	1.61	0.89	2.60	3.71	1.01	1.10	0.39	1.37	0.59
MB	−0.07	−0.39	0.09	0.91*	−0.28	−0.02	−0.30	0.05	0.89*	−0.16
PMB	−2.12	−7.06	3.13	17.45*	−2.76	−0.76	−5.38	1.85	17.10*	−1.53
R	−0.39	0.10	0.24	0.04	0.22	−0.08	0.20	0.52*	0.83*	0.96*
NSE	−0.74	0.38	0.05	1.59	0.68	−0.75	0.82	0.62	0.39	0.91

* Significant at 0.05 significant level.

RMSE, root-mean-square-error; MB, mean bias; PMB, percentage mean bias; NSE, Nash-Sutcliff efficiency; r, correlation coefficient.

speeds ≤3.84 m s^{-1} (Ohunakin et al. 2011; Wen et al. 2021). In general, the findings suggested that ERA produces reliable sea surface wind speed suitable for the assessment of wind energy potential (as done in this present study) and climate impact studies (to be carried out in future works) over the study area.

9.3.2 Buoy-Station Observations and ERA Simulations of Wind Characteristics

The estimated Weibull shape (k) and scale (c) factors at 100 m hub-height for both OBS and ERA are shown in Figure 9.3. The results showed that the wind speed characteristics obtained with ERA compared well with those of the OBS. The sea wind speed characteristics for both OBS and ERA were adequately captured by the Weibull shape (k) and scale factors (c).

The values of k for the OBS ranged between 5.5 ± 0.1 (at Forcados) and 9.8 ± 0.15 (at Bonny), while those of the ERA ranged between 4.5 ± 0.4 (at Bonga) and 9.2 ± 0.15 (at Bonny) (Figure 9.3a). The disparities between ERA and OBS followed the patterns previously obtained for the sea surface wind. As previously established in the literature, the values of k range between 1 and 10 over the study area (Ohunakin 2011). The pattern of the spatial distribution of c at the chosen hub height perfectly mimics that of the sea surface wind speed (Figure 9.3b). The value of c for the OBS at Agbami (17.5 ± 2.0 m s^{-1}) was found to be the highest and the lowest (6.6 ± 0.5 m s^{-1}) at Bonny. Similarly, c for ERA ranged between 17.2 ± 2.0 m s^{-1} at Agbami and 7.0 ± 0.5 m s^{-1} at Bonny (Figure 3b). As expected, the results demonstrated that the offshore wind speeds increase with height up to the chosen hub height.

9.3.3 Spatial and Temporal Variations in Observed and Simulated Offshore Wind Power Density

Table 9.2 describes the variations in WPD based on OBS and ERA at the selected offshore stations. The indicators used are minimum WPD, maximum WPD, CV, MVI, annual trend, the distance of the station from the coast, and the episode of extreme wind speed at 100 m height (≥16 m s^{-1}). Regarding their distances from the coast, Bonny is the closest, while Agbami is the farthest (Table 9.2). The results demonstrated that the wind power potential in the GoG varied with

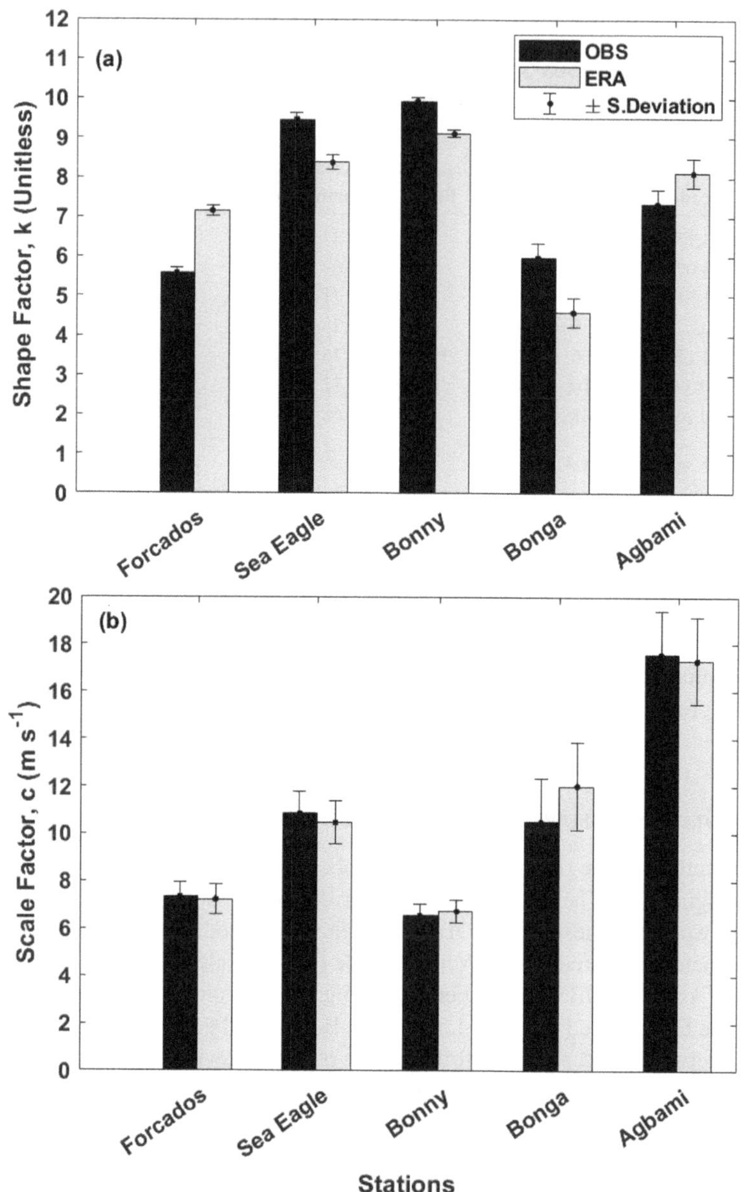

FIGURE 9.3 Observed (OBS) and simulated (ERA) means and standard deviations of the Weibull (a) Shape factor k, and (b) Scale factor, c of wind speeds at 100 m-Hub height at the selected offshore stations in the Gulf of Guinea.

the distance from the coast with Agbami showing the best potential for wind power resources. The maximum monthly mean WPD varied from 150.9 (Bonny) to 3092.6 W m^{-2} (Agbami) for the OBS dataset and 164.0 (Bonny) to 2943.2 W m^{-2} (Agbami) for the ERA simulation. On the other hand, the minimum monthly mean WPD varied from 150.8 to 3003.5 W m^{-2} for OBS and 163.4 to 2873.3 W m^{-2} for ERA (Table 9.2). However, typical of an offshore site, as documented in Wen et al. (2021) and Kamranzad and Lin (2020), the results of the present study showed low ranges (0.13–89.5 W m^{-2} for the OBS and 0.61–69.89 W m^{-2} for ERA) in the monthly WPD over

TABLE 9.2
Values of Indicators of Variations in the Observed (OBS) and Simulated (ERA) Wind Power Density at the Selected Stations

	Stations									
	OBS					ERA				
Indicators of Variations	Forcados	Sea Eagle	Bonny	Bonga	Agbami	Forcados	Sea Eagle	Bonny	Bonga	Agbami
Maximum WPD (Wm^{-2})	213.35	707.21	150.93	638.95	3092.61	203.13	632.66	164.00	971.11	2943.19
Minimum WPD (Wm^{-2})	211.56	695.09	150.80	622.03	3003.46	201.44	620.37	163.39	938.83	2873.30
Range of WPD (Wm^{-2})	1.79	12.12	0.13	16.92	89.15	1.69	12.29	0.61	32.28	69.89
CV	0.001958	0.004216	0.000212	0.006029	0.006291	0.001766	0.004528	0.000759	0.008427	0.005700
MVI	0.005585	0.011739	0.000581	0.017944	0.019985	0.005555	0.013275	0.002521	0.022731	0.016426
Annual trend (Wm^{-2} $year^{-1}$)	0.000984	−0.00426	0.000697	−0.01069	−0.04516	0.00139	−0.00333	0.00030	−0.01589	−0.04332
Distance from the coast (referenced to the coast of Warri in nautical miles)	On coast	49	On coast	72	131	On coast	49	On coast	72	131
Extreme wind speed (%)	–	–	–	–	13	–	–	–	–	12

* Significant at 0.05 significant level.

WPD, wind power density; CV, coefficient of variation; MVI, monthly variability index.

the offshore locations (Table 9.2). The findings demonstrate that both CV and MVI values were very low. For example, the CV at the highest wind-producing location (i.e., Agbami) ranged from 0.005700 to 0.006291, while MVI varied from 0.016426 to 0.019985 for ERA and OBS datasets, respectively (Table 9.2). The annual trends in WPD were found to be generally low and insignificant. In addition, the patterns and signs of the trends were the same but with slightly varying magnitudes for both OBS and ERA. It was negative at Sea Eagle, Bonga, and Agbami but positive at Forcados and Bonny. For example, the annual trends in the observed WPD were −0.00426 (Sea Eagle), −0.01069 (Bonga), −0.04516 (Agbami), 0.000984 (Forcados), and 0.000697 W m^{-2} year^{-1} (Bonny). The episodes of extreme wind speed were very low with 13% (OBS) and 12% (ERA) at Agbami during the SOP, and none in other locations. It has been previously established that the best offshore location should have the least variation indicators with the best accessibility (ANRC-AfDB 2021). Low variability and intermittency obtained (in terms of trends, range, CV, MVI, and extreme wind speed episodes) are all indications of good wind power availability at the sites. In addition, relatively large mean monthly and annual WPD obtained over a few offshore stations are positive indications of high potential for wind energy resources. The experience of Nigeria with existing offshore oil and gas exploitations and industries is a very strong factor that makes the buoy-stations accessible for efficient offshore wind power development, as suggested in ANRC-AfDBG (Ayala et al. 2020). In summary, the results of the present study suggest that a few locations in the GoG are highly viable for offshore wind energy generation. We found that Agbami has the highest potential for offshore wind energy among the selected offshore locations. Hence, future investment in offshore wind energy projects in the GoG is expected to significantly improve the energy supply to meet the ever-increasing energy demand occasioned by the population explosion and industrial revolution in West Africa.

Figure 9.4 illustrates the daily mean and annual total WPD at 100 m hub height at the selected offshore stations in the GoG. The highest daily mean WPD (2.9 ± 0.5 and 2.8 ± 0.5 kW m^{-2} for OBS and ERA, respectively) and annual total WPD (1.05 ± 0.17 and 1.02 ± 0.17 MW m^{-2} for OBS and ERA, respectively) were obtained at Agbami, while Bonny recorded the least daily mean WPD (~0.2 ± 0.05 kW m^{-2} for both OBS and ERA) and annual total WPD (~0.05 ± 0.002 MW m^{-2} for both OBS and ERA). These results confirmed the potential for offshore wind energy resources among the study stations. We found the farthest station, Agbami, to have the highest potential for wind power generation in the GoG.

9.4 CONCLUSIONS

This present study has utilized buoy-station measurements of wind speed to validate the use of high-resolution RegCM4 regional climate model simulations forced by ERA for the assessment of wind energy potentials of some selected synoptic offshore stations from 1979 to 2015 in the GoG. It further adopted a series of standardized criteria, such as WPD, annual trends, coefficient of variation (CV), MVI, accessibility, extreme wind speed, and distance to the coast, to identify the most suitable sites for offshore wind energy over the study area. The study demonstrates that ERA wind speed fits the field measurements at most of the selected stations fairly well. The findings reveal that ERA produces reliable sea surface wind speed suitable for assessment of wind energy potential and climate impact studies in the GoG. The wind power potential varied with the distance from the coast with Agbami buoy-station showing the best potential for wind power generation. There was evidence of the high potential of wind energy resources and very low and insignificant variability/ intermittency (in terms of trends, range, CV, MVI, and frequency of extreme wind speed episodes) at the selected locations, particularly at Agbami. It was also established that the offshore locations were very viable for future offshore wind power generation. The use of very few offshore buoy-stations that appear to cluster around a narrow coastline, due to paucity of observational wind data is perceived to be a probable limitation of this present study. The findings have implications for

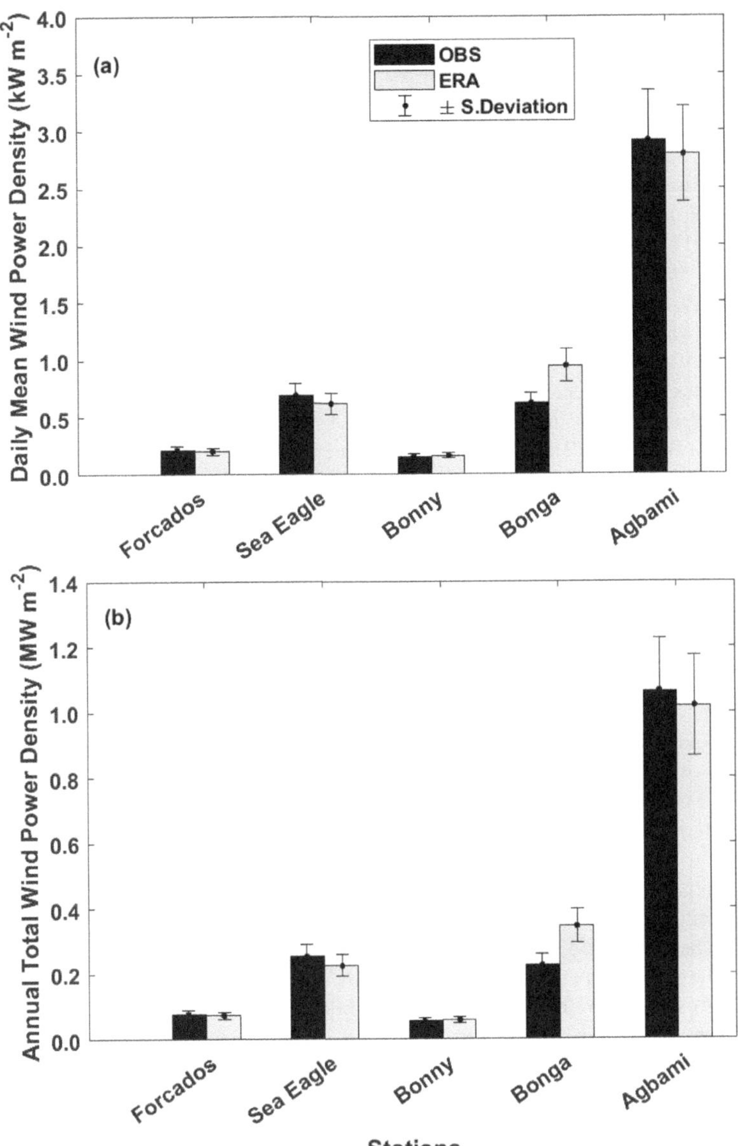

FIGURE 9.4 Observed, OBS and simulated, ERA, for (a) daily mean wind power density in kW m⁻², and (b) annual total wind power in MW m⁻² at 100 m hub height at the selected offshore stations in the Gulf of Guinea.

offshore wind energy resources in the GoG as a dependable and efficient alternative energy source under a future warming climate. It is recommended that further investigation should be carried out on the cost of investing in offshore wind energy projects in the GoG.

ACKNOWLEDGEMENTS

The authors wish to acknowledge the support offered by Covenant University, Ota, Lagos State, Nigeria, in the actualization of this research work.

REFERENCES

Abiodun, B. J., Salami, A. T., Matthew, O. J., Odedokun, S., 2012. Potential impact of afforestation on climate change and extreme events in Nigeria. *Climate Dynamics* 41, 277–293. https://dx.doi.org/10.1007/s00 382-012-1523-9

Aboobacker, V. M., Shanas, P. R., Veerasingam, S., Al-Ansari, E. M. A. S., Sadooni, F. N., Vethamony P., 2021. Long-term assessment of onshore and offshore wind energy potentials of Qatar. *Energies* 14, 1178. https://dx.doi.org/10.3390/en14041178

Adeniyi, M. O., 2017. Modeling the impact of changes in Atlantic sea surface temperature on the climate of West Africa. *Meteorology and Atmospheric Physics* 129, 187–210.

Adeniyi M. O., 2019. Sensitivity of two dynamical cores in RegCM4.7 to the 2012 intense rainfall events over West Africa with focus on Lau, Nigeria. *International Journal of Modelling and Simulation* 40 (1), 1–11. DOI: 10.1080/02286203.2019.1641777

African Natural Resources Center-African Development Bank Group, ANRC-AfDBG, 2021. *Assessing the potential of Offshore Renewable Energy in Africa: A Backround Paper*. Published by African Development Bank (AfDB), Abidjan, Côte d'Ivoire, pp. 92.

Ayala, A. I., Moras, S., Pierson, D. C., 2020. Simulations of future changes in thermal structure of Lake Erken (Sweden): Proof of concept for ISIMIP2b lake sector local simulation strategy. *Hydrology and Earth System Sciences* 24, 3311–3330.

Bennington, V., Notaro, M., Holman K. D., 2014. Improving climate sensitivity of deep lakes within a regional climate model and its impact on simulated climate. *Journal of Climate* 27, 2886–2911.

Chen, C. Q, Zheng, L., Zhou, J. L., Zhao, H., 2017. Persistence and risk of antibiotic residues and antibiotic resistance genes in major mariculture sites in Southeast China. *Science of the Total Environment* 580, 1175–1184. https://doi.org/10.1016/j.scitotenv.2016.12.075

Chou, J. S., Ou, Y. C., Lin, K. Y., 2019. Collapse mechanism and risk management of wind turbine tower in strong wind. *Journal of Wind Engineering & Industrial Aerodynamics* 193, 103962. https://doi.org/10.1016/ j.jweia.2019.103962

Elsner, P., 2019. Continental-scale assessment of the African offshore wind energy potential: Spatial analysis of an under-appreciated renewable energy resource. *Renewable and Sustainable Energy Reviews* 104, 394–407.

Emeksiz, C., Demirci, B., 2019. The determination of offshore wind energy potential of Turkey by using novelty hybrid site selection method. *Sustain Energy Technol Assessments* 33(36), 100562. https://doi.org/ 10.1016/j.seta.2019.100562

Esteban, M. D., Diez, J. J., López, J. S., Negro, V., 2011. Why offshore wind energy? *Renewable Energy* 36, 444–450. https://doi.org/10.1016/j.renene.2010.07.009

Gallagher, S., Tiron, R., Whelan, E., Gleeson, E., Dias, F., McGrath, R., 2016. The nearshore wind and wave energy potential of Ireland: A high resolution assessment of availability and accessibility. *Renew Energy* 88, 494–516.

Giorgi, F., Coppola, E., Solmon, F., Mariotti, L., Sylla, M., Bi, X., Elguindi, N., Diro, G. T., Nair, V., Giuliani, G., Turuncoglu, U. U., Cozzini, S., Güttler, I., O'Brien, T. A., Tawfik, A. B., Shalaby, A., Zakey, A. S., Steiner, A. L., Stordal, F., Sloan, L. C., Brankovic, C., 2012. RegCM4: model description and preliminary tests over multiple CORDEX domains. *Climate Research* 52, 7–29.

Grist, J. P., Nicholson, S. E., 2001. A study of the dynamic factors influencing the rainfall variability in the West African Sahel. *Journal of Climate* 14, 1337–1359.

Hostetler, S. W., Bates, G. T., Giorgi, F., 1993. Interactive coupling of a lake thermal model with a regional climate model. *Journal Geophysical Research* 98, 5045–5057.

Jurasz, J., Mikulik, J., Dabek, P. B., Guezgouz, M., Kazmierczak, B., 2021. Complementarity and 'resource droughts' of solar and wind energy in Poland: An ERA5-based analysis. *Energies* 14, 1118. https://doi. org/10.3390/en14041118

Kamranzad, B., Lin, P., 2020. Sustainability of wave energy resources in the South China Sea based on five decades of changing climate. *Energy* 210 (1) 118604. https://doi.org/10.1016/ j.energy.2020.118604

Karnauskas, K. B., Lundquist, J. K., Zhang, L., 2018. Southward shift of the global wind energy resource under high carbon dioxide emissions. *Nature Geoscience* 11(1), 38.

Kronenberg, R., Barfus, K., Franke, J., Bernhofer, C., 2013. On the downscaling of meteorological fields using recurrent networks for modelling the water balance in a meso-scale catchment area of Saxony, Germany. *Atmospheric and Climate Sciences* 3, 552–561.

Matthew, O. J., 2022. Estimation of diurnal patterns of global solar radiation, temperature, relative humidity and wind speed from daily datasets at a humid tropical location. *Agricultural and Forest Meteorology* 322(2022), 109003. https://doi.org/10.1016/j.agrformet.2022.109003

Moriasi, D. N., Arnold, J. G., Van Liew, M. W., Bingner, R. L., Harmel, R. D., Veith, T. L., 2007. Model evaluation guidelines for systematic quantification of accuracy in watershed simulations. *Transactions of the American Society of Agricultural and Biological Engineers* 50, 885–900.

Nie, B., Li, J., 2018. Technical potential assessment of offshore wind energy over shallow continent shelf along China coast. *Renewable Energy* 128, 391–399. https://doi.org/10.1016/j.renene.2018.05.081

Ohunakin O. S., 2011. Wind resource evaluation in six selected high altitude locations in Nigeria. *Renewable Energy*, 36 (12), 3273–3281.

Ohunakin, O. S., Adaramola, M. S., Oyewola O. M., 2011. Wind energy evaluation for electricity generation using WECS in seven selected locations in Nigeria. *Applied Energy* 88, 3197–3206.

Ohunakin O. S., Akinnawonu O. O., 2012. Assessment of wind energy potential and the economics of wind power generation in Jos, Plateau State, Nigeria. *Energy for Sustainable Development*, 16 (1), 78–83.

Ohunakin, O. S., Matthew, O. J., Adaramola, M. S., Atiba, O. E., Adelekan, D. S., Aluko, O. O., Henry, E. U., Ezekiel, V. U. 2023. Techno-economic assessment of offshore wind energy potential at selected sites in the Gulf of Guinea. *Energy Conversion and Management* 288(2023), 117110. https://doi.org/10.1016/j.enconman.2023.117110

Peng, X., She, J., Zhang, S., Tan, J., Li, Y., 2019. Evaluation of multi-reanalysis solar radiation products using global surface observations. *Atmos* 10, 42. https://doi.org/10.3390/atmos10020042

Sant'Anna de Sousa Gomes M. S., de Paiva J. M. F., Moris V. A. d.-S., Nunes A. O., 2019. Proposal of a methodology to use offshore wind energy on the southeast coast of Brazil. *Energy* 185, 327–336. https://doi.org/10.1016/j.energy.2019.07.057

Sawadogo W., Reboita M. S., Faye A., da Rocha R. P.,Odoulami R. C., Olusegun C. F., Adeniyi M. O., Abiodun B. J., Sylla M. B., Diallo I., Coppola E., Giorgi F., 2020. Current and future potential of solar and wind energy over Africa using the RegCM4 CORDEX-CORE ensemble. *Climate Dynamics* 26. https://doi.org/10.1007/s00382-020-05377-1

Soukissian T. H., Karathanasi F. E., Zaragkas D. K., 2021. Exploiting offshore wind and solar resources in the Mediterranean using ERA5 reanalysis data. *Energy Conversion & Management* 237 (2021), 114092. https://doi.org/10.1016/j.enconman.2021.114092

Teichmann, C., Jacob, D., Remedio, A. R., Coppola, E., et al., 2020. *Assessing Mean Climate Change Signals in the Global CORDEX-CORE Ensemble*, A paper delivered at EGU General Assembly 2020 on 4th of May 2020.

Tian Y.-C., Kou L., Han Y.-D., Yang X., Hou T.-T., Zhang W.-K., 2021. Evaluation of offshore wind power in the China Sea. *Energy Exploration & Exploitation* 0(0), 1–14. https://doi.org/10.1177/0144598721992268

Wen, Y., Kamranzad, B., Lin, P., 2021. Long-term assessment of offshore wind energy in the south and southeast coasts of China based on a 55-year dataset. *Energy*. https://doi.org/10.1016/j.energy.2021.120225

10 Neural Network-Based Wind Turbine Power Curve Models Using Several Wind Farms' Influencing Parameters and Topography

Olayinka S. Ohunakin, Emerald U. Henry,
Victor U. Ezekiel, Olaniran J. Matthew,
Damola S. Adelekan, and Mutali Nepfumbada

10.1 INTRODUCTION

The demand for energy rises astronomically as countries develop [1,2]. Industrialized nations make up only a fourth part of the world's population and utilize four-fifths of the world's energy. In most developed countries, fossil fuels are the primary source of energy; nevertheless, the usage of renewable energy has increased, and there are ongoing plans to replace fossil fuels with renewable energy. Due to environmental consequences associated with fossil fuel utilization (greenhouse gas emissions) and rising energy demands with the ever-increasing population, the need to switch to alternative energy sources is inevitable [3–5]. Actions must also be taken to improve the energy efficiency of production and consumption patterns, as well as to encourage the adoption of low-carbon technologies. The deployment of renewable energy technologies has been found to be a significant alternative for reducing Green House Gases (GHG) emissions, which contribute to global warming [6]. They currently meet 14% of the total global energy demand [7].

Wind energy is one of the fastest-growing renewable sources of energy. In the last decade, the number of wind farms increased significantly [8]. The global installed wind energy capacity has improved from 1.29GW in 1995 [9] to 837GW at the end of 2022 [10]. This progressive increase in deployment of the technology, calls for the need for an efficient method of wind energy assessment.

The theoretical power output (P) of a wind turbine (WT) is given by the mathematical expression in Equation (10.1):

$$P = \frac{1}{2} C_P (\lambda, \beta) \rho A v^3 \tag{10.1}$$

where $C_P (\lambda, \beta)$ is the power coefficient, λ and β are the tip-speed ratio and blade pitch angle, respectively, ρ is the air density, A is the rotor swept area, and v is the wind speed [11]. The main factor influencing the power generated is the wind speed, which exhibits a cubic relationship with the power and varies significantly with height. Other factors may include wind direction, blade pitch angle, rotor dimensions, air density, etc. The wind turbine power curve (WTPC) (Figure 10.1) is a

DOI: 10.1201/9781032651958-10

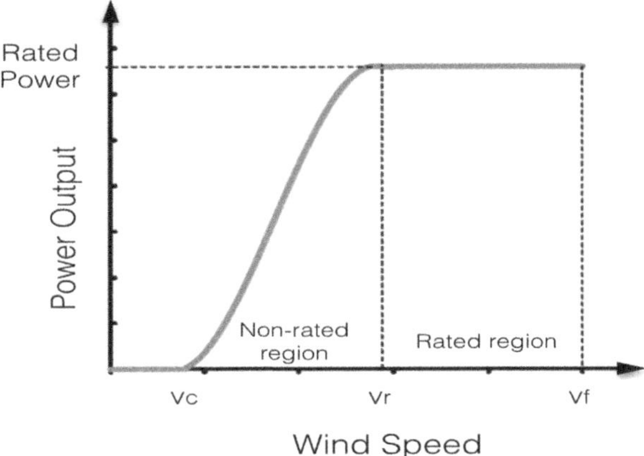

FIGURE 10.1 A typical wind turbine power curve as defined by the manufacturer. It is represented by a single line and usually not an appropriate representation of the wind-to-power relationship experienced in wind farms.

graph that shows the relationship between a given wind speed and the electrical power generated for a specific wind turbine; the corresponding electrical power delineates the performance of the wind turbine.

The theoretical power curve of a wind turbine, as provided by the manufacturer, is calculated under the ideal, ignoring actual field conditions such as climate variability, the effect of neighbouring wind turbines, and turbine inertia [11]. The actual power curve differs from a single-line power curve due to the variables associated with power production on the field. Some of these variables are turbulence intensity (TI), wake effect, ambient temperature, atmospheric pressure, and wind direction. WTPC modelling can be utilized for the choice of wind turbines, wind energy assessment and prediction, monitoring and troubleshooting, and predictive control and optimization. Details about these can be further studied in the following references [8,12–18].

10.1.1 Approaches to WTPC Modelling

There are generally four approaches to WTPC modelling. These include discrete approach, stochastic approach, non-parametric or physics-based approach, and parametric or data-driven approach.

1. Discrete methods consist of modelling a continuous process by making discrete approximations [19]. One method is dividing wind speed into intervals and obtaining a generic power value for each interval. Within this category falls the IEC-61400-12-1 standard. It discretizes the wind speed range into bins of 0.5 m/s in size, all the υ and P for each data point within a bin is averaged to obtain the (υ, P) pair for each bin [20]. In this evaluation technique, air density is implicitly considered as input with wind speed and power being the output. This method takes into account the nonlinear relationship between υ and P, but a large number of measured data is required for improved accuracy [21]. In Llombart et al. [22], modifications were made to the IEC 61400-12 bins method with the aim of improving the accuracy; when compared to the initial bins method, their work achieved improved accuracy.

2. The Markov chain theory is the prevailing stochastic model for the analysis of a wind turbine power output. In Gottschall and Peinke [23], the dynamic behaviour of the wind turbine was analysed with respect to a stochastic signal comprising wind speed and turbulence intensity, thereby yielding power curves independent of turbulence intensity. Another advantage is that the power curve can be obtained within a few days; its disadvantage is that no parameters other than wind speed and turbulence intensity are considered [19].

3. Parametric models or physics-based models are developed from a set of mathematical equations that includes a set of parameters that must dynamically adapt through a set of continuous data. They are generally developed from linear, nonlinear, polynomial, and differential equations [19]. In Shokrzadeh [24], a penalized spline regression was used in modelling the WTPC, and it resulted in a better performance than the other tested models. In Marčiukaitis [25], nonlinear regression functions are used to model the WTPC; the model's input is the wind speed, while the wind direction is used to vary the curve. Kusiak and Verma [26], in their work scale this method for implementation in the entire wind farm by the utilization of wind direction-specific operational curves. In Pinson et al. [27], a simple and adjustable double exponential function was used in modelling WTPC. A three-parameter [28], four-parameter [29], five-parameter [30], and six-parameter [31] logistic functions have been used in modelling WTPC. The accuracy of a logistic function was found to improve with increasing parameters. A nine-parameter hyperbolic tangent function has also been employed by Taslimi-Renani et al. [32] in a bid to improve accuracy. Though parametric models may contain many free parameters that can fit the data, they have a few limitations: they have a fixed form and are rigid, they are complex to use, and they form less accurate wind power curve models compared to their non-parametric counterparts [11].

4. Due to the availability of supervisory control and data acquisition (SCADA) data that enable access to a tremendous amount of data, models that consider the nonlinear relationship between features regarded as input parameters and the output power can thus be developed. Hence, rather than a mathematical approach, a data-driven approach is required. There are currently a few data approaches available to researchers. From available research, the cubic spline method [33] and computational intelligent (CI) approaches such as the Gaussian process [34], Bayesian-based methods [35], support vector regression methods [9], hybrid relevant vector methods [36], monotonic regression [37], heuristic and metaheuristic methods [30,38], and artificial neural network (ANN) methods [11,19,21,34]. Generally, non-parametric models are more flexible, precise, and are computationally expensive. Quite a large number of hybrid models have been implemented due to the need for outlier filtration before utilizing the dataset for model creation because non-parametric models are only as good as the available data [34].

Least square and cubic spline interpolation methods were utilized in modelling the actual power of a wind turbine utilizing wind speed as input in Thapar et al. [33]. This method achieves comparably moderate accuracy and neglects other input parameters that influence the actual power curve. In Manobel et al. [34], a hybrid model based on Gaussian filtering and ANN was employed in modelling the actual power curve. In general, ANN models yield comparably higher accuracies than other CI models because they try to reduce prediction losses through backpropagation; however, only wind speed is used as an input parameter, and the utilization of hybrid models invariably increase computational cost. The Bayesian method and multi-kernel regression were employed by Wang et al. [35]. The aim of the work was to achieve both a deterministic and probabilistic power output. A merger of both the deterministic and probabilistic power curve is the ideal power curve for power prediction; however, only the wind speed was utilized in the model creation. In Pandit and Infield [9], a dataset filtered by binning was fed into a support vector regression model with moderate accuracy. Monotonic regression was utilized for WTPC modelling in Mehrjoo et al. [37]. A temperature-considerate power curve was obtained in Rodríguez-López et al. [6], utilizing fuzzy logic and ANN. The input parameters are wind speed and ambient temperature, thereby yielding a more realistic power curve; other influencing parameters were not considered. In Xu et al. [21], a quantile-based power curve is obtained by using ANN. Power curves for nine user-defined quantiles are obtained and plotted, with the collective defining the possible range of power output. This method is relevant for wind farm monitoring but not suitable for energy assessment and prediction. It also only takes wind speed as input. A Tabu search non-symmetric fuzzy mean (TS-NSFM) approach combined with a radial basis function (RBF) neural network was used in modelling the WTPC in the work of Karamichailidou et al. [11]. Four input parameters were utilized (wind speed, wind direction, blade pitch angle, and ambient temperature), and cluster centres and weights were obtained from the TS-NSFM algorithm and then fed into the RBF network. Modelling accuracy is greatly improved; however, the triple hybrid model is extremely computationally expensive.

We found all models utilized in literature to generally develop power curves without considerations to wind farm topography and field conditions. However, several factors have been found to influence the power output of wind turbines, including climate variability, effect of neighbouring wind turbines, turbulence intensity, wake effect, ambient temperature, atmospheric pressure, wind direction, and terrain. Therefore, this work aims to develop a power curve model that accurately represents the actual field conditions of wind turbines on wind farms by considering all the mentioned influencing factors. This work considers climate variability by analysing a time series plot of wind speed over four years to capture all climate conditions. The effect of neighbouring wind turbines, turbulence intensity and wake effect, is achieved by analysing the relative position of turbines on the wind farms to capture the various flow characteristics, while an assessment of the terrain map was carried out to account for disparate land formation. Further, other than using wind speed as the only variable for the WTPC model (as presented in past works), we utilized several other variables alongside wind speed, including ambient temperature, wind direction, and atmospheric pressure to develop the model in this work. We further provided an assessment of the accuracy and sensitivity of the two non-parametric approaches, including RBF and multi-layer perceptron (MLP), that we utilized for the WTPC modelling. The developed model is wind farm specific, generic to all turbines within the wind farm regardless of local orography, and invariant with geographical seasons. Lastly, the accuracy of our models was compared with various parametric and non-parametric models in literature [1,12,35] to validate their performances.

10.2 MATERIAL AND METHODS

10.2.1 DATA DESCRIPTION

Two wind farms were considered for this work including Kelmarsh wind farm located near Haselbach, Northamptonshire in the United Kingdom at latitude 52°24′5.8″, longitude −0°56′34.6″ (comprising six 2.05 MV Senvion MM92 turbines [39] [see supplementary material [A], for the dataset]), and Penmanshiel wind farm located near Grantshouse in the Scottish Borders, the United Kingdom at latitude 55°52′16.2″, longitude −2°21′16.2″ (comprising 14 Senvion MM82' turbines [40] [see Supplementary material [B], for the dataset]). A 10-minute SCADA data, ranging from mid-2016 to mid-2021, was obtained for six turbines from the two wind farms (three turbines from each wind farm) for modelling. The turbines were selected based on their relative position on the wind farms: top, midway, and bottom, and on the terrain in which they fall. Six different turbine SCADA data were utilized for modelling.

A total of four parameters were required for modelling. These include density normalized wind speed, wind direction, blade pitch angle (three input parameters), and output power (one output parameter). The effect of turbulence intensity (TI) and wake effect was a primary consideration in our turbine selection process. Turbines situated midway and at the bottom of a wind farm will usually experience incoming turbulence due to wind exiting from preceding wind turbines. However, this phenomenon depends on the topological spacing of wind turbines on the wind farm. The terrain of a wind farm also influences the nature of on-coming experienced by each turbine. Turbines situated in hilly sites will experience more laminar wind at higher speeds, while those situated at valley sites will experience less laminar winds at lower speeds. Due to this asymmetry, care must be taken in the selection process to include different terrains.

10.2.2 Artificial Neural Networks

ANN are one of the most essential computational intelligence tools, and they have been utilized for a number of tasks like function approximation and pattern recognition [11]. ANNs are inspired by biological neural systems. They are made up of a collection of simple processing pieces known as neurons or nodes that are connected together by links, each having a value regarded as their weight. Each neuron can receive information from the neurons that are input to them and any other external input; its output information is received as input information into another neuron. Using information inputted into each neuron, they create an output, which is a linear combination of the inputs. Consider Equation (10.2), y_i is the output of the neuron i, x_j is the input value to the output neuron, indexed as j, w_{ij} is the weight of the link between the ith and jth neurons, θ_i is the bias of the neurons, and f_i is the activation function. This can be linear or nonlinear. A nonlinear activation function (sigmoid, Gaussian, etc.) is preferable as it allows the network to act more universally [6].

$$y_i = f_i \sum\nolimits_{j=0}^{n} w_{ij} \cdot x_j - \theta_i \tag{10.2}$$

Neural networks are divided into feed-forward neural networks (unidirectional) or backpropagation (recursive) neural networks. For the feed-forward type, neurons are connected in a single direction from input to output. In a backpropagation neural network, the link between the neurons is in both directions. The arrangement of the neurons determines the structure of the neural network. Neurons are arranged in a sequence regarded as layers. A typical neural network will comprise an input layer, 1 to ∞ number of hidden layers, and an output layer. The most peculiar property of a neural network is its ability to learn from the data. The speed at which this learning occurs (steps taken to arrive at a local minimum) is influenced by the learning rate. Learning means adjusting the weights of the connection between neurons in order to improve accuracy [6].

10.2.3 Radial Basis Function Neural Networks

The RBF neural network architecture was first proposed by Broom head, Lowe in their work entitled 'Multivariate Functional Interpolation and Adaptive Networks' 1988 [41]. RBF neural networks consist of three layers by design: the input layer, the hidden layer, and the output layer [42]. Figure 10.2

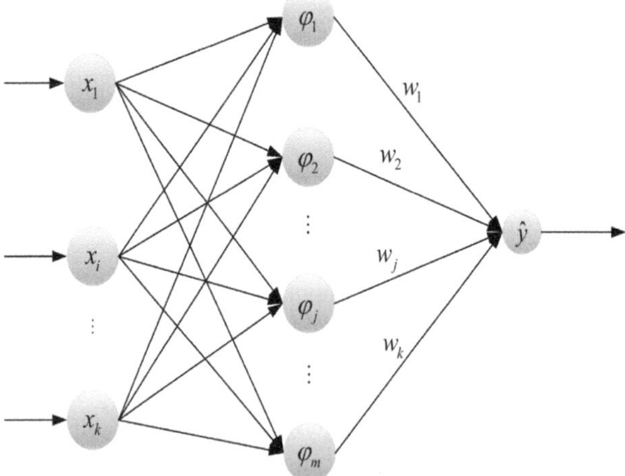

FIGURE 10.2 Typical Structure of a Radial Basis Function Network. This type of consist of one hidden layer with weight and bias nodes.

represents the typical structure of a RBF network with a single output node for single-value tasks (e.g., regression, etc.).

The input node distributes the k input variables to the m nodes of the hidden layer. In the hidden layer, each node has a centre with the same dimensions as the number of input variables. The hidden layer applies a nonlinear transformation to the input space, transforming it into a higher-dimensional space [43]. The activity $\mu_l\left(\mathbf{x}(f)\right)$ of the lth node is the Euclidean normal of the difference between the fth input vector and the node centre, and given in Equation (10.3) as:

$$\mu_l\left(\mathbf{x}(f)\right) = \left\|\mathbf{x}(f) - \dot{\mathbf{x}}_l\right\| = \sqrt{\sum_{i=1}^{k}\left(\mathbf{x}(f) - \dot{\mathbf{x}}_{l,i}\right)^2}, \qquad f = 1,\dots,f \tag{10.3}$$

where f is the total number of available data, $\mathbf{x}^T(f) = \left[x_1(f), x_2(f),\dots,x_k(f)\right]$ is the input vector, and $\mathbf{x}_l^T = \left[\dot{\mathbf{x}}_{1l,}\ \dot{\mathbf{x}}_{2\,l,\dots,\dot{\mathbf{x}}_{kl,}}\right]$ is the centre of the lth node.

The activation function for each node is a radially symmetric function. In this work, we employed the sigmoid function (Equation (10.4)) because the sigmoid activation function introduces non-linearity into the output of a neuron, allowing the model to learn thereby modelling more complex relationships in the data.

$$g(\mu) = \frac{1}{1 + e^{-\mu}} \tag{10.4}$$

The hidden node response is denoted by $\mathbf{z}(f)$ (Equation (10.5)):

$$\mathbf{z}(f) = \left[g\left(\mu_1\left(\mathbf{x}(f)\right)\right), g\left(\mu_2\left(\mathbf{x}(f)\right)\right),\dots, g\left(\mu_m\left(\mathbf{x}(f)\right)\right)\right] \tag{10.5}$$

The output of an RBF network contains y unit, where y is the singular possible output value. The numerical output $y(f)$ is produced by a linear combination of the hidden nodes' response (Equation (10.6)):

$$y(f) = \mathbf{z}(f).\mathbf{w}_n = \sum_{i=1}^{m} w_{l,n} g\left(\mu_l\left(\mathbf{x}(f)\right)\right) \tag{10.6}$$

where $\mathbf{w}_n = \left[w_{1,n}, w_{2,n},\dots, w_{m,n}\right]^T$ is a vector containing the synaptic weights corresponding to the output n.

The synaptic weights are commonly determined using linear regression of the hidden layer outputs to the real measured output after the RBF centres and nonlinearities in the hidden layer have been fixed. In most cases, linear least squares in matrix form can be used to solve the regression problem [43].

$$\mathbf{W} = \left(\mathbf{Z}^T.\mathbf{Z}\right)^{-1}.\mathbf{Z}^T.\mathbf{Y} \tag{10.7}$$

where $\mathbf{Z} = \left[z(1), z(2),\dots, z(F)\right]^T$ is a matrix containing the hidden layer responses for all input vectors. $W = \left[w_1, w_2,\dots, w_n\right]$ is a matrix containing all the synaptic weights for the output layer and converges to a scalar containing the target vector. The target vector $y(f)$ carries the information of the value predicted by the fth input vector [43].

10.2.4 MULTI-LAYER PERCEPTRON (MLP) NEURAL NETWORKS

A MLP is a supplement of a feed-forward neural network. It consists of the basic three layers of a neural network, just as in the RBF neural network [44]. They are utilized for generic approximation because they can model any continuous function. A perceptron receives n features as input $x = x_1, x_1, \ldots, x_n$, each of these features has an associated weight. The features inputted into the network must be numeric; all non-numeric features must be first converted into numbers before being inputted into the network. The input features are passed on to an input function u, the function u computes the weighted sum of the input features [45].

$$u(x) = \sum_{i=1}^{n} w_i x_i \tag{10.8}$$

The result $u(\mathbf{x})$ is passed onto an activation function f, this function assists in producing the output of the perceptron. The activation function utilized in this step is a Rectifed Linear Unit (RELU).

$$y(x) = MAX(\mathbf{0}, x) \tag{10.9}$$

$$y(x) = \begin{cases} 0 \ for \ x < 0 \\ x \ for \ x \geq 0 \end{cases} \tag{10.10}$$

Learning in MLPs, as in RBFs, consist of adjusting the weights in order to reduce the error in predicting the training data. Learning is a backpropagation task achieved by a backpropagation algorithm (optimizer), that attempts to minimize the loss in predicting the ground truth.

10.2.5 OPTIMIZERS

These are methods and algorithms utilized by a neural network for backpropagation; they assist in adjusting the weights based on a user-defined learning rate with the aim of reducing prediction losses. There are varieties of optimizers. We will focus on three types [46]:

1. **Stochastic Gradient Descent (SGD):** This is a revised version of gradient descent, where the model parameters are updated on every iteration. Gradient descent is an optimization algorithm for finding the local minimum of differentiable functions. The basic representation is $\theta_j \leftarrow \theta_j - \alpha \dfrac{\delta}{\delta \theta_j} j(\theta)$. In gradient decent, we take all the data for each iteration. However, in SGD, we randomly select batches of data. The procedure is to select the initial parameters v and learning rate η, and thereafter randomly shuffle the data at each iteration to reach an approximate minimum [47], $Q_i(v)$ is the derivative of Q_i with respect to v_n.

$$v := v - \eta \nabla Q_i(v) \tag{10.11}$$

2. **Adam:** It is also called the Adaptive Moment Estimation algorithm. It combines root-mean-square propagation and momentum-based gradient descent to achieve better optimization. It introduces two hyper-parameters β_1 and β_2; their optimal values have been found to be 0.9 and 0.99, respectively [47]. The root-mean-square prop $\left[V_{dw} = \dfrac{V_{dw}}{\left(1 - \beta_1^t\right)} \right]$ is combined with the corrected momentum-based gradient descent, $\left[S_{dw} = \dfrac{S_{dw}}{\left(1 - \beta_2^t\right)} \right]$ to yield the Adam optimizer.

$$W^{new} = W - \eta \frac{V_{dw}}{\sqrt{S_{dw}} + \varepsilon} \tag{10.12}$$

3. **RMSProp**: This is short for root-mean-square propagation. This is an adaptive learning rate method proposed by Geoffrey Hinton. The algorithm focuses on accelerating the optimization process by decreasing the number of function evaluations required to reach the local minima. It retains the moving average of squared gradients for every weight and divides the gradients by the square root of the mean square [48].

$$v(w,t) := \gamma v(w,t-1) + (1-\gamma)\left(\nabla Q_i(w)\right)^2 \tag{10.13}$$

where gamma is the forgetting factor, i.e., an optimal value determined experimentally to be 0.95 weights, and is updated by the expression in Equation (10.14).

$$w := w - \frac{\eta}{\sqrt{v(w,t)}} \nabla Q_i(w) \tag{10.14}$$

10.2.6 EVALUATION METRICS

These metrics statistically compare the models' output distribution with the ground truth distribution. In this work, we have considered a mean absolute error (MAE), root-mean-square error (RMSE), and coefficient of determination (R^2) as given in Equations (10.15), (10.16), and (10.17), respectively:

$$\text{MAE} = \frac{1}{n}\sum_{i=1}^{N}\left(|\acute{p}_i - P_i|\right) = \frac{1}{n}\sum_{i=1}^{N} AE(i) \tag{10.15}$$

$$\text{RMSE} = \sqrt{\frac{1}{N}\sum_{i=1}^{N}\left(\acute{p}_i - P_i\right)^2} = \sqrt{\frac{1}{N}\sum_{i=1}^{N} AE^2(i)} \tag{10.16}$$

$$R^2 = \frac{(a\sum P) + (b\sum XP) - n\acute{P}^2}{\left(\sum P^2\right) - n\acute{P}^2} \tag{10.17}$$

10.2.7 METHODS AND ALGORITHMS

A quantile defines a particular part of a data set by assigning each point within a distribution to be either above or below a certain limit [49]. Figure 10.3 shows the distribution of q-quantile plots for all values $a \epsilon S$; the probability that x falls within quantile q is given by $P[X < x] \leq k / q$ (where x is a kth q-quantile for a variable X), while the probability that x falls without the quantile q is given by $P[X < x] \geq 1 - \frac{k}{q}$, considering also that x is the kth q-quantile for a variable X. The distribution is represented mathematically in Equation (10.18).

$$P[X < x] = \frac{1}{\sqrt{2\pi}}\int_{\infty}^{x} e^{-\frac{t^2}{2}} dt \tag{10.18}$$

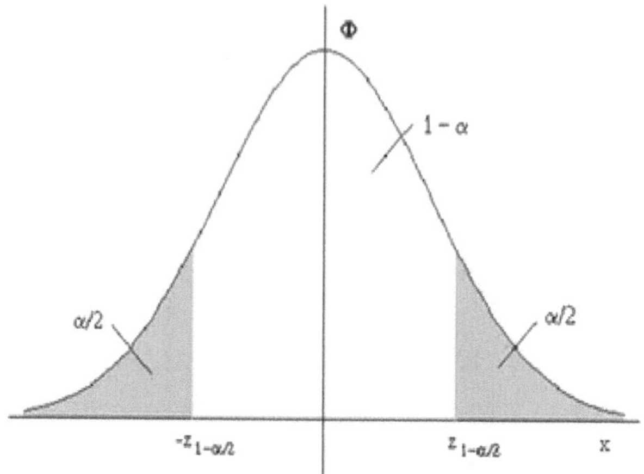

FIGURE 10.3 The quantile specification on a normal distribution. Every Z-score value defines an inference threshold. In this case, it separates faulty data from normal data.

The αth quantile $\theta_\gamma(\propto), 0 < \propto < 1$ of a finite population vector $y = (y_1,\ldots,y_N)$ is defined as

$$\theta_\gamma(\propto) = \inf\{t : F_\gamma(t) \geq \alpha\} \tag{10.19}$$

where $F_\gamma(t)$ is the distribution function γ. In case $\dot{F}_\gamma(t)$, an estimator of $F_\gamma(t)$ is a monotonic non-decreasing function of t, the customary estimator of $\theta_\gamma(\propto)$ is obtained as

$$\theta_\gamma(\propto) = \inf\{t : \dot{F}_\gamma(t) \geq \alpha\} \tag{10.20}$$

Let $\dot{F}_\gamma(t)$ be the customary estimator of $F_x(t)$. In case the population αth quantile $\theta_x(\propto)$ of x is known, the ratio estimator of $\theta_\gamma(\propto)$ is given by

$$\ddot{\theta}_{r\gamma}(\propto) = \frac{\ddot{\theta}_\gamma(\propto)}{\ddot{\theta}_x(\propto)}\theta_x(\propto) \tag{10.21}$$

Similarly, a difference estimator of $\theta_\gamma(\propto)$ is given by:

$$\ddot{\theta}_{d\gamma}(\propto) = \ddot{\theta}_\gamma(\propto) - R\{\ddot{\theta}_x(\propto) - \theta_x(\propto)\} \tag{10.22}$$

where $R = \dfrac{\sum_{i \in S} \dfrac{y_i}{\pi_i}}{\sum_{i \in S} \dfrac{x_i}{\pi_i}}$ is a consistent estimator of the population ratio $R = Y / X$.

Both estimators $\ddot{\theta}_{r\gamma}(\propto)$ and $\ddot{\theta}_{d\gamma}(\propto)$ reduce to $\theta_\gamma(\propto)$ if $y_i \propto x_i \forall i \in U$. In this case, the variance becomes zero [50]. The quantile filtering is as given in Algorithm 10.1.

ALGORITHM 10.1 Quantile Filtering

Input: $[x_1, x_2, x_3]$: from a dataset of initial size, in a Data Frame.
$[y]$: from a dataset of initial size, in the same Data Frame.

Output: $[x_1, x_2, x_3]$: returned Data Frame of filtered size.
$[y]$: returned in the same Data Frame, of filtered size.

1. Divide Data Frame into sub-Frames up to 50.
2. Define a single Data Frame, and set the power equal to the max power.
3. Define the distribution.
4. Apply quantiles to the distribution for each sub-Frame to remove outliers.
5. Merge all Data Frames.
6. END

10.2.8 NETWORK ARCHITECTURE

The base parameters of the neural networks are specified for the network creation. A total of three neural networks are developed and compared for the best-performing networks. It comprises of a RBF network with a single hidden layer, two variations of MLP network, i.e., one with four-hidden layers of varying nodes, and the other with six-hidden layers of varying nodes. A link to the code is provided for model visualization [see supplementary material [C]]. Figure 10.4 articulates the entire modelling process. The cleaned data is passed into a neural network containing an optimizer that assists in the backpropagation of the model; the neural network evaluates the output of the forward process by some user-defined metrics, and thereafter employs the user-specified activation function in an attempt to reduce loss.

10.3 CASE STUDY

10.3.1 DATA SELECTION

A temporal 10-minute interval dataset from mid-2016 to mid-2022 was acquired from Kelmarsh and Penmanshiel wind farms in the United Kingdom (see supplementary material [A] and [B] for the dataset). Out of these years, only a single year's data (dynamic year), is used in developing the model. The wind speed is plotted against time for all available years to aid visual identification of the most dynamic year (Figure 10.5). The choice of the year to be used as the dynamic year depends on how randomly distributed wind speed to time appeared for that year. The year with the highest standard deviation of wind speed found to be 2020 is thus taken to be the most dynamic year (Figure 10.5a). Figure 10.5b shows the variation of wind speed to time for the year 2020.

It was discussed in Karamichailidou et al. [11] that the variations in turbulence intensity (TI) due to field conditions, such as the relative position of turbines, local orography, and wake effects of neighbouring wind turbines, have a significant effect on the power output. Hence, for a generic wind turbine model, the effect of TI must be considered. We utilized SCADA data from three turbines selected from each of the two wind farms. The three turbines were selected with the aim of capturing power variations resulting from TI as influenced by the relative positions of the turbines and topography of the respective farms. Based on the most prevalent wind direction for the year, turbines at the top of the wind farm will experience little or no TI, thus resulting in limited/no reduction in power output, while wind turbines at the midway of the wind farm will experience the most power reduction due to TI and wake effect, irrespective of the wind direction. However, wind turbines at the bottom of the wind farm will experience some power reductions due to TI. An area view of the

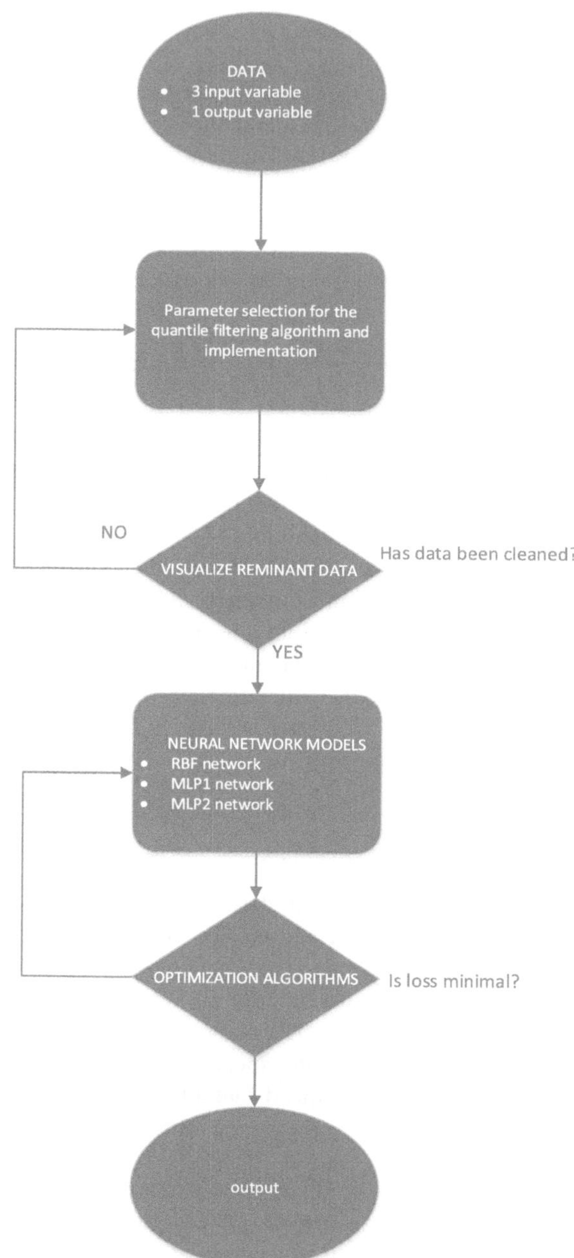

FIGURE 10.4 The QF-NN algorithm, which comprises data filtration by a quantile-based filtration approach and three neural network architectures for modelling the power curve.

wind farms (by satellite) and turbine numbers used in the selection of the three turbines is given in Figure 10.6.

Terrain variation is also an important factor to be considered in wind turbine model development. As height increases, wind speeds tend to be faster and more laminar. At hill sites, the wind will be

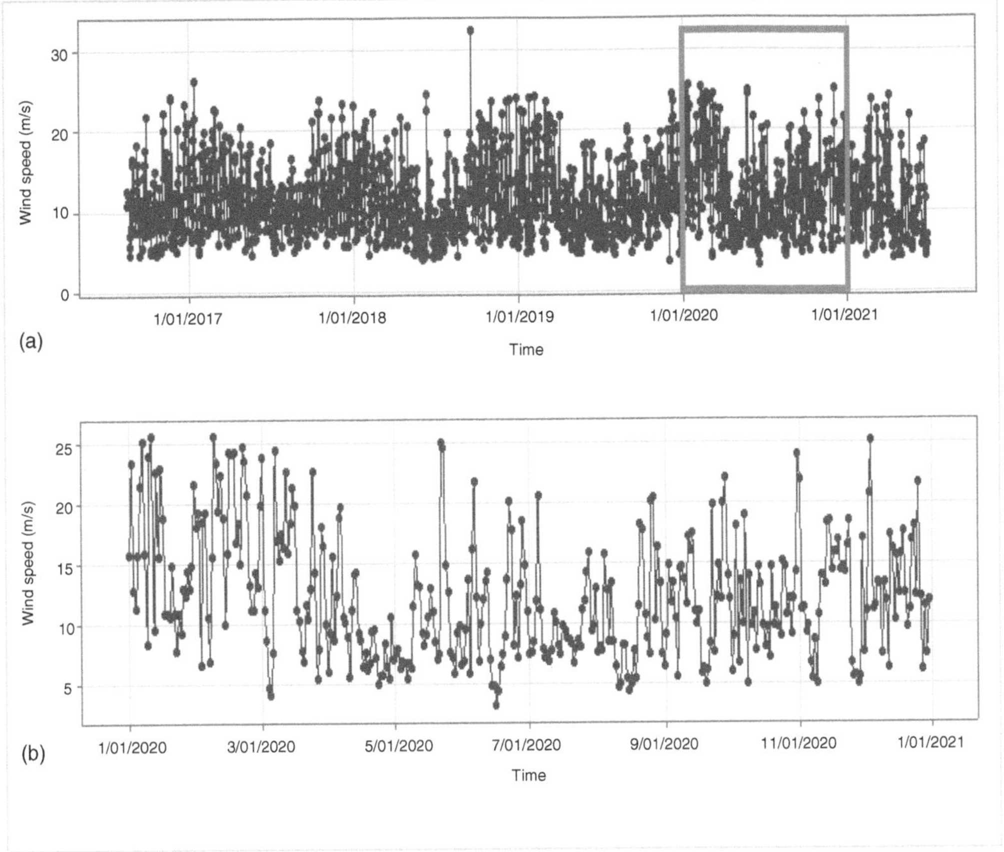

FIGURE 10.5 Variation in wind speed with time for: (a) various years, and (b) selected year.

laminar because there will be little obstruction to air flow, whereas at valley sites wind speed tends to be slower and more turbulent. Hence, topographic consideration is critical to selecting the best turbine data from SCADA that is needed for modelling. Figure 10.7 represents the terrain maps of the selected wind farms, indicating hilly and valley portions needed for the dataset selection process. It can be seen that the contour maps in Figure 10.7 show very little variation in land topography. The topographical variations will constitute only a slight variation in wind speed from turbine to turbine within the selected wind farms.

10.3.2 Data Visualization

A plot of wind speed to power generated is the basis for this analysis. Figure 10.8 shows a scatter plot of wind speed to power output for the six turbines (from the two wind farms) used for model development. It is evident from Figure 10.8 that the actual curve of these wind turbines deviates from the standard manufacturer's power curve. Generally, the actual power curve of a wind turbine will contain abnormal values (outliers) such as non-operational periods, invalid, missing due to sensor and pitch malfunction, etc., [51]. Hence, to aid the development of an accurate model, a data-cleaning process must be employed to filter out all erroneous data.

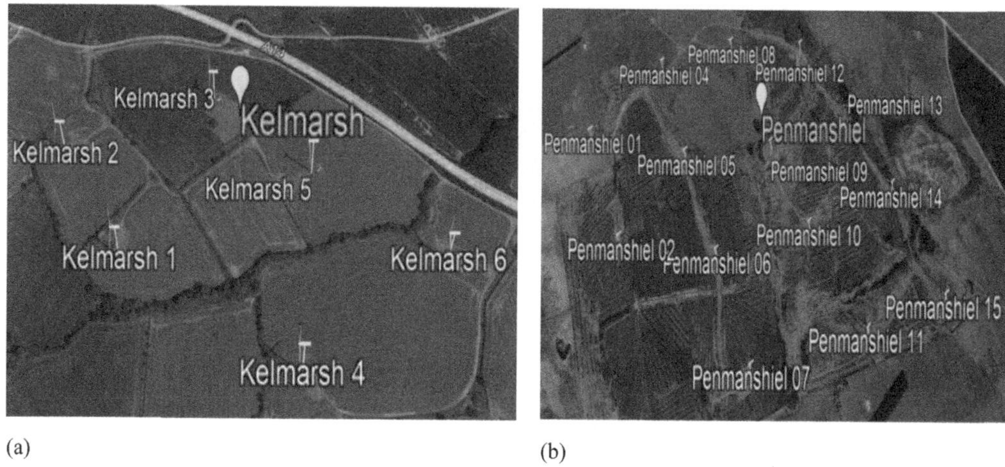

(a)   (b)

FIGURE 10.6 Satellite image of (a) Kelmarsh and (b) Penmanshiel wind farms showing the turbine numbers. (Source: Ref. [36,37].)

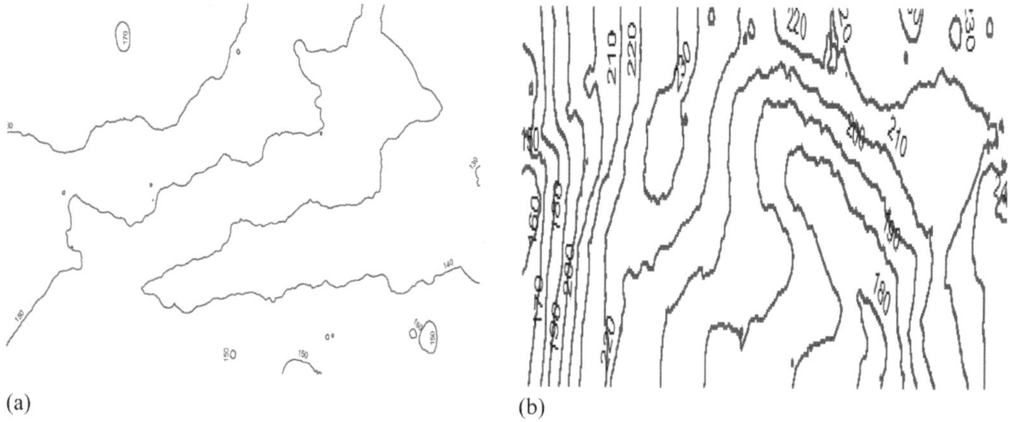

(a)   (b)

FIGURE 10.7 The developed contour maps for (a) Kelmarsh, and (b) Penmanshiel wind farms, with the aim of highlighting disparate terrains.

10.3.3 DATA CLEANING

Several methods are employed in the literature for the filtration of erroneous data. Most of these methods are computationally expensive due to their utilization of complex mathematics. In this work, we proposed a filtering approach which is based on quantiles set on a distribution. This is achieved by dividing data into wind speed bins and setting user-defined quantiles. Quantile values are obtained experimentally and may differ from one wind farm to another. A link to the database containing the code has been provided to aid the visualization of the set quantiles [see supplementary material [C]]. The output (filtered power curve) from the filtering algorithm is shown in Figure 10.9.

A test of the efficiency of any filtering algorithm is the quantity of data points it regards as outliers in its process of cleaning the dataset. A more efficient filtering algorithm will filter out fewer data points in order to achieve a clean dataset, while a less efficient filtering algorithm will filter out more

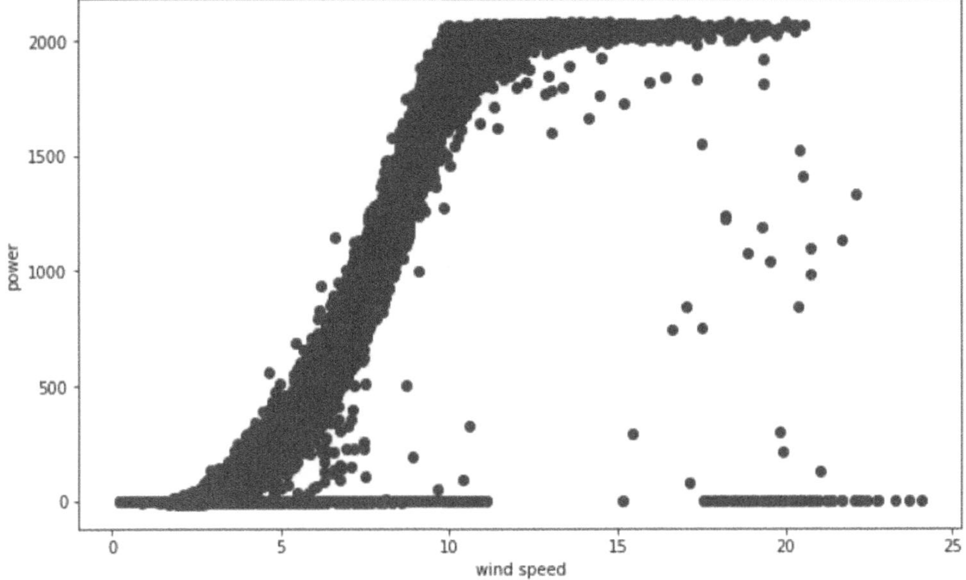

a)

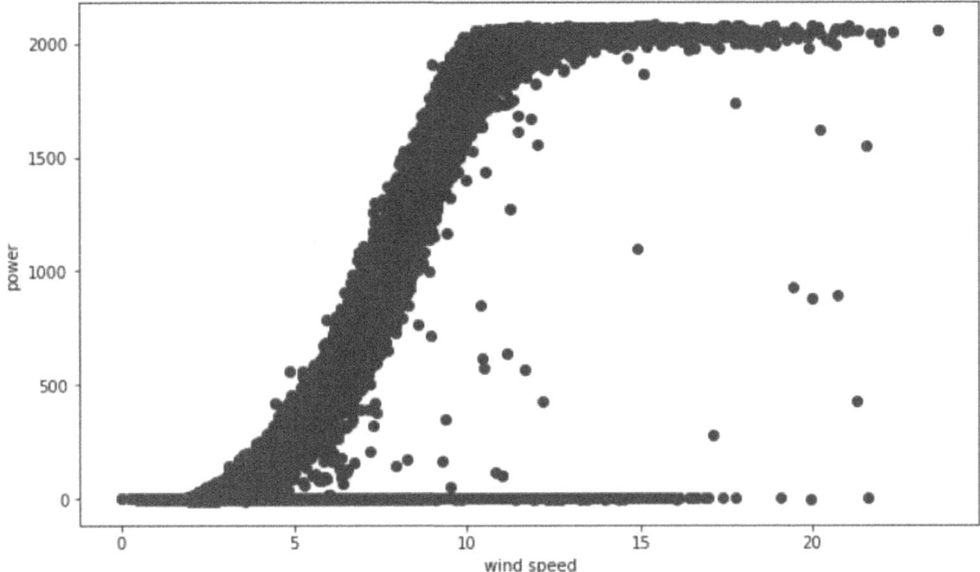

b)

FIGURE 10.8 The unfiltered plot of wind speed and output power for all turbines used in this study: (a) Turbine 1, (b) Turbine 2, (c) Turbine 3, (d) Turbine 4, (e) Turbine 5, and (f) Turbine 6.

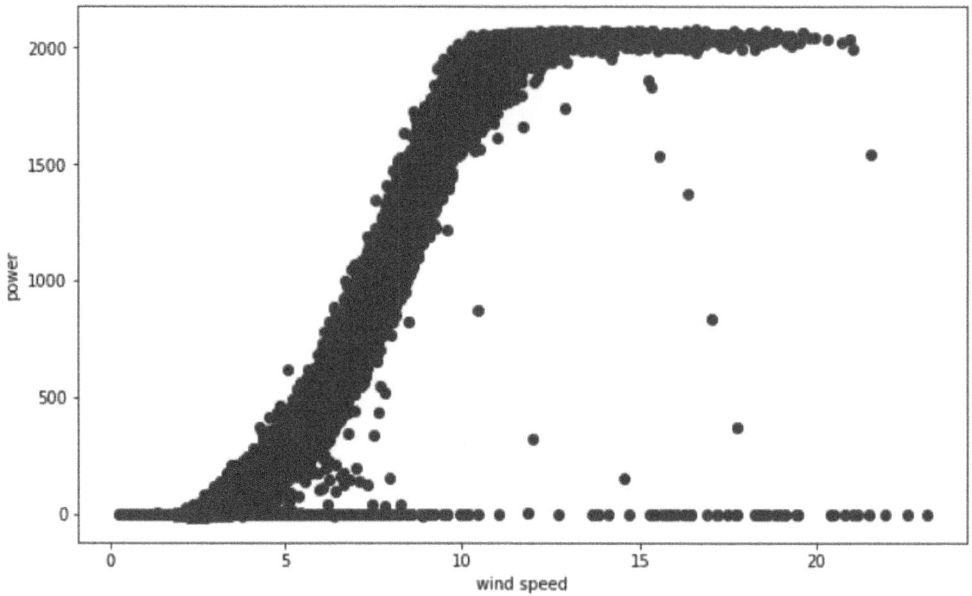

c)

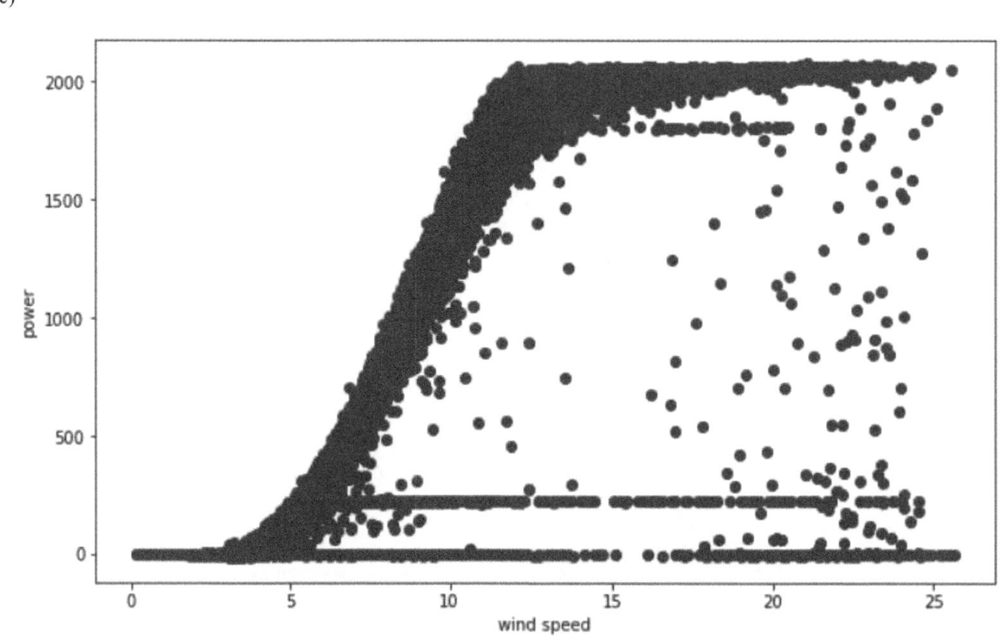

d)

FIGURE 10.8 (Continued)

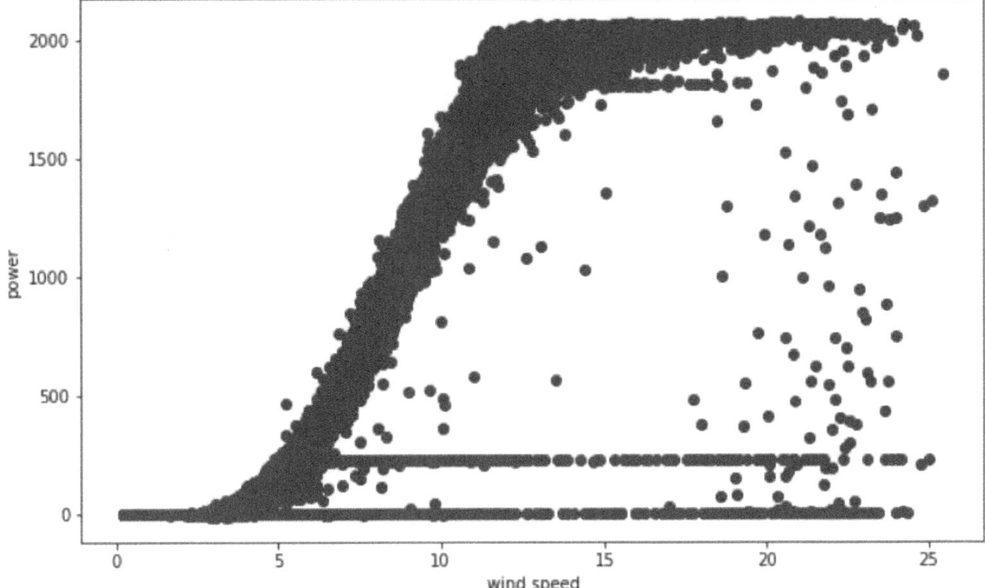

e)

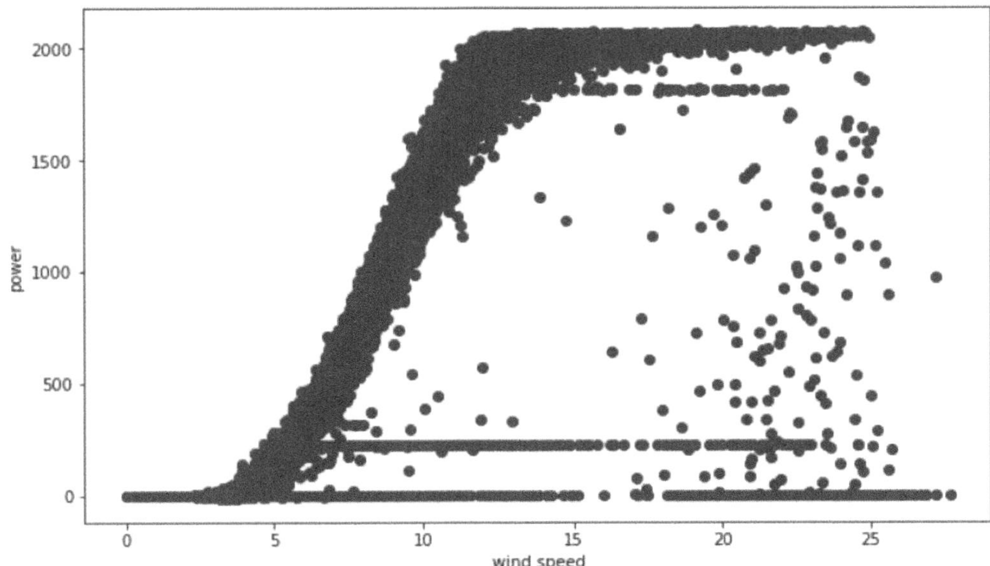

f)

FIGURE 10.8 (Continued)

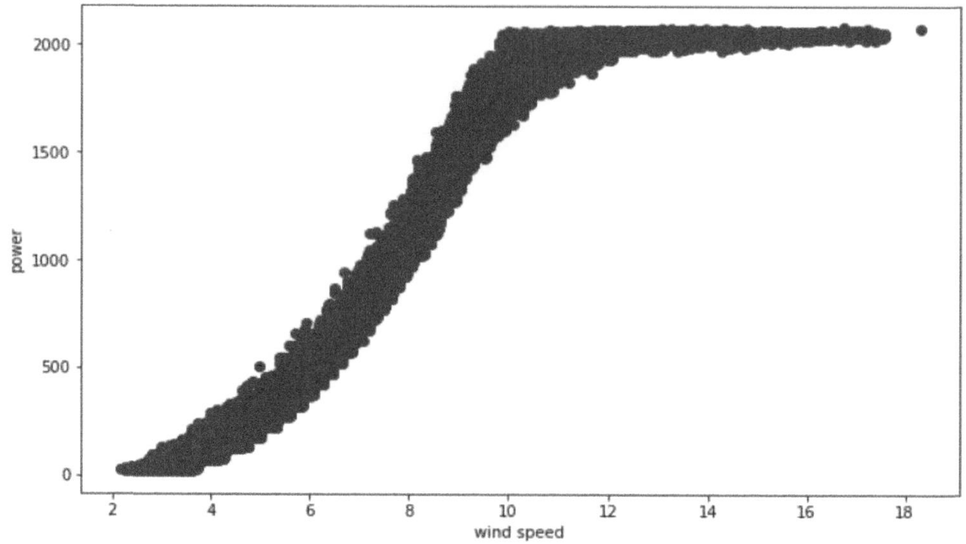

a)

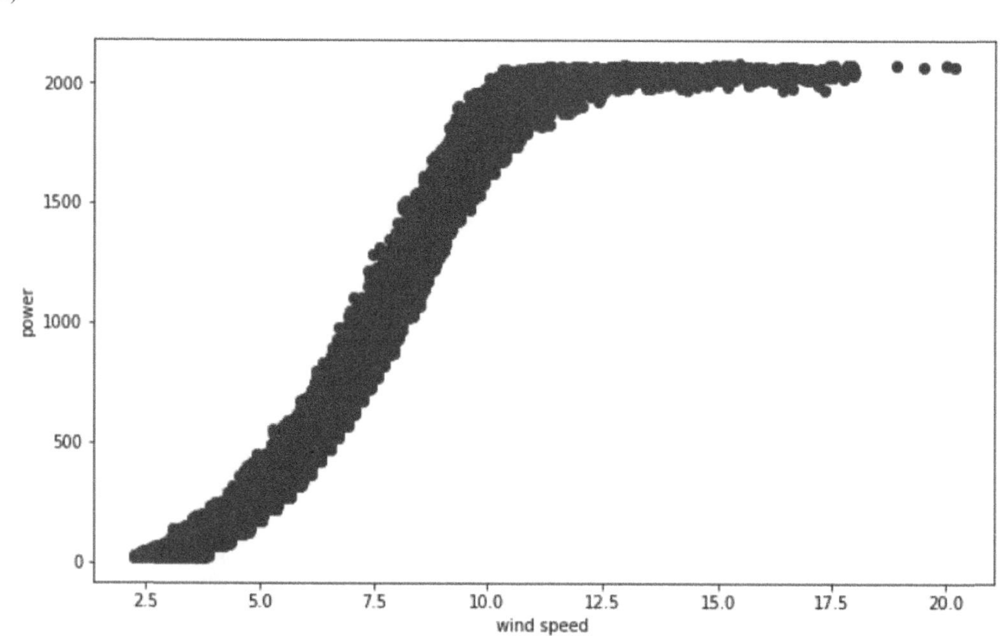

b)

FIGURE 10.9   The plot of all turbines used in this study after filtration was performed using the quantile-based filtration approach: (a) Turbine 1, (b) Turbine 2, (c) Turbine 3, (d) Turbine 4, (e) Turbine 5, and (f) Turbine 6.

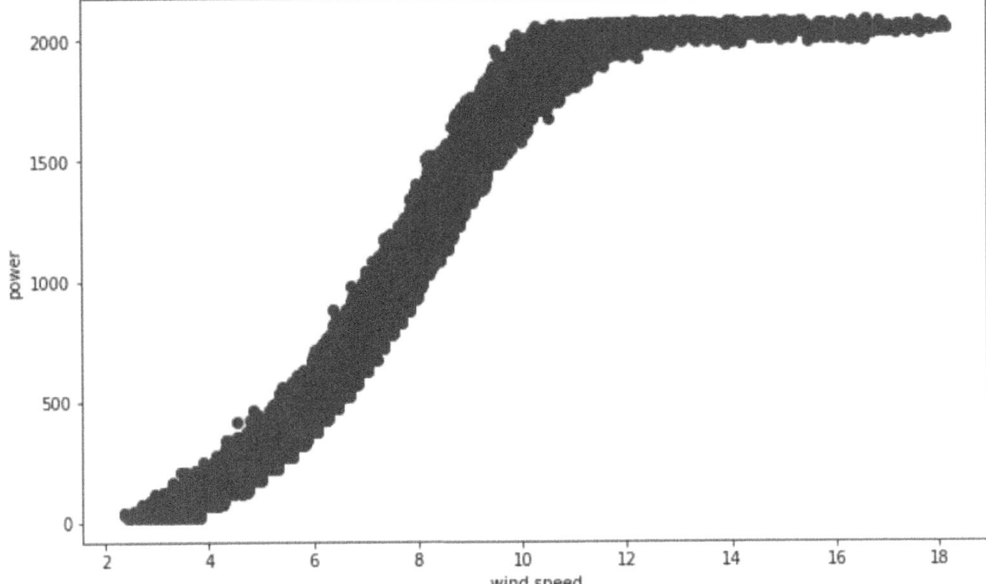

c)

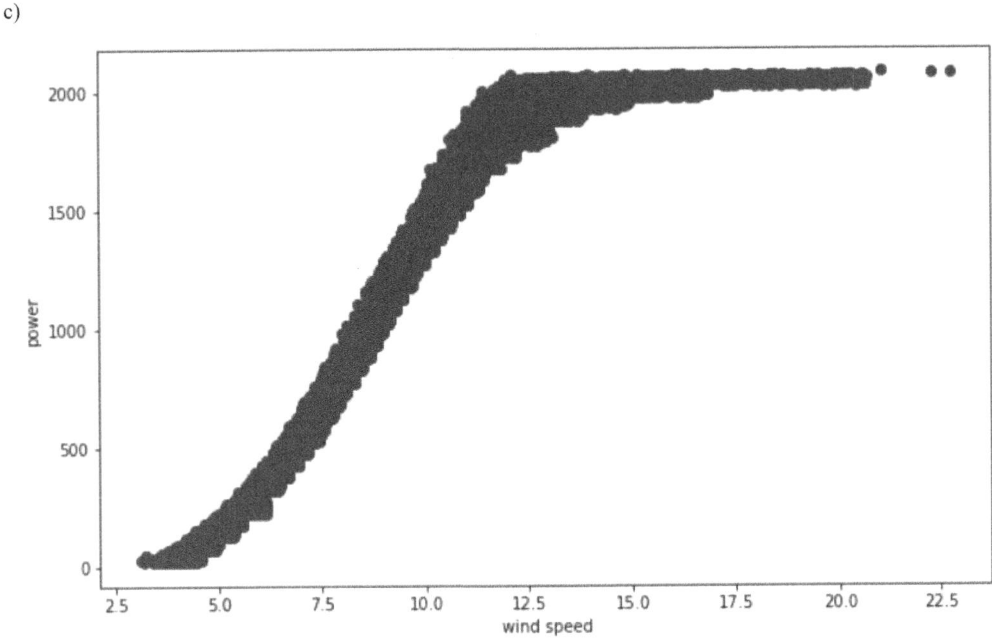

d)

FIGURE 10.9 (Continued)

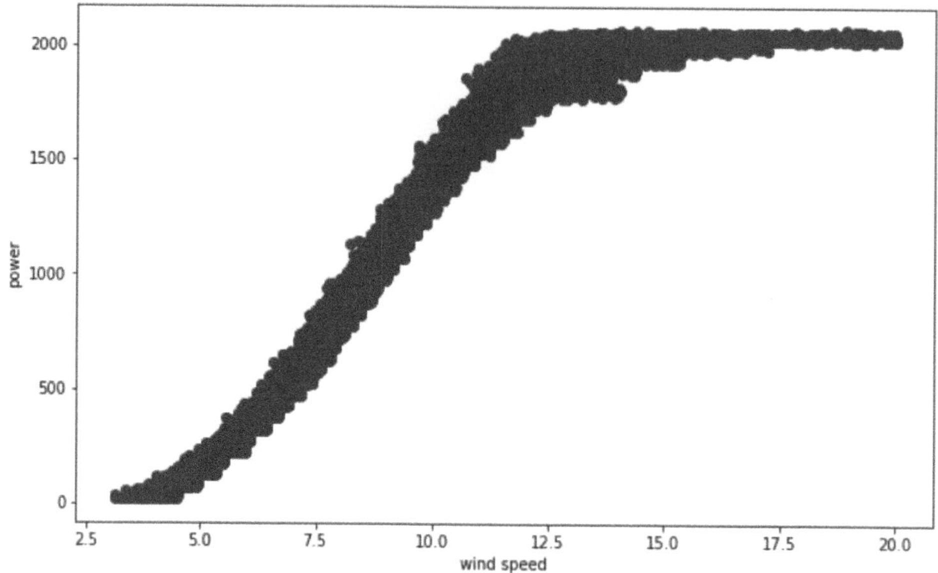

e)

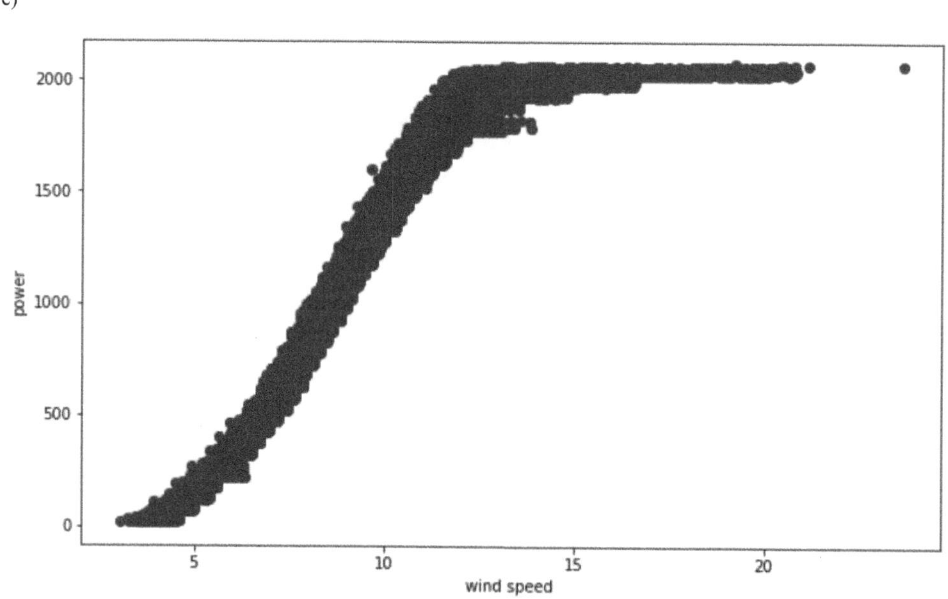

f)

FIGURE 10.9   (Continued)

TABLE 10.1
Percent Data Loss Due to Filtering

	Before Cleaning	After Cleaning	Percent Data Loss (%)
Dataset 1	52,704	43,425	17.6
Dataset 2	52,704	41,196	21.8
Dataset 3	52,704	41,278	21.6
Dataset 4	52,704	37,247	28.1
Dataset 5	52,704	35,282	29.5
Dataset 6	52,704	37,760	27.9

TABLE 10.2
Division of Dataset

	Training Size	Validation Size	Test Size
Dataset 1	28,226	10,856	8685
Dataset 2	26,777	10,299	8239
Dataset 3	26,830	10,319	8255
Dataset 4	24,210	9311	7449
Dataset 5	22,933	8820	7056
Dataset 6	24,544	9440	7552

data points in order to achieve a clean dataset. In the literature, models utilized for data cleaning are static, in that upon implementation, the data points filtered out are constant. However, in this work, outputs of the quantile filtering (QF) method vary the amount of filtered data points contingent on user pre-set quantiles. The number of data points cleaned out based on the quantile set developed in this research is shown in Table 10.1.

After filtering, the dataset passes a stage of batch normalization. This standardizes the dataset between 0 and 1. The efficiency and speed of a neural network is enhanced when we convert the variable data to values between 0 and 1 for the fitting process. Normalized data are then split into training, testing, and validation datasets. In this research, the dataset from each turbine was divided randomly using 65% for training, 25% for validation, and 20% for testing. This is shown in Table 10.2.

10.3.4 IMPLEMENTATION OF ALGORITHM

The visual analysis of the actual wind speed to power plot of a wind turbine is the basis for the choice of the most optimum quantile for filtering. An assessment of the distribution of the dataset must be carried out during the specification of quantiles. A skewed distribution of the dataset will constitute poor filtering even with the most generic quantile; it thus requires a rather unique quantile values that should be selected based on the nature of its distribution. The filtered data comprising three input parameters (density normalized wind speed, wind direction, and blade pitch angle) are fed into three neural networks, with the aim of accurately predicting the power output. These input parameters were found to improve the modelling accuracy of power curve models, as

TABLE 10.3
Architecture of the Developed Models

	Model Name	Input Layer	Hidden Layer	Output Layer	Activation	Loss Function
Model 1	RBF	1	1	1	*sigmoid*	*Binary crossentropy*
Model 2	MLP1	1	4	1	*relu*	*Binary crossentropy*
Model 3	MLP2	1	6	1	*relu*	*Binary crossentropy*

TABLE 10.4
Test Result for Model 1 (RBF Model)

	Optimizer	MAE (KW)	RMSE (KW)	R^2	Computation time (sec)
	adam	21.7	30	0.9924	145.94
Data 1	rmsprop	21.9	30	0.9923	202.69
	sgd	33.0	42.6	0.9833	142.50
	adam	23.7	32.1	0.9899	133.99
Data 2	rmsprop	23.6	32.7	0.9897	142.96
	sgd	35.2	46.6	0.9789	142.48
	adam	21.1	28.7	0.9917	142.59
Data 3	rmsprop	21.2	29.6	0.9913	142.475
	sgd	31.9	42.2	0.9815	142.511
	adam	**16.8**	**22.8**	**0.9955**	114.999
Data 4	rmsporp	17.1	23.3	0.9951	124.435
	sgd	23.9	30.1	0.9920	111.959
	adam	17.3	23.8	0.9950	142.571
Data 5	rmsprop	17.8	24.3	0.9933	142.663
	sgd	28.1	36.6	0.9880	112.231
	adam	18.5	25.6	0.9945	142.561
Data 6	rmsprop	17.3	24.7	0.9948	142.676
	sgd	25.7	32.5	0.9911	142.507

Note: The results in bold signifies the best performance.

each constitutes an influence on the power output of a wind turbine [52,53]. A RBF network of one-hidden layers, a MLP network of four-hidden layers, and a MLP network of six-hidden layers were used for modelling. There is an increase in the number of hidden layers for the second MLP network; this was done with the aim of inspecting the performance of MLP network with more hidden layers. The architectures are shown in Table 10.3.

10.4   RESULTS AND DISCUSSION

From Table 10.4, data 1–3 are obtained from the Kelmarsh wind farm, while data 4–6 are from the Penmanshiel wind farm. The differences in the accuracies between datasets from the same wind farm are due to the variation of the data distribution and TI, while the differences recorded between datasets from different wind farms are a result of the differences in field conditions at both farm sites. An analysis of the data distribution was performed.

The field conditions at the Kelmarsh wind farm are generally more dynamic than those at the Penmanshiel wind farm. Hence, the fitting accuracy is slightly less for the Kelmarsh wind farm than

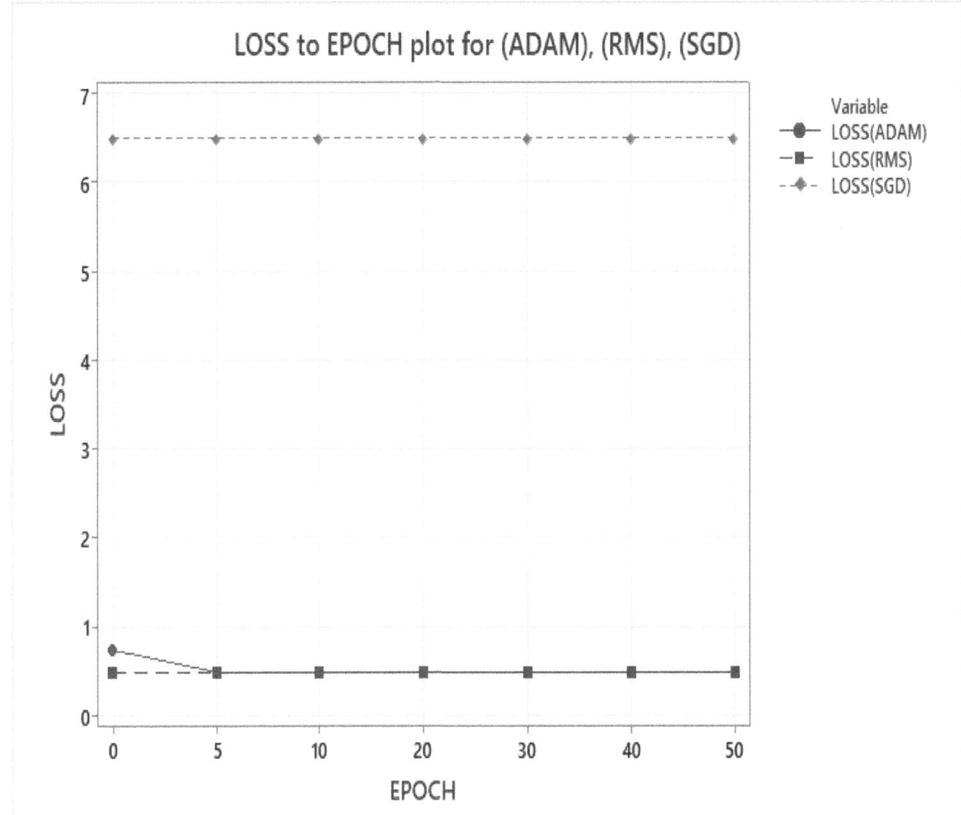

FIGURE 10.10   The plot of loss to epoch of the RBF network.

the Penmanshiel wind farm. The test for a good model is its generalization ability. The standard deviation of the datasets from both wind farms is minimal, thus indicating that the model generalizes appropriately. There is a slight variation in performance when various optimizers are applied on the RBF network. The best optimizer for this network is *adam* as it has shown better performance on all the datasets compared to the other optimizers (though not far behind). All optimizers performed generally well on the RBF network. In this work, the RBF network is generally appropriate for modelling WTPC regardless of field conditions. Data 1 and 4 are obtained from wind turbines at the top of the wind farms where TI has a negligible effect on power output. Data 2 and 5 are obtained from wind turbines midway through wind farms and are mostly affected by TI and the wake effect. Data 3 and 6 are obtained from wind turbines at the bottom of the wind farms; these are somewhat influenced by TI and the wake effect. In this work, it is worth noting that we obtained similar fitting accuracies regardless of the position of the wind turbine.

We ran the models on Google Collaboratory (12.63GB GPU), for 50 epochs. Figure 10.10 depicts the loss to epoch plot for each optimizer. The model arrived at a minimum loss within just a few epochs. Hence, more time could be saved by reducing the number of epochs. The MLP model is generally sensitive to optimizers; only the *adam* optimizer performed appropriately. It can be seen from Figure 10.11 that the *rmsprop* and *sgd* optimizers worked to increase the loss rather than to decrease it. Therefore, care should be taken when choosing the right optimizer for an MLP model.

The variations found in the dataset due to field conditions do not affect the accuracy of the MLP model. Taking the values of the *adam* optimizer as a case study, the values of both MAE and R^2 vary

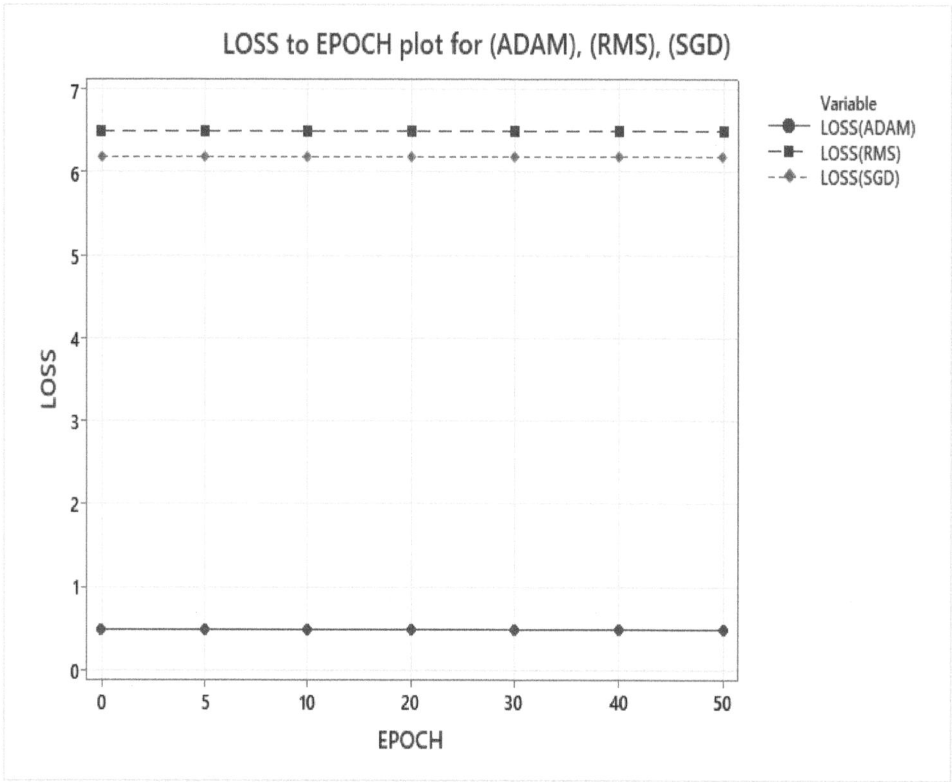

FIGURE 10.11　The plot of loss to epoch of the four-hidden layer MLP network.

only slightly across data 1–6. This is an indication that the MLP network generalizes properly for all turbines within a wind farm and for turbines in multiple wind farms. The best-performing case of the four-input MLP network outperforms all cases in the RBF network. Generally, an increase in the number of hidden layers of an MLP network reduces the sensitivity of the model to change in the optimizer. However, it also slightly increases the models' sensitivity to dataset distribution. This can be observed in Table 10.5 with the standard deviation of the six-hidden layers of the MLP network being relatively more than that of the four-input MLP network. The six-hidden layers MLP network generally requires more time to train, but results in higher fitting accuracy than the other models. From Figure 10.12, it can be seen that the network accommodates the use of the *rmsprop* optimizer but not the *sgd* optimizer.

Table 10.6 summarizes the results of the application of QF-NN algorithm in comparison with the best performing parametric and non-parametric models found in literature, and with comparison to the manufacturer's power curve [54].

From Table 10.7, QF-NN models outperform all other parametric and non-parametric models in terms of fitting accuracy. Generally, the standard deviation of the QF-NN's fitting accuracy is found to be relatively small.

It is also evident from Table 10.7 that hybrid non-parametric models utilizing computational intelligence (CI) generally perform better than parametric or other non-parametric models. However, they are computationally expensive and will take a longer time to compute. QF-NN proves to be an excellent alternative to other hybrid non-parametric models, in that it outperforms them in terms of fitting accuracy and also computes in very little time regardless of the size of the data. This is important if these models are to be implemented for real-time power prediction and forecasting. The

TABLE 10.5
Test Result for Model 2 (QF-MLP With 4-Hidden Layer)

	Optimizer	MAE (KW)	RMSE (KW)	R^2	Computation Time (sec)
Data 1	adam	21.6	28.6	0.9932	202.680
	rmsprop	50.2	60.8	−2.2444	166.041
	sgd	45.7	56.8	−1.8431	140.157
Data 2	adam	23.4	31.5	0.9902	142.671
	rmsprop	38.2	50.1	−1.4062	202.954
	sgd	98.8	100.3	−9.3963	143.103
Data 3	adam	21.7	28.8	0.9915	138.004
	rmsprop	35.4	46.77	−1.2681	159.599
	sgd	122.6	125.8	−15.0733	134.258
Data 4	**adam**	**16.1**	**22.1**	**0.9960**	142.637
	rmsporp	42.6	55.3	−1.4986	142.953
	sgd	51.3	63.2	−2.2674	115.142
Data 5	adam	16.7	23.1	0.9953	121.813
	rmsprop	36.7	49.4	−1.1796	140.560
	sgd	144.8	148.3	−19.0171	116.485
Data 6	adam	19.4	27.7	0.9940	125.826
	rmsprop	46.1	58.4	−1.7645	202.962
	sgd	49.5	61.1	−2.0359	121.506

Note: The results in bold signifies the best performance.

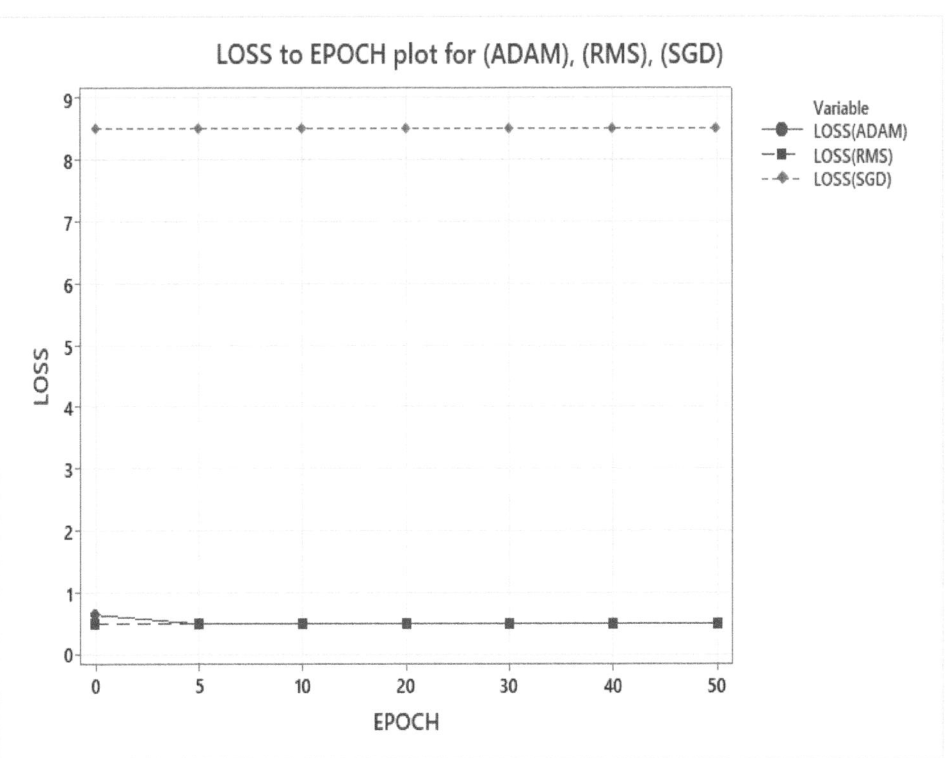

FIGURE 10.12   The plot of loss to epoch of the six-hidden layer MLP network.

TABLE 10.6
Test Result for Model 3 (QF-MLP With 6-Hidden Layer)

	Optimizer	MAE (KW)	RMSE (KW)	R^2	Computation Time (sec)
	adam	25.6	35.0	0.7495	202.733
Data 1	rmsprop	43.8	55.06	−1.6645	181.260
	sgd	742.5	743.4	−486.80	202.629
	adam	40.9	52.1	−1.6036	202.737
Data 2	rmsprop	24.0	32.2	0.9898	174.694
	sgd	38.2	49.8	−1.373	140.108
	adam	22.5	30.5	0.9905	202.761
Data 3	rmsprop	34.4	46.4	−1.2357	203.162
	sgd	35.6	47.1	−1.3049	137.383
	adam	**15.3**	**21.5**	**0.9962**	142.753
Data 4	rmsporp	31.5	40.0	0.9871	151.888
	sgd	33,755	3755	−11,680	122.132
	adam	28.1	44.8	0.9825	142.739
Data 5	rmsprop	16.9	23.1	0.9925	203.155
	sgd	47.4	57.9	−1.9906	120.360
	adam	17.5	24.2	0.9953	142.737
Data 6	rmsprop	417.3	543.5	−1.4034	203.118
	sgd	421.0	547.6	−1.423	125.839

Note: The results in bold signifies the best performance.

TABLE 10.7
Comparison of Results of Parametric and Non-Parametric Models

	Hidden Layers	MAE (KW)	R^2 score	Computational Time(s)
QF-RBF(adam)	1	**16.8**	**0.995**	**114.999**
QF-RBF(rmsprop)	1	**17.1**	**0.995**	**124.656**
QF-MLP1-(adam)	4	**16.1**	**0.996**	**142.637**
QF-MLP2-(adam)	6	**15.3**	**0.996**	**142.753**
TS-NSFM	1	18.935	0.994	4896
NSFM	1	46.823	0.955	−
SFM	1	19.103	0.992	987
Cubic Spline	−	32.7123	0.985	1.50
SVM	−	27.345	0.989	80
GP	−	23.497	0.990	45.43
Logistic Function	−	29.956	0.986	1.89
Parametric	−	33.545	0.984	1.46
Modified IEC-BINS	−	29.453	0.989	1.32
Manufacturer power curve	−	40.234	0.979	−

Note: The results in bold signifies the best performance.

best-performing model is the QF-MLP with four-hidden layer using *adam* optimizer, having MAE of 15.3 and R^2 of 0.996. Though it is the best performing model, other NN models are also found to be close in performance. The importance of a better model includes better power forecasting, accurate instantaneous power prediction, and efficient performance monitoring of wind turbines, leading to better energy management that will result in a reduction in the total cost of energy.

An overall analysis of the QF-NN algorithm is necessary in order to justify the research objectives. The input variables of the models are density, wind speed, temperature, and blade pitch angle. In the literature, very few works have utilized more than one input variable, this was mainly due to the methods employed for power curve modelling that accommodates just one input variable. It is worth noting that models developed under such conditions cannot accurately represent actual power curves at field conditions. In this research, the four variables considered are those that majorly contribute to the power output of a wind turbine.

Quantile filtering only considers wind speed and power output in the cleaning process because it is a two-dimensional cleaning process requiring two parameters to define the plane. The two-dimensional cleaning is sufficient to filter out outliers and it does that in less time than a higher-dimensional cleaning algorithm. Passing the cleaned data into a shallow neural network harnesses the power of the network to efficiently fit the input parameters to the output power in a supervised framework. We have also compared the accuracy and sensitivity of RBF and MLP networks used for the WTPC modelling.

10.5 CONCLUSION

In this work, we introduced a new method for obtaining a very accurate WTPC model, by taking into account several variables (found to affect power output from a wind turbine), other than wind speed only. These variables are density, wind speed, temperature, and blade pitch angle. The datasets were obtained from Kelmarsh and Penmanshiel wind farms in Haselbach and Grantshouse, the United Kingdom, respectively. Since wind turbine performance vary with TI, we also considered the topography of the locations to aid the selection of appropriate turbines to be selected for the modelling.

We used a quantile filtering approach, a data cleaning method where a set point or quantile is used to distinguish relevant data points from outliers (that are thereafter removed). The filtered data was then applied to three models, including RBF, QF-MLP with four-hidden layers, and QF-MLP with six-hidden layers. The results of these models were compared to determine the best performer for modelling WTPC using shallow neural networks. The performance of the new model was also compared with those of parametric and non-parametric models found in the literature [Ref. 1,11,34]. The results showed that the QF-MLP with six-hidden layers was the best performer and the most sensitive to dataset distribution. The four-layer and six-layer QF-MLP networks are sensitive to the chosen optimizer. The RBF network was the most stable, having the smallest standard deviation between datasets. All three models outperformed parametric and non-parametric models in terms of fitting accuracy. They also computed much faster than other hybrid models (such as TS-NSFM and SFM).

We have developed a WTPC model that is generic to all turbines in wind farms, taking topography and other field conditions into consideration during the dataset selection and modelling. The developed model in this work could aid in better power forecasting, accurate instantaneous power prediction and efficient performance monitoring of wind turbines, thereby fostering efficient energy management that leads to a reduction in the total cost of energy.

DECLARATION OF COMPETING INTEREST

The authors declare that there are no known competing interests.

SUPPLEMENTARY MATERIALS

Supplementary material for this chapter can be found at:

[A]   Kelmarsh wind farm data: https://zenodo.org/record/5841834#.YsGnSHbMLIU.
[B]   Penmanshiel Wind Farm Data: https://zenodo.org/record/5946808#.YsGnSXbMLIU.
[C]   Link to the code: https://github.com/henrii1/wind-turbine-power-curve-modelling.git.

REFERENCES

[1] Wang Y, Hu Q, Li L, Foley AM, Srinivasan D. Approaches to wind power curve modeling: A review and discussion. *Renew. Sustain. Energy Rev.* 2019; 116 (109422). DOI: 10.1016/j.rser.2019.109422.

[2] Ohunakin O. S., Adaramola M. S., Oyewola O. M. Wind energy evaluation for electricity generation using WECS in seven selected locations in Nigeria. *Appl. Energy* 2011; 88 (9): 3197–3206.

[3] Ohunakin O. S., Akinnawonu O. O. Assessment of wind energy potential and the economics of wind power generation in Jos, Plateau State, Nigeria. *Energy Sustain. Develop.* 2012; 16 (1): 78–83.

[4] Ohunakin O. S. Wind resource evaluation in six selected high-altitude locations in Nigeria. *Renew. Energy* 2011; 36 (12): 3273–3281.

[5] Bölük, G., Mert, M. Fossil & renewable energy consumption, GHGs (greenhouse gases) and economic growth: Evidence from a panel of EU (European Union) countries. *Energy* 2014; 74: 439–446.

[6] Rodríguez-López M, Cerdá E, del Rio P. Modeling wind-turbine power curves: Effects of environmental temperature on wind energy generation. *Energies* 2020; 13 (18): 1–21.

[7] United Nations Development Programme. *Energy and The Challenge of Sustainability*. 2000. Available online: www.undp.org/sites/g/files/zskgke326/files/publications/World%20Energy%20Assessment-2000.pdf. Accessed [03/07/2022].

[8] Mehrjoo M. *Using Machine Learning Methods for Wind Turbine Power Curve Modeling*. 2021. Available online: https://mspace.lib.umanitoba.ca/handle/1993/35697. Accessed [03/07/2022].

[9] Pandit R. Infield D. Comparative analysis of binning and support vector regression for wind turbine rotor speed based power curve use in condition monitoring. *Proc. – 2018 53rd Int. Univ. Power Eng. Conf. UPEC 2018*; 1–6. DOI: 10.1109/UPEC.2018.8542057.

[10] Global Wind Energy Council. *Global Wind Report* 2022. Available online: https://gwec.net/global-wind-report-2022:text:Total global wind power capacity,carbon emission of South America. Accessed [03/03/2022].

[11] Karamichailidou, D. V., Kaloutsa, V., Alexandridis, A. Wind turbine power curve modeling using radial basis function neural networks and tabu search. *Renew. Energy* 2021; 163: 2137–2152.

[12] Lydia, M., Kumar, S. S., Selvakumar, A. I., Prem Kumar, G. E. A comprehensive review on wind turbine power curve modeling techniques. *Renew. Sustain. Energy Rev.* 2014; 30: 452–460.

[13] Manwell, J. F., McGowan, J. G., Rogers, A. L. *Wind Energy Explained (Theory, Design and Application)*. Second Edition. 2010. ISBN: 978-0-470-01500-1.

[14] Jangamshetti, S. H., Guruprasada Rau, V. Normalized power curves as a tool for identification of optimum wind turbine generator parameters. *IEEE Trans. Energy Convers.* 2001; 16 (3): 283–288.

[15] Norgaard, P., Holttinen, H. A multi-turbine power curve approach. *Nord. Wind Power Conf.* 2004; 1–2. Available Online: http://scholar.google.com/scholar?hl=en&btnG=Search&q=intitle:A+Multi-Turbine+Power+Curve+Approach#0.

[16] Ohunakin, O. S., Henry, E. U., Matthew O. J., Ezekiel V. U., Adelekan, D. S., Oyeniran, A. T. Conditional monitoring and fault detection of wind turbines based on Kolmogorov – Smirnov non-parametric test. *Energy Rep.* 2024; 11: 2577–2591.

[17] Kusiak, A., Zheng, H., Song Z. On-line monitoring of power curves. *Renew. Energy* 2009; 34 (6): 1487–1493.

[18] Uluyol, O., Parthasarathy, G., Foslien, W., Kim, K. Power curve analytic for wind turbine performance monitoring and prognostics. *Proc. Annu. Conf. Progn. Heal. Manag. Soc.* 2011; 435–442.

[19] Pelletier, F., Masson, C., Tahan A. Wind turbine power curve modelling using artificial neural network. *Renew. Energy* 2015; 89: 207–214.

[20] IEC 61400-12-1:2017 RLV. *Wind Energy Generation Systems – Part 12-1: Power Performance Measurements of Electricity Producing Wind Turbines*. 2005; 1–8.

[21] Xu, K., Yan, J., Zhang, H., Zhang, H., Han, S., Liu, Y. Quantile based probabilistic wind turbine power curve model. *Appl. Energy* 2021; 296: 116913. DOI: 10.1016/j.apenergy.2021.116913.

[22] Llombart, A., Watson, S. J., Llombart, D., Fandos, J. M. Power curve characterization I: Improving the bin method. *J. Renew. Energy Power Qual.* 2005; 1 (3): 367–371.

[23] Gottschall, J., Peinke J. Stochastic modelling of a wind turbine's power output with special respect to turbulent dynamics. *J. Phys. Conf. Ser.* 2007; 75 (012045): 1–8.

[24] Shokrzadeh, S., Jafari Jozani, M., Bibeau E. Wind turbine power curve modeling using advanced parametric and nonparametric methods. *IEEE Trans. Sustain. Energy* 2014; 5 (4): 1262–1269.

[25] Marčiukaitis, M., Žutautaitė, I., Martišauskas, L., Jokšas, B., Gecevičius, G., Sfetsos, A. Non-linear regression model for wind turbine power curve. *Renew. Energy* 2017; 113: 732–741.

[26] Kusiak, A., Verma A. Monitoring wind farms with performance curves. *IEEE Trans. Sustain. Energy* 2013; 4 (1): 192–199.

[27] Pinson, P., Nielsen, H. A., Madsen H. Robust estimation of time-varying coefficient functions – Application to the modeling of wind power production. *Math. Model* 2007; 57 (1) 1–36. Available Online: www2.compute.dtu.dk/~ppin/docs/robust_IntProgProject_v4.pdf.

[28] Hernández-Escobedo, Q., Manzano-Agugliaro, F., Zapata-Sierra, A. The wind power of Mexico. *Renew. Sustain. Energy Rev.* 2010; 14 (9): 2830–2840.

[29] Sohoni, V., Gupta, S., Nema, R. A comparative analysis of wind speed probability distributions for wind power assessment of four sites. *Turkish J. Elec. Eng. Comput. Sci.* 2016; 24 (6): 4724–4735.

[30] Lydia, M., Selvakumar, A. I., Kumar, S. S., Kumar G. E. P. Advanced algorithms for wind turbine power curve modeling. *IEEE Trans. Sustain. Energy* 2013; 4(3): 827–835.

[31] Villanueva, D., Feijóo, A. Comparison of logistic functions for modeling wind turbine power curves. *Elec. Power Syst. Res.* 2018; 155: 281–288.

[32] Taslimi-Renani, E., Modiri-Delshad, M., Elias, M. F. M., Rahim, N. A. Development of an enhanced parametric model for wind turbine power curve. *Appl. Energy* 2016; 177: 544–552.

[33] Thapar, V., Agnihotri, G., Sethi, V. K. Critical analysis of methods for mathematical modelling of wind turbines. *Renew. Energy* 2011; 36(11): 3166–3177.

[34] Manobel, B., Sehnke, F., Lazzús, J. A., Salfate, I., Felder, M., Montecinos, S. Wind turbine power curve modeling based on Gaussian Processes and Artificial Neural Networks. *Renew. Energy* 2018; 125: 1015–1020.

[35] Wang, Y., Hu, Q., Meng, D., Zhu, P. Deterministic and probabilistic wind power forecasting using a variational Bayesian-based adaptive robust multi-kernel regression model. *Appl. Energy* 2017; 208: 1097–1112.

[36] Jing, B., Qian, Z., Wang, A., Chen, T., Zhang F. Wind turbine power curve modelling based on hybrid relevance vector machine. *J. Phys. Conf. Ser.* 2020; (1) 1659. DOI: 10.1088/1742-6596/1659/1/012034.

[37] Mehrjoo, M., Jafari Jozani, M., Pawlak, M. Wind turbine power curve modeling for reliable power prediction using monotonic regression. *Renew. Energy* 2020; 147: 214–222.

[38] Goudarzi, A., Davidson, I. E., Ahmadi, A., Venayagamoorthy, G. K. *Intelligent Analysis of Wind Turbine Power Curve Models.* IEEE Symp. Comput. Intell. Appl. Smart Grid, CIASG. 2015. DOI: 10.1109/CIASG.2014.7011548.

[39] Plumley, C. *Kelmarsh Wind Farm Data (0.0.3) [Data set].* 2022. DOI: 10.5281/zenodo.5841834.

[40] Plumley, C. *Penmanshiel Wind Farm Data (0.0.2) [Data set].* 2022. DOI: 10.5281/zenodo.5946808.

[41] Irwine, S. J. C., Ward, M. C., Mullin, J. B. Royal signals and radar establishment. *Emerg. Technol. Situ Proc.* 2012; 139 (4148): 221.

[42] Yang, Y., Wang, P., Gao, X. A novel radial basis function neural network with high generalization performance for nonlinear process modelling. *Processes* 2022; 10 (1): 1–16.

[43] Alexandridis, A., Chondrodima, E. A medical diagnostic tool based on radial basis function classifiers and evolutionary simulated annealing. *J. Biomed. Inform.* 2014; 49: 61–72.

[44] Abirami, S., Chitra, P. Chapter fourteen-energy-efficient edge based real-time healthcare support system. *Adv. Comput.* 2020; 117 (1): 339–368.

[45] Menzies, T., Kocagüneli, E., Minku, L., Peters, F., Turhan, B. Using goals in model-based reasoning. *Shar. Data Model. Softw. Eng.* 2015; 1: 321–353. DOI: 10.1016/b978-0-12-417295-1.00024-2.

[46] www.kdnuggets.com/2020/12/optimization-algorithms-neural-networks.html#:~:text=Optimizersarealgorithmsor methods,problemsbyminimizingthefunction.

[47] https://medium.com/analytics-vidhya/this-blog-post-aims-at-explaining-the-behavior-of-different-algorithms-for-optimizing-gradient-46159a97a8c1.

[48] www.analyticsvidhya.com/blog/2021/10/a-comprehensive-guide-on-deep-learning-optimizers/.

[49] www.statista.com/statistics-glossary/definition/356/quantile/#:~:text=Aquantiledefinesaparticular,)andpercentiles(hundredth).

[50] Arnab, R. Estimation of distribution functions and quantiles. *Surv. Sampl. Theory Appl.* 2017; 747–772. DOI: 10.1016/b978-0-12-811848-1.00023-6.

[51] Park, J. Y., Lee, J. K., Oh, K. Y., Lee, J. S. Development of a novel power curve monitoring method for wind turbines and its field tests. *IEEE Trans. Energy Convers.* 2014; 29 (1): 119–128.

[52] Alexandridis, A., Chondrodima, E., Giannopoulos, N., Sarimveis, H. A fast and efficient method for training categorical radial basis function networks. *IEEE Trans. Neural Networks Learn. Syst.* 2017; 28(11): 2831–2836.

[53] Shetty, R. P., Sathyabhama, A., Pai, P. S. Comparison of modeling methods for wind power prediction: A critical study. *Front. Energy* 2020; 14 (2): 347–358.

[54] Li, T., Liu, X., Lin, Z., and Morrison, R. Ensemble offshore wind turbine power curve modelling – An integration of isolation forest, fast radial basis function neural network, and metaheuristic algorithm. *Energy*, vol. 239, p. 122340, 2022, DOI: 10.1016/j.energy.2021.122340.

11 Review of Solar Cells Deposition Techniques for the Global South

George Chukwuebuka Enebe, Kingsley Ukoba, and Tien-Chien Jen

11.1 INTRODUCTION

The global south refers to regions of countries struggling with development. Top among the challenges experienced in this region is epileptic power supply and poor internet connection. Power supply and internet access are key to the industrialization of this region. Fossil fuel has been utilized in this region causing climate issues. This is despite the presence of natural resources such as sunlight, water, and wind. Solar energy is one of the technologies that can solve the problem of climate change, sustainable energy security, and industrialization in the global south. Solar has not been utilized fully because there are limited funds to purchase high-power-consuming solar cell deposition equipment. This review examines deposition techniques that can be used in the global south base on cost, power consumption, and sustainability.

Spanning from the revelation years of 1839 through 1904, this is 175 years. Adams and Day discovered the Photovoltaic (PV) effect in coagulated selenium in the year 1877, and Hallwachs created a semiconductor-junction solar cell along with copper and copper oxide in 1904 [1,2]. Unfortunately, this was only a moment of discovery, with no genuine understanding of the principles underlying the performance of these early PV systems. Solar cells are electrical appliances that convert solar electricity into direct current (DC) using photovoltaic (PV) panels or mirrors. Bell Labs scientists first created a functioning silicon solar cell in 1954, which produced electrical currents when exposed to sunlight [3–5]. The direct current (DC) produced by solar energy is transformed into alternating current (AC), which is used to power all home appliances. Solar cell electricity is now affordable in many regions, and photovoltaic systems are being used on a big scale to power the electric grid. Solar cells are connected in series or parallel circuits to form larger units known as solar panels or modules, which are referred to as building blocks of solar panels/modules [6–9]. PV modules and arrays can be linked to the electrical grid to form a larger PV system, allowing them to meet any electric power requirement. PV modules and arrays are merely a component of the PV system, with the solar panel increasing, directing, and protecting power [10–13].

The deposition process is an essential component in the creation of innovative thin-film materials, as it dictates virtually all the thin film's characteristics and has the potency to be used to alter/modify existing qualities [14,15]. Deposition procedures must be cautiously chosen based on their application since not all procedures give the same qualities, such as surface morphology, corrosion, electrical, hardness, optical, microstructure, tribological, and biocompatibility. Thin film deposition may be accomplished in a variety of ways. This study discusses and reports the numerous deposition processes and resultant systems for effective thin film deposition. The various deposition techniques confront several problems that impede the creation of high-performance solar cells. These difficulties include, to name a few, the process's operating parameters, the existence of contaminants, nanoparticles produced by homogenous processes, precursors working conditions, the kind of solvent used, and sluggish deposition rates. Yet, the Supplementary materials of this study include further data on the various deposition techniques and their compatibility with solar cell deposition. For thin film solar cell depositions, various deposition procedures can be considered with regard to its affordability in the global south. These deposition techniques include chemical bath deposition (CBD) [16–19], dip coating, electrodeposition [20–24], screen printing [25–32], sol–gel process [33–46], spray pyrolysis [47–54], pulsed laser deposition [55–64], thermal evaporation [65–70], sputtering [71–77], spin coating [78–84].

The focus of this study is on thin film deposition with regards to the global south and the economic viabilities of the deposition process; hence, this report has been cautiously chosen to concentrate more on the various cheap deposition processes used for solar cell deposition and thin film growth suited for desired optimizations.

11.2 SOLAR CELLS

A solar cell is a device that generates electricity when a packet of sunlight falls on it. It is a physical and chemical phenomenon that exhibits a photovoltaic effect because the voltage, current, and resistance vary when sunlight falls on it. Solar cells' operation is based on three basic attributes viz: the absorption of sunlight, exciton (bounding of electron–hole pairs), and plasmons (unbound electron–hole pairs) [85]. Secondly, there is the charge carrier separation, and lastly, there is the extraction of the charge carrier to the external circuit [85,86].

11.2.1 History of Solar Cells

A 19-year-old French physicist, Becquerel experimentally demonstrated a photovoltaic phenomenon inside his father's laboratory [87]. An English electrical engineer discovered the photoconductivity of selenium immediately, and an electric current was sent through it in 1873. This led to the discovery of solar cells, which were used in the early generation television system. Charles Fritts, in 1883, developed the first solid-state solar cells by coating selenium with gold to form a p–n junction. Fritts device achieved a 1% efficiency. Russian scientist Stoletov developed the first cell using the photoelectric effect in 1888. However, the first practical cell was built by Julius Elster and Hans Geitel in 1904. The explanation of the photoelectric effect was made by Albert Einstein in 1905 and later won the Nobel Prize in Physics in 1921. It was in 1941 that the p–n junction was invented by Vadim Lashkaryov in Ag_2S and CuO protocells. Russel Ohl patented the semiconductor solar cells' modern junction culminating in the discovery of transistors in 1946. The first peer-review journal publication was made in 1948 by Kurt Lehovec, entitled "Introduction to the World of Semiconductor States" in *Journal Physical Review*. Bell laboratories' inventors (Daryl Chapin, Pearson Gerald, and Calvin Fuller) demonstrated the first solar cells in 1954. The fourth artificial earth-orbiting satellite, entitled Vanguard 1, incorporated solar cells in 1958, resulting in huge prominence for solar cells.

11.3 SOLAR CELLS DEPOSITION TECHNIQUES

Figure 11.1 illustrates deposition techniques classified as solution-based, physical, and chemical vapour [47].

Gas-phase and solution deposition are two types of chemical deposition. Gas-phase processes include atomic layer epitaxy, chemical vapour deposition (CVD) [88] and atomic layer deposition (ALD) [89]. Solution deposition techniques include spray pyrolysis [47], sol–gel [6], spin coaters [81,83,84,90], and dip-coating [91]. Physical techniques include molecular beam epitaxy [92], pulsed laser deposition [94,95], physical vapour deposition (PVD) [93], and magnetron sputtering

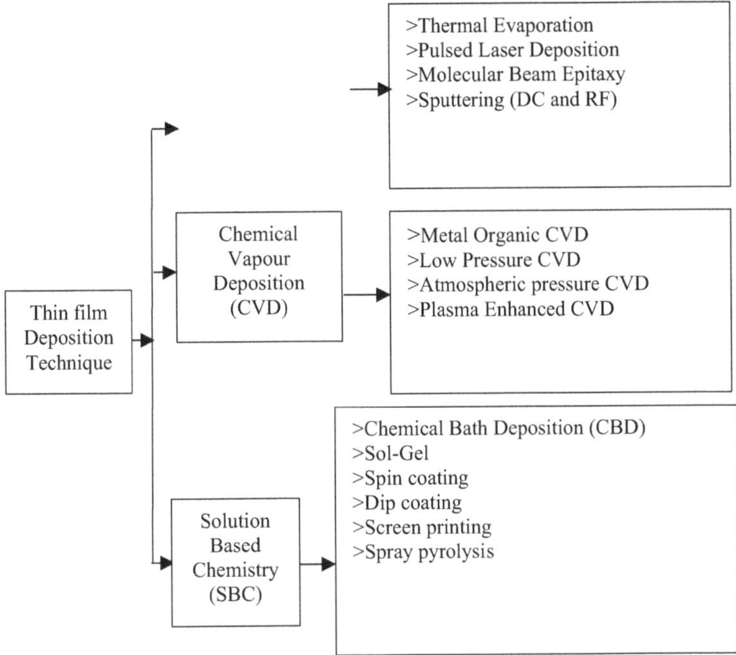

FIGURE 11.1   Classification of thin film deposition methods based on the state of deposition and nature of deposition [47].

TABLE 11.1
Solar Cell Generations

Generation	Features
First Generation	Crystalline Silicon: average efficiency and standard cost Production
Second Generation	Thin film: fewer materials, affordable substrates, cost-effective fabrication equipment, and reduced efficiency.
Third Generation	Photovoltaic solar cells, dye-sensitized solar cells (DSSCs), perovskite solar cells, quantum solar cells, copper zinc tin sulphide (CZTS): emerging technologies with less commercial advancement.

[96,97]. Other methods include anodic oxidation [98–104], advanced reactive gas deposition [105–109], electron beam evaporation [110–117], vacuum evaporation [118], and CBD [119–121]. Table 11.1 shows the key parameters of the various deposition methods for solar cells, as described in Table 11.2.

11.4   DEPOSITION TECHNIQUES FOR GLOBAL SOUTH

The global south is a region struggling with development. At the peak of this is the epileptic power supply and the high cost of products caused by imports from the developed world. This has resulted in a lag in the fabrication of solar cell devices in the global south. On this basis, the following criteria of cost-effectiveness, low power consumption, sustainability, and eco-friendliness were used to discuss deposition techniques for solar cells.

TABLE 11.2
A Summary of Deposition Method [14,15]

Deposition Type	Description	Merit	Parameters	Demerit
		Solution Based		
Chemical Bath Deposition	It is a process of depositing thin film that utilizes an aqueous precursor solution to produce solids from a solution or gas. Chemical Bath Deposition generates homogenous thin films of metal chalcogenides (primarily oxides, sulphides, and selenides)	CBD is generally affordable, well-situated for wide-range deposition, and has the capacity to tune thin film properties by modifying and controlling it.	The temperature of the solution, stirring time, precursor material concentration, and substrate type	There is a waste of solution, and the substrate should be clean.
Dip-coating	It is the process of slowly dipping a substrate into a solution. The substrate is then allowed with a solid thin covering once the liquid component of this solution has dried.	It is affordable and the thickness of the deposited layer can be finetuned easily	viscous drag, the forces of inertia, surface tension, and gravitational force	The process is slow and screen blocking resulting in lower efficiency
Electrodeposition	The method of controlled deposition of material on conducting surfaces using electric current from a solution containing ionic species is known as electrodeposition or electroplating. it has been widely utilized to create thin films with no binder. Electrodeposition can be divided into three types: electroplating, electrophoretic deposition, and underpotential deposition.	the capacity to wrap items with very thin metal layers to enhance their appearance	Content, current density and temperature. Step by step, the electrodeposition rate increases as the current density rises.	Lack of thick shell
Screen Printing	Screen printing otherwise called serigraphy and serigraph printing) is a deposition technique whereby the ink (or dye) is deposited onto a substrate using a mesh, with the exception of regions designated to be impermeable to the ink by a blocking stencil. After dragging a blade or squeegee across the screen to fill the open mesh apertures with ink, a reverse stroke causes the screen to briefly touch the substrate along a line of contact. The ink wets the substrate and is pulled out through the mesh holes as the screen springs back after the blade has passed. Because each color is printed independently, a multi-colored picture or pattern may be created using numerous screens.	The process is simple, costs are inexpensive, and the deposition area is large.	The paste's viscosity, the number of mesh screens, the snap-off distance amidst the substrate, screen and pressure	Low output amount, inadequate ink assistance, and lengthy time for drying are all issues that must be addressed.

Sol–gel	Sol–gel method is a wet-chemical approach for creating solid materials from small molecules. Transforming monomers into a colloidal solution (sol) acts as a precursor for producing an integrated network (or gel) of distinct particles or network polymers.	Simplest, most homogeneous, least expensive, most reliable, most reproducible	amount of water, temperature, pH, solvent, and precursor are all factors to consider.	Precursor costs are high, the process takes a long time, monolith synthesis is difficult, and the process chemistry is difficult to regulate and repeat.
Spray Pyrolysis	Spray pyrolysis is a technique for covering large regions with even layers of thin film. Spray coating is the technique of propelling printing ink through a nozzle, producing a tiny aerosol.	Cost-effectiveness, versatility, thickness range, and processing speed		Non-uniformity of deposited films. Extra resources and cost are incurred to improve the deposited films
Physical Vapour Deposition				
Pulsed laser Deposition	In this physical vapour deposition process, a high-power pulsed laser beam is targeted to contact a target of the appropriate composition. The material is then vaporized and placed as a thin layer on a substrate facing the target.	The main benefit is that it is conceptually simple and versatile.	maximum pulse energy, wavelength	Slow deposition rate, does not support large area deposition, target fragmentation in depositing some materials
Thermal Evaporation	It's one of the most basic types of PVD, and it involves using resistive heat to form a thin film, a solid material is evaporated in a vacuum. The material is then heated in a high vacuum chamber until vapor pressure is reached..	Good purity because of low pressure	The main parameter is the source material's vapour pressure at the evaporation temperature	Firstly, pollution from the boat's outgassing or evaporation, as well as the heating circuits used during the high-temperature evaporation process. Secondly, obtaining a suitable boat material is difficult, especially when evaporating refractory materials. Also, Poor step coverage, harder forming of alloys, and decreased throughput due to low vacuum
Sputtering	Sputtering is a phenomenon that occurs when tiny particles of a solid object are expelled off its surface after being struck by strong plasma or gas particles. Since it happens naturally in space, it can be an unfavourable source of wear in precision components. Yet, because of its capacity to operate on extremely thin layers of material, it is used in research and industry to do precise etching, perform analytical procedures, and deposit thin film layers in the manufacture of optical coatings, semiconductor devices, and nanotechnology goods. It's a type of physical vapour deposition.	Uniform covering across a large area with little contaminants	Operating pressure, direct current power, and substrate temperature are all factors to consider.	The bombardment's target region is too tiny, and the rate of deposition is often poor.

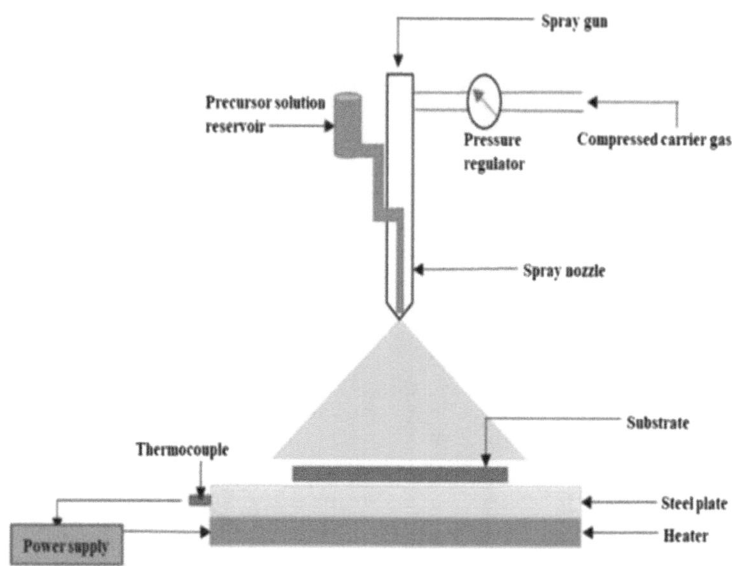

FIGURE 11.2 Representations of a chemical spray pyrolysis setup [14].

11.4.1 SPRAY PYROLYSIS

Spray pyrolysis involves coating large areas with films of very thin layers of identical thickness. Spray coating is the technique of propelling printing ink through a nozzle, producing a tiny aerosol (Figure 11.2). Numerous studies have been conducted using this deposition method for solar cell deposition [122–131].

11.4.2 SCREEN PRINTING

Screen printing (otherwise acknowledged as serigraphy printing and serigraph) comprises a printing technique where the ink (or dye) is deposited on a substrate by the use of a mesh, with the exception of regions made impenetrable to the ink by a blocking stencil [29]. A reverse stroke enables the screen to momentarily touch the substrate along a line of contact after sliding a blade or squeegee over the screen to fill the mesh pores with ink. As the screen springs back after the blade has passed, the ink wets the substrate and is drawn out through the mesh holes. Because each colour is printed independently, a multi-coloured picture or pattern may be created using numerous screens [29,32]. Solar cell investigation shows that several studies have contributed towards solar cell modifications using the screen-printing deposition method [132–141].

11.4.3 CHEMICAL BATH DEPOSITION

It is a thin-film deposition process that uses an aqueous precursor solution to produce solids from a solution or gas. CBD generates homogenous thin films of metal chalcogenides (primarily oxides, sulphides, and selenides) [18,19].

11.4.3.1 Chemical Bath Deposition

It is also known as chemical solution deposition [16]. It is a technique for thin film deposition that uses an aqueous precursor solution to produce solids from a solution or gas [142]. CBD generates homogenous thin films of metal chalcogenides (primarily selenides, sulphides, and oxides) and

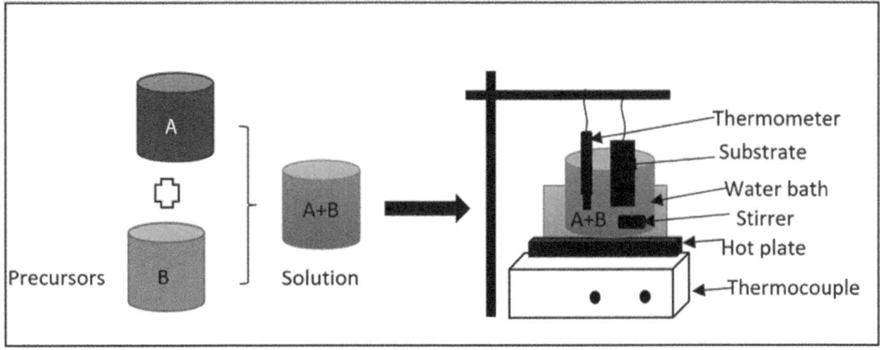

FIGURE 11.3 Representation of the chemical bath deposition process.

a smaller amount of common ionic compounds utilizing heterogeneous nucleation (deposition or adsorption of water ions onto a solid substrate) [143,144]. CBD dependably generates films with a simple technique that requires little equipment at a reduced temperature (100°C) at a reasonable cost [145–149]. CBD can also be used for batch processing or continuous deposition over a vast area [150–153]. CBD films are commonly employed in semiconductors, supercapacitors, and solar cells, which has attracted a growing interest in the area of nanomaterials studies/research. Figure 11.3 shows the representation of the CBD process.

11.4.4 Spin Coating

A major deposition technique used for depositing thin films onto a substrate. It is a centrifugal force method for applying a uniform layer to a solid surface that requires a liquid–vapour interface [81,83]. A thin layer (a few nm to a few μm) is applied by spreading a solution of the desired substance in a solvent uniformly across the surface of a substrate (for example, an "ink") while the substrate is spinning [79]. To put it another way, by creating a thin coating of a solid substance, such as a polymer, a liquid solution is deposited onto a rotating substrate. When the substrate is spun at an increased speed (usually >10 revolutions per second = 600 rpm), the centripetal force integrates with the surface tension of the solution, forcing the liquid coating to form a uniform layer. During this period, the solvent evaporates, producing an even coating of the targeted material on the substrate. When the material and solvent solution are spun at faster rotation per minute (RPM), surface tension and the centripetal force of the liquid combine to generate a homogeneous coating [79]. A small amount of coating material is often applied to the substrate's core, which rotates slowly or not at all. The coating substance is subsequently disseminated by centrifugal force while the substrate is rotated at speeds up to 10,000 rpm. A spin coater, or simply a spinner, is a machine used for spin coating [81,83,154]. When the fluid spins off the edges of the substrate, the spinning remains till the desired film thickness is attained. The solvent used is generally volatile and evaporates simultaneously. Parameters such as the viscosity and concentration of the solution, as well as the solvent, influence the film thickness.

Deposition, spin-up, spin-off, and evaporation are the four basic phases in this process. Firstly, the solution is cast onto the substrate with a pipette. The solution will then be dispersed throughout the substrate by centrifugal motion, the substrate is then allowed to spin (dynamic spin coating) or after deposition (static spin coating).

The substrate is then adjusted to be at a suitable pace, either instantaneously or following the slow spread step. At this time, most of the solution has been expelled from the substrate. Initially, the fluid will spin at a higher pace than the substrate; however, as soon as drag equalizes rotational

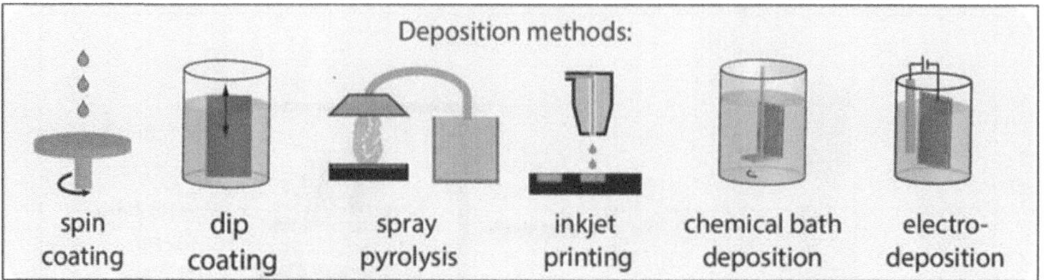

FIGURE 11.4   Schematic summary of various deposition methods.

accelerations, rotation velocity will finally balance, leading to the fluid being uniform. When viscous forces now dominate the fluid, it starts to form thin [155]. When the fluid spills off, the film usually changes colour due to conflicting effects. When the colour stops shifting, this signals that the film is dry. Since the fluid must form droplets at the edge to be flung off, edge effects could occur. Ultimately, fluid-flowing ends and solvent evaporation dominates thinning [150]. The pace at which the solvent evaporation is influenced by the solvent's volatility, surrounding environment, and vapour pressure. Inconsistencies in the evaporation rate, such as along the substrate's edge, will lead to inconsistencies in the film [79,82,90,155].

11.4.5 Dip Coating

It is the process of slowly dipping a substrate into a solution. The substrate is left with a solid thin covering once the liquid component of this solution has dried.

Figure 11.4 is the schematic summary of the selected deposition technique of solar cells for use in the global south.

11.5 CONCLUSION

The need for cheap and earthly abundant material for solar cell fabrication necessitates the use of affordable deposition techniques to accommodate and satisfy the growing demand for power generation. Thin films have had a significant influence on modern technologies, providing the foundation for sophisticated applications in a variety of disciplines. The structure and stability of the deposits are significant issues for all functions, and deposition processes can be used to deposit thin layers. Adapting and tailoring these surface qualities to fulfil a specific requirement for improved performance is possible and has been widely used in several aspects of life. This may be accomplished by covering the surface using thin film deposition. This study conducts a literature evaluation on several solar cell deposition processes utilized for surface modification. The two key fields of interest mentioned in this paper are solar cells and affordable deposition processes. This study observed that the most suitable deposition techniques for the global south include the chemical bath technique, dip coating, screen printing, spray pyrolysis, and spin coating. These deposition techniques were considered for the global south due to their affordability, accessibility, and ease of the process.

ACKNOWLEDGEMENT

Authors acknowledge the funding of the Commonwealth University of Johannesburg, the National Research Foundation of South Africa (NRF), and the University Research Council (URC) of the University of Johannesburg.

REFERENCES

[1] Fraas, L. M., & Fraas, L. M. (2014). History of solar cell development. *Low-Cost Solar Electric Power*, 1–12.

[2] Kim, H., Nam, S., Jeong, J., Lee, S., Seo, J., Han, H., & Kim, Y. (2014). Organic solar cells based on conjugated polymers: History and recent advances. *Korean Journal of Chemical Engineering, 31*, 1095–1104.

[3] Heller, A. (1981). Conversion of sunlight into electrical power and photoassisted electrolysis of water in photoelectrochemical cells. *Accounts of Chemical Research, 14*(5), 154–162.

[4] Bihouix, P. (2020). *The Age of Low Tech: Towards a Technologically Sustainable Civilization.* Policy Press.

[5] Brusso, B. C. (2019). A brief history of the energy conversion of light [history]. *IEEE Industry Applications Magazine, 25*(4), 8–13.

[6] Lukong, V. T., Mouchou, R. T., Enebe, G. C., Ukoba, K., & Jen, T. C. (2022). Deposition and characterization of self-cleaning TiO_2 thin films for photovoltaic application. *Materials Today: Proceedings, 62*, S63–S72.

[7] Enebe, G. C., Lukong, V. T., Mouchou, R. T., Ukoba, K. O., & Jen, T. C. (2022). Optimizing nanostructured TiO_2/Cu_2O pn heterojunction solar cells using SCAPS for fourth industrial revolution. *Materials Today: Proceedings, 62*, S145–S150.

[8] Enebe, G. C., Ukoba, K., & Jen, T. C. (2019). Numerical modeling of effect of annealing on nanostructured CuO/TiO 2 pn heterojunction solar cells using SCAPS. *AIMS Energy, 7*(4), 527–538.

[9] Ukoba, K. O., Eloka-Eboka, A. C., & Inambao, F. L. (2018). Review of nanostructured NiO thin film deposition using the spray pyrolysis technique. *Renewable and Sustainable Energy Reviews, 82*, 2900–2915.

[10] El-Dein, M. S., Kazerani, M., & Salama, M. M. A. (2012). Optimal photovoltaic array reconfiguration to reduce partial shading losses. *IEEE Transactions on Sustainable Energy, 4*(1), 145–153.

[11] Glavin, M. E., & Hurley, W. G. (2012). Optimisation of a photovoltaic battery ultracapacitor hybrid energy storage system. *Solar Energy, 86*(10), 3009–3020.

[12] Köntges, M., Kunze, I., Kajari-Schröder, S., Breitenmoser, X., & Bjørneklett, B. (2011). The risk of power loss in crystalline silicon based photovoltaic modules due to micro-cracks. *Solar Energy Materials and Solar Cells, 95*(4), 1131–1137.

[13] Woyte, A., Nijs, J., & Belmans, R. (2003). Partial shadowing of photovoltaic arrays with different system configurations: literature review and field test results. *Solar Energy, 74*(3), 217–233.

[14] Emeka, N. C., Imoisili, P. E., & Jen, T. C. (2020). Preparation and characterization of Nb x O y thin films: A review. *Coatings, 10*(12), 1246.

[15] Nwanna, E. C., Imoisili, P. E., & Jen, T. C. (2020). Fabrication and synthesis of SnO X thin films: A review. *The International Journal of Advanced Manufacturing Technology, 111*, 2809–2831.

[16] Hodes, G. (2007). Semiconductor and ceramic nanoparticle films deposited by chemical bath deposition. *Physical Chemistry Chemical Physics, 9*(18), 2181–2196.

[17] Lokhande, C. D., Ennaoui, A., Patil, P. S., Giersig, M., Diesner, K., Muller, M., & Tributsch, H. (1999). Chemical bath deposition of indium sulphide thin films: Preparation and characterization. *Thin Solid Films, 340*(1–2), 18–23.

[18] Nair, P. K., Nair, M. T. S., García, V. M., Arenas, O., Peña, Y., Castillo, A., ... & Rincon, M. E. (1998). Semiconductor thin films by chemical bath deposition for solar energy related applications. *Solar Energy Materials and Solar Cells, 52*(3–4), 313–344.

[19] Cheng, J., Fan, D., Wang, H., Liu, B., Zhang, Y., & Yan, H. (2003). Chemical bath deposition of crystalline ZnS thin films. *Semiconductor Science and Technology, 18*(7), 676.

[20] Lincot, D. (2005). Electrodeposition of semiconductors. *Thin Solid Films, 487*(1–2), 40–48.

[21] Brenner, A. (2013). *Electrodeposition of Alloys: Principles and Practice.* Elsevier.

[22] Gamburg, Y. D., & Zangari, G. (2011). *Theory and Practice of Metal Electrodeposition.* Springer Science & Business Media.

[23] Schwarzacher, W. (2006). Electrodeposition: a technology for the future. *Electrochemical Society Interface, 15*(1), 32–35.

[24] Pandey, R. K., Sahu, S. N., & Chandra, S. (2017). *Handrook of Semiconductor Electrodeposition.* CRC Press.

[25] Kim, Y. D., & Hone, J. (2017). Screen printing of 2D semiconductors. *Nature, 544*(7649), 167–168.

[26] Kazani, I., Hertleer, C., De Mey, G., Schwarz, A., Guxho, G., & Van Langenhove, L. (2012). Electrical conductive textiles obtained by screen printing. *Fibres & Textiles in Eastern Europe, 20*(1), 57–63.

[27] Jabbour, G. E., Radspinner, R., & Peyghambarian, N. (2001). Screen printing for the fabrication of organic light-emitting devices. *IEEE Journal of Selected Topics in Quantum Electronics, 7*(5), 769–773.

[28] Riemer, D. E. (1989). The theoretical fundamentals of the screen printing process. *Microelectronics International, 6*(1), 8–17.

[29] Adam, R., & Robertson, C. (2003). *Screen Printing*. Thames & Hudson.

[30] Zavanelli, N., & Yeo, W. H. (2021). Advances in screen printing of conductive nanomaterials for stretchable electronics. *ACS Omega, 6*(14), 9344–9351.

[31] Bull, C. L., Gleeson, D., & Knight, K. S. (2003). Determination of B-site ordering and structural transformations in the mixed transition metal perovskites La_2CoMnO_6 and La_2NiMnO_6. *Journal of Physics: Condensed Matter, 15*(29), 4927.

[32] Riemer, D. E. (1989). The theoretical fundamentals of the screen printing process. *Microelectronics International, 6*(1), 8–17.

[33] Hart, J. N., Menzies, D., Cheng, Y. B., Simon, G. P., & Spiccia, L. (2006). TiO_2 sol–gel blocking layers for dye-sensitized solar cells. *Comptes Rendus Chimie, 9*(5–6), 622–626.

[34] Oh, H., Krantz, J., Litzov, I., Stubhan, T., Pinna, L., & Brabec, C. J. (2011). Comparison of various sol–gel derived metal oxide layers for inverted organic solar cells. *Solar Energy Materials and Solar Cells, 95*(8), 2194–2199.

[35] Lee, S. C., Lee, J. H., Oh, T. S., & Kim, Y. H. (2003). Fabrication of tin oxide film by sol–gel method for photovoltaic solar cell system. *Solar Energy Materials and Solar Cells, 75*(3–4), 481–487.

[36] Baik, D. G., & Cho, S. M. (1999). Application of sol-gel derived films for ZnO/n-Si junction solar cells. *Thin Solid Films, 354*(1–2), 227–231.

[37] Zhu, Z., Bai, Y., Zhang, T., Liu, Z., Long, X., Wei, Z., ... & Yang, S. (2014). High-performance hole-extraction layer of sol–gel-processed NiO nanocrystals for inverted planar perovskite solar cells. *Angewandte Chemie International Edition, 53*(46), 12571–12575.

[38] Rani, S., Suri, P., Shishodia, P. K., & Mehra, R. M. (2008). Synthesis of nanocrystalline ZnO powder via sol–gel route for dye-sensitized solar cells. *Solar Energy Materials and Solar Cells, 92*(12), 1639–1645.

[39] Sun, Y., Seo, J. H., Takacs, C. J., Seifter, J., & Heeger, A. J. (2011). Inverted polymer solar cells integrated with a low-temperature-annealed sol-gel-derived ZnO film as an electron transport layer. *Advanced Materials, 23*(14), 1679–1683.

[40] Dimitriev, Y., Ivanova, Y., & Iordanova, R. (2008). History of sol-gel science and technology. *Journal of the University of Chemical Technology and Metallurgy, 43*(2), 181–192.

[41] Brinker, C. J., & Scherer, G. W. (2013). *Sol-gel Science: The Physics and Chemistry of Sol-Gel Processing*. Academic press.

[42] Pierre, A. C. (2020). *Introduction to Sol-Gel Processing*. Springer Nature.

[43] Hench, L. L., & West, J. K. (1990). The sol-gel process. *Chemical Reviews, 90*(1), 33–72.

[44] Sakka, S., & Kozuka, H. (Eds.). (2005). *Handbook of Sol-Gel Science and Technology. 1. Sol-Gel Processing* (Vol. 1). Springer Science & Business Media.

[45] Hench, L. L., & West, J. K. (1990). The sol-gel process. *Chemical Reviews, 90*(1), 33–72.

[46] Brinker, C. J., & Scherer, G. W. (2013). *Sol-Gel Science: The Physics and Chemistry of Sol-Gel Processing*. Academic press.

[47] Ukoba, K. O., Eloka-Eboka, A. C., & Inambao, F. L. (2018). Review of nanostructured NiO thin film deposition using the spray pyrolysis technique. *Renewable and Sustainable Energy Reviews, 82*, 2900–2915.

[48] Filipovic, L., Selberherr, S., Mutinati, G. C., Brunet, E., Steinhauer, S., Köck, A., ... & Schrank, F. (2014). Methods of simulating thin film deposition using spray pyrolysis techniques. *Microelectronic Engineering, 117*, 57–66.

[49] Ashour, A., Kaid, M. A., El-Sayed, N. Z., & Ibrahim, A. A. (2006). Physical properties of ZnO thin films deposited by spray pyrolysis technique. *Applied Surface Science, 252*(22), 7844–7848.

[50] Perednis, D., & Gauckler, L. J. (2005). Thin film deposition using spray pyrolysis. *Journal of Electroceramics, 14*, 103–111.

[51] Kaid, M. A., & Ashour, A. (2007). Preparation of ZnO-doped Al films by spray pyrolysis technique. *Applied Surface Science, 253*(6), 3029–3033.

[52] Perednis, D., & Gauckler, L. J. (2005). Thin film deposition using spray pyrolysis. *Journal of Electroceramics, 14*, 103–111.

[53] Kamoun, N., Bouzouita, H., & Rezig, B. J. T. S. F. (2007). Fabrication and characterization of Cu_2ZnSnS_4 thin films deposited by spray pyrolysis technique. *Thin Solid Films, 515*(15), 5949–5952.

[54] Patil, P. S. (1999). Versatility of chemical spray pyrolysis technique. *Materials Chemistry and Physics, 59*(3), 185–198.

[55] Chrisey, D. B., & Hubler, G. K. (Eds.). (1994). *Pulsed Laser Deposition of Thin Films. Handbook of Laser Technology and Applications* (CRC Press/Taylor and Francis Group.

[56] Greer, J. A. (2013). History and current status of commercial pulsed laser deposition equipment. *Journal of Physics D: Applied Physics, 47*(3), 034005.

[57] Blank, D. H., Dekkers, M., & Rijnders, G. (2013). Pulsed laser deposition in Twente: from research tool towards industrial deposition. *Journal of Physics D: Applied Physics, 47*(3), 034006.

[58] Lowndes, D. H., Geohegan, D. B., Puretzky, A. A., Norton, D. P., & Rouleau, C. M. (1996). Synthesis of novel thin-film materials by pulsed laser deposition. *Science, 273*(5277), 898–903.

[59] Smagorinsky, J. (1963). General circulation experiments with the primitive equations: I. The basic experiment. *Monthly Weather Review, 91*(3), 99–164.

[60] Aziz, M. J. (2008). Film growth mechanisms in pulsed laser deposition. *Applied Physics A, 93*, 579–587.

[61] Christen, H. M., & Eres, G. (2008). Recent advances in pulsed-laser deposition of complex oxides. *Journal of Physics: Condensed Matter, 20*(26), 264005.

[62] Aziz, M. J. (2008, Nov.). Film growth mechanisms in pulsed laser deposition. *Applied Physics A: Materials Science & Processing, 93*(3), 579–587. DOI: 10.1007/S00339-008-4696-7

[63] Eason, R. (Ed.). (2007). *Pulsed Laser Deposition of Thin Films: Applications-Led Growth of Functional Materials*. John Wiley & Sons.

[64] Chrisey, D. B., & Hubler, G. K. (Eds.). (1994). *Pulsed Laser Deposition of Thin Films. Handbook of Laser Technology and Applications* (CRC Press /Taylor and Francis Group.

[65] Dai, Z. R., Pan, Z. W., & Wang, Z. L. (2003). Novel nanostructures of functional oxides synthesized by thermal evaporation. *Advanced Functional Materials, 13*(1), 9–24.

[66] Indirajith, R., Srinivasan, T. P., Ramamurthi, K., & Gopalakrishnan, R. (2010). Synthesis, deposition and characterization of tin selenide thin films by thermal evaporation technique. *Current Applied Physics, 10*(6), 1402–1406.

[67] Mohammed, M. A., Abdulridha, W. M., & Abd, A. N. (2018). Thickness effect on some physical properties of the Ag thin films prepared by thermal evaporation technique. *Journal of Global Pharma Technology, 10*(3), 613–619.

[68] Pan, C. A., & Ma, T. P. (1980). High-quality transparent conductive indium oxide films prepared by thermal evaporation. *Applied Physics Letters, 37*(2), 163–165.

[69] Timoumi, A., Bouzouita, H., & Rezig, B. (2008). Synthesis and characterization of In_2S_3: Na thin films prepared by vacuum thermal evaporation technique for photovoltaic applications. *The European Physical Journal-Applied Physics, 42*(3), 187–191.

[70] Pan, C. A., & Ma, T. P. (1980). High-quality transparent conductive indium oxide films prepared by thermal evaporation. *Applied Physics Letters, 37*(2), 163–165. DOI: 10.1063/1.91809

[71] Depla, D., Mahieu, S., & Greene, J. E. (2010). Sputter deposition processes. In *Handbook of Deposition Technologies for Films and Coatings* (pp. 253–296). William Andrew Publishing.

[72] Rossnagel, S. (2001). Sputtering and sputter deposition. In *Handbook of Thin Film Deposition Processes and Techniques* (pp. 319–348). William Andrew Publishing.

[73] Greene, J. E. (2017, Sep.). Review article: Tracing the recorded history of thin-film sputter deposition: From the 1800s to 2017. *Journal of Vacuum Science & Technology A: Vacuum, Surfaces, and Films, 35*(5), 05C204. DOI: 10.1116/1.4998940

[74] Windischmann, H. (1992, Jan.). Intrinsic stress in sputter-deposited thin films. *Critical Reviews in Solid State and Materials Sciences, 17*(6), 547–596. DOI: 10.1080/10408439208244586

[75] Rossnagel, S. M. (1999). Sputter deposition for semiconductor manufacturing. *IBM Journal of Research and Development, 43*(1.2), 163–179.

[76] Wasa, K., & Hayakawa, S. (2023). *Handbook of sputter deposition technology, 1992.* Accessed: March 22, 2023. [Online]. Available: www.osti.gov/biblio/7105337

[77] Thornton, J. A. (1986, Nov.). The microstructure of sputter-deposited coatings. *Journal of Vacuum Science & Technology A: Vacuum, Surfaces, and Films, 4*(6), 3059–3065. DOI: 10.1116/1.573628

[78] Bornside, D. E. Macosko, C. W., & Scriven, L. E. (1989). Spin coating: One-dimensional model. *Journal of Applied Physics, 66*(11), 5185–5193. DOI: 10.1063/1.343754

[79] Scriven, L. E. (1988). Physics and applications of dip coating and spin coating. *MRS Online Proceedings Library (OPL), 121,* 717.

[80] Tyona, M. D. (2013). A theoritical study on spin coating technique. *Advances in Materials Research, 2*(4), 195.

[81] Larson, R. G., & Rehg, T. J. (1997). Spin coating. *Liquid Film Coating,* 709–734. DOI: 10.1007/978-94-011-5342-3_20

[82] Sahu, N., Parija, B., & Panigrahi, S. (2009, Apr.). Fundamental understanding and modeling of spin coating process: A review. *Indian Journal of Physics, 83*(4), 493–502. DOI: 10.1007/S12648-009-0009-Z

[83] Birnie, D. P. (2004). Spin coating technique. *Sol-Gel Technologies for Glass Producers and Users,* 49–55. DOI: 10.1007/978-0-387-88953-5_4

[84] Bornside, D. E., Macosko, C. W., & Scriven, L. E. (1987). Modeling of SPIN coating. *Journal of Imaging Technology, 13*(4), 122–130.

[85] Fonash, S. (2012). *Solar Cell Device Physics.* Elsevier.

[86] Kaushika, N., Mishra, A., & Rai, A. (2018). *Solar Photovoltaics.* Accessed: March 22, 2023. [Online]. Available: https://link.springer.com/content/pdf/10.1007/978-3-319-72404-1.pdf

[87] Fayeez, A. T. I., Gannapathy, V. R., Isa, I. S. M., Nor, M. K., & Azyze, N. L. (2015). Literature review of battery-powered and solar-powered wireless sensor node. *Asian Research Publishing Network-Journal of Engineering and Applied Sciences, 10*(2), 671–677.

[88] Patra, C., & Das, D. (2022). Room temperature synthesized highly conducting B-doped nanocrystalline silicon thin films on flexible polymer substrates by ICP-CVD. *Applied Surface Science, 583,* 152499.

[89] Oviroh, P. O., Akbarzadeh, R., Pan, D., Alfred, R., Coetzee, M., & Jen, T.-C. (2019 Dec.). New development of atomic layer deposition: Processes, methods and applications. *Taylor & Francis, 20*(1), 465–496. DOI: 10.1080/14686996.2019.1599694

[90] Shafi, M. A., Bouich, A., Fradi, K., Guaita, J. M., Khan, L., & Mari, B. (2022). Effect of deposition cycles on the properties of ZnO thin films deposited by spin coating method for CZTS-based solar cells. *Optik, 258,* 168854.

[91] Zaremba, O. T., Goldt, A. E., Khabushev, E. M., Anisimov, A. S., & Nasibulin, A. G. (2022). Highly efficient doping of carbon nanotube films with chloroauric acid by dip-coating. *Materials Science and Engineering: B, 278,* 115648.

[92] Richards, R. D., Bailey, N. J., Liu, Y., Rockett, T. B. O., & Mohmad, A. R. (2022, Feb.). GaAsBi: From molecular beam epitaxy growth to devices. *Physica Status Solidi (B): Basic Research, 259*(2). DOI: 10.1002/PSSB.202100330

[93] Dan, A., Bijalwan, P. K., Pathak, A. S., & Bhagat, A. N. (2022, Mar.). A review on physical vapor deposition-based metallic coatings on steel as an alternative to conventional galvanized coatings. *Journal of Coatings Technology and Research, 19*(2), pp. 403–438, DOI: 10.1007/S11998-021-00564-Z

[94] Albu, D. F., Lungu, J., Popescu-Pelin, G., Mihăilescu, C. N., Socol, G., Georgescu, A., ... & Mihailescu, I. N. (2022). Thin film fabrication by pulsed laser deposition from TiO_2 Targets in O_2, N_2, He, or Ar for dye-sensitized solar cells. *Coatings, 12*(3), 293.

[95] Bjelajac, A., Petrović, R., Stan, G. E., Socol, G., Mihailescu, A., Mihailescu, I. N., ... & Janaćković, D. (2022). C-doped TiO_2 nanotubes with pulsed laser deposited Bi_2O_3 films for photovoltaic application. *Ceramics International, 48*(4), 4649–4657.

[96] Li, J., Ren, G. K., Chen, J., Chen, X., Wu, W., Liu, Y., ... & Shi, Y. (2022). Facilitating complex thin film deposition by using magnetron sputtering: A review. *JOM, 74*(8), 3069–3081.

[97] Si, Y., Wang, G., Wen, M., Tong, Y., Wang, W., Li, Y., ... & Ren, P. (2022). Corrosion and friction resistance of TiVCrZrWNx high entropy ceramics coatings prepared by magnetron sputtering. *Ceramics International, 48*(7), 9342–9352.

[98] Stergiopoulos, T., Ghicov, A., Likodimos, V., Tsoukleris, D. S., Kunze, J., Schmuki, P., & Falaras, P. (2008). Dye-sensitized solar cells based on thick highly ordered TiO_2 nanotubes produced by controlled anodic oxidation in non-aqueous electrolytic media. *Nanotechnology, 19*(23), 235602.

[99] Du, Y., Cai, H., Wen, H., Wu, Y., Huang, L., Ni, J., ... & Zhang, J. (2016). Novel combination of efficient perovskite solar cells with low temperature processed compact TiO_2 layer via anodic oxidation. *ACS Applied Materials & Interfaces, 8*(20), 12836–12842.

[100] Chen, C. C., Chung, H. W., Chen, C. H., Lu, H. P., Lan, C. M., Chen, S. F., ... & Diau, E. W. G. (2008). Fabrication and characterization of anodic titanium oxide nanotube arrays of controlled length for highly efficient dye-sensitized solar cells. *The Journal of Physical Chemistry C, 112*(48), 19151–19157.

[101] Fortin, E., & Masson, D. (1982). Photovoltaic effects in Cu_2O Cu solar cells grown by anodic oxidation. *Solid-State Electronics, 25*(4), 281–283.

[102] Yang, D. J., Park, H., Cho, S. J., Kim, H. G., & Choi, W. Y. (2008). TiO_2-nanotube-based dye-sensitized solar cells fabricated by an efficient anodic oxidation for high surface area. *Journal of Physics and Chemistry of Solids, 69*(5–6), 1272–1275.

[103] Li, H. H., Chen, R. F., Ma, C., Zhang, S. L., An, Z. F., & Huang, W. (2011). Titanium oxide nanotubes prepared by anodic oxidation and their application in solar cells. *Acta Physico-Chimica Sinica, 27*(5), 1017–1025.

[104] Chen, C. C., Chung, H. W., Chen, C. H., Lu, H. P., Lan, C. M., Chen, S. F., ... & Diau, E. W. G. (2008). Fabrication and characterization of anodic titanium oxide nanotube arrays of controlled length for highly efficient dye-sensitized solar cells. *The Journal of Physical Chemistry C, 112*(48), 19151–19157.

[105] Reyes, L. F., Saukko, S., Hoel, A., Lantto, V., & Granqvist, C. G. (2004). Structure engineering of WO_3 nanoparticles for porous film applications by advanced reactive gas deposition. *Journal of the European Ceramic Society, 24*(6), 1415–1419.

[106] Reyes, L. F., Hoel, A., Saukko, S., Heszler, P., Lantto, V., & Granqvist, C. G. (2006). Gas sensor response of pure and activated WO_3 nanoparticle films made by advanced reactive gas deposition. *Sensors and Actuators B: Chemical, 117*(1), 128–134.

[107] Solis, J. L., Hoel, A., Kish, L. B., Granqvist, C. G., Saukko, S., & Lantto, V. (2001). Gas-sensing properties of nanocrystalline WO_3 films made by advanced reactive gas deposition. *Journal of the American Ceramic Society, 84*(7), 1504–1508.

[108] Cindemir, U., Topalian, Z., Granqvist, C. G., Österlund, L., & Niklasson, G. A. (2019). Characterization of nanocrystalline-nanoporous nickel oxide thin films prepared by reactive advanced gas deposition. *Materials Chemistry and Physics, 227*, 98–104.

[109] Jilani, A., Abdel-Wahab, M. S., & Hammad, A. H. (2017). Advance deposition techniques for thin film and coating. *Modern Technologies for Creating the Thin-film Systems and Coatings, 2*(3), 137–149.

[110] Agarwal, D. C., Chauhan, R. S., Kumar, A., Kabiraj, D., Singh, F., Khan, S. A., ... & Satyam, P. V. (2006). Synthesis and characterization of ZnO thin film grown by electron beam evaporation. *Journal of Applied Physics, 99*(12), 123105.

[111] George, J., & Menon, C. S. (2000). Electrical and optical properties of electron beam evaporated ITO thin films. *Surface and Coatings Technology, 132*(1), 45–48.

[112] Zhou, W., Zhang, J., Liu, Y., Li, X., Niu, X., Song, Z., ... & Feng, S. (2008). Characterization of anti-adhesive self-assembled monolayer for nanoimprint lithography. *Applied Surface Science, 255*(5), 2885–2889.

[113] Shamala, K. S., Murthy, L. C. S., & Rao, K. N. (2004). Studies on tin oxide films prepared by electron beam evaporation and spray pyrolysis methods. *Bulletin of Materials Science, 27*(3), 295–301. DOI: 10.1007/BF02708520

[114] Tanuševski, A., & Poelman, D. (2003). Optical and photoconductive properties of SnS thin films prepared by electron beam evaporation. *Solar Energy Materials and Solar Cells, 80*(3), 297–303.

[115] Kuroyanagi, A. (1989). Properties of aluminum-doped ZnO thin films grown by electron beam evaporation. *Japanese Journal of Applied Physics, 28*(2R), 219–222. DOI: 10.1143/JJAP.28.219/META

[116] Al-Kuhaili, M. F., Durrani, S. M. A., & Khawaja, E. E. (2004). Characterization of hafnium oxide thin films prepared by electron beam evaporation. *Journal of Physics D: Applied Physics, 37*(8), 1254.

[117] Lorenz, P., Zieger, G., Dellith, J., & Schmidt, H. (2022). Electron beam co-deposition of thermoelectric BiSb thin films from two separate targets. *Thin Solid Films, 745,* 139082.

[118] Mahana, D., Mauraya, A. K., Pal, P., Singh, P., & Muthusamy, S. K. (2022). Comparative study on surface states and CO gas sensing characteristics of CuO thin films synthesised by vacuum evaporation and sputtering processes. *Materials Research Bulletin, 145,* 111567.

[119] Carrillo-Castillo, A., Rivas-Valles, B. G., Castillo, S. J., Ramirez, M. M., & Luque-Morales, P. A. (2022). New formulation to synthetize semiconductor Bi_2S_3 thin films using chemical bath deposition for optoelectronic applications. *Symmetry, 14*(12), 2487. doi.org/10.3390/sym14122487

[120] Fazal, T., Iqbal, S., Shah, M., Bahadur, A., Ismail, B., Abd-Rabboh, H. S., ... & Qayyum, M. A. (2022). Deposition of bismuth sulfide and aluminum doped bismuth sulfide thin films for photovoltaic applications. *Journal of Materials Science: Materials in Electronics,* 1–12.

[121] Fazal, T., Iqbal, S., Shah, M., Ismail, B., Shaheen, N., Alharthi, A. I., ... & Ibrahium, H. A. (2022). Correlation between structural, morphological and optical properties of Bi_2S_3 thin films deposited by various aqueous and non-aqueous chemical bath deposition methods. *Results in Physics, 40,* 105817.

[122] Pavan, M., Rühle, S., Ginsburg, A., Keller, D. A., Barad, H. N., Sberna, P. M., ... & Fortunato, E. (2015). TiO_2/Cu_2O all-oxide heterojunction solar cells produced by spray pyrolysis. *Solar Energy Materials and Solar Cells, 132,* 549–556.

[123] Ryo, T., Nguyen, D. C., Nakagiri, M., Toyoda, N., Matsuyoshi, H., & Ito, S. (2011). Characterization of superstrate type $CuInS_2$ solar cells deposited by spray pyrolysis method. *Thin Solid Films, 519*(21), 7184–7188.

[124] Isac, L., Duta, A., Kriza, A., Manolache, S., & Nanu, M. (2007). Copper sulfides obtained by spray pyrolysis—Possible absorbers in solid-state solar cells. *Thin Solid Films, 515*(15), 5755–5758.

[125] Ortega-Lopez, M., & Morales-Acevedo, A. (1998). Characterization of $CuInS_2$ thin films for solar cells prepared by spray pyrolysis. *Thin Solid Films, 330*(2), 96–101.

[126] Krunks, M., Katerski, A., Dedova, T., Acik, I. O., & Mere, A. (2008). Nanostructured solar cell based on spray pyrolysis deposited ZnO nanorod array. *Solar Energy Materials and Solar Cells, 92*(9), 1016–1019.

[127] Aranovich, J., Ortiz, A., & Bube, R. H. (1979). Optical and electrical properties of ZnO films prepared by spray pyrolysis for solar cell applications. *Journal of Vacuum Science & Technology, 16*(4), 994–1003. DOI: 10.1116/1.570167

[128] Yuksel, S. A., Gunes, S., & Guney, H. Y. (2013). Hybrid solar cells using CdS thin films deposited via spray pyrolysis technique. *Thin Solid Films, 540,* 242–246.

[129] Tomar, M. S., & Garcia, F. J. (1981). Spray pyrolysis in solar cells and gas sensors. *Progress in Crystal Growth and Characterization, 4*(3), 221–248.

[130] Vijayan, K., Vijayachamundeeswari, S. P., Sivaperuman, K., Ahsan, N., Logu, T., & Okada, Y. (2022). A review on advancements, challenges, and prospective of copper and non-copper based thin-film solar cells using facile spray pyrolysis technique. *Solar Energy, 234,* 81–102.

[131] Kobayashi, H., Mori, H. Ishida, T., & Nakato, Y. (1995). Zinc oxide/n-Si junction solar cells produced by spray-pyrolysis method. *Journal of Applied Physics, 77*(3), 1301–1307. DOI: 10.1063/1.358932

[132] Marchetti, F., Moroni, E., Pandini, A., & Colombo, G. (2021). Machine learning prediction of allosteric drug activity from molecular dynamics. *The Journal of Physical Chemistry Letters, 12*(15), 3724–3732.

[133] Rong, Y., Ming, Y., Ji, W., Li, D., Mei, A., Hu, Y., & Han, H. (2018). Toward industrial-scale production of perovskite solar cells: screen printing, slot-die coating, and emerging techniques. *The Journal of Physical Chemistry Letters, 9*(10), 2707–2713.

[134] Tepner, S., Wengenmeyr, N., Linse, M., Lorenz, A., Pospischil, M., & Clement, F. (2020 Oct.). The link between Ag-paste rheology and screen-printed solar cell metallization. *Advanced Materials Technologies, 5*(10) 2000654, 1–9. DOI: 10.1002/ADMT.202000654

[135] Mette, A. G. W. A., Schetter, C., Wissen, D., Lust, S., Glunz, S. W., & Willeke, G. (2006, May). Increasing the efficiency of screen-printed silicon solar cells by light-induced silver plating. In *2006 IEEE 4th World Conference on Photovoltaic Energy Conference* (vol. 1, pp. 1056–1059). IEEE.

[136] Lorenz, A., Linse, M., Frintrup, H., Jeitler, M., Mette, A., Lehner, M., ... & Clement, F. (2018). Screen printed thick film metallization of silicon solar cells—recent developments and future perspectives. In *35th European Photovoltaic Solar Energy Conference and Exhibition* (pp. 819–824).

[137] Gatz, S., Hannebauer, H., Hesse, R., Werner, F., Schmidt, A., Dullweber, T., ... & Brendel, R. (2011). 19.4%-efficient large-area fully screen-printed silicon solar cells. *Physica Status Solidi (RRL) – Rapid Research Letters*, 5(4), 147–149.

[138] Tepner, S., Ney, L., Linse, M., Lorenz, A., Pospischil, M., & Clement, F. (2020). Studying knotless screen patterns for fine-line screen printing of Si-solar cells. *IEEE Journal of Photovoltaics*, 10(2), 319–325.

[139] Shaheen, S. E., Radspinner, R., Peyghambarian, N., & Jabbour, G. E. (2001). Fabrication of bulk heterojunction plastic solar cells by screen printing. *Applied Physics Letters*, 79(18), 2996–2998.

[140] Bull, C. L., Gleeson, D., & Knight, K. S. (2003). Determination of B-site ordering and structural transformations in the mixed transition metal perovskites La_2CoMnO_6 and La_2NiMnO_6. *Journal of Physics: Condensed Matter*, 15(29), 4927.

[141] Krebs, F. C., Jørgensen, M., Norrman, K., Hagemann, O., Alstrup, J., Nielsen, T. D., ... & Kristensen, J. (2009). A complete process for production of flexible large area polymer solar cells entirely using screen printing—First public demonstration. *Solar Energy Materials and Solar Cells*, 93(4), 422–441.

[142] De Guire, M. R., Bauermann, L. P., Parikh, H., & Bill, J. (2013 Dec.). Chemical bath deposition. *Chemical Solution Deposition of Functional Oxide Thin Films*, 9783211993118, 319–339. DOI: 10.1007/978-3-211-99311-8_14

[143] Nair, P. K., Nair, M. T. S., García, V. M., Arenas, O., Pena, Y., Castillo, A., ... & Rincon, M. E. (1998). Semiconductor thin films by chemical bath deposition for solar energy related applications. *Solar Energy Materials and Solar Cells*, 52(3–4), 313–344.

[144] Nair, M. T. S., Guerrero, L., & Nair, P. K. (1998). Conversion of chemically deposited CuS thin films to and by annealing. *Semiconductor Science and Technology*, 13(10), 1164.

[145] Sun, J., Lu, J., Li, B., Jiang, L., Chesman, A. S., Scully, A. D., ... & Jasieniak, J. J. (2018). Inverted perovskite solar cells with high fill-factors featuring chemical bath deposited mesoporous NiO hole transporting layers. *Nano Energy*, 49, 163–171.

[146] Cheng, H. C., Chen, C. F., & Lee, C. C. (2006). Thin-film transistors with active layers of zinc oxide (ZnO) fabricated by low-temperature chemical bath method. *Thin Solid Films*, 498(1–2), 142–145.

[147] Roy, P., Ota, J. R., & Srivastava, S. K. (2006). Crystalline ZnS thin films by chemical bath deposition method and its characterization. *Thin Solid Films*, 515(4), 1912–1917.

[148] Cheng, S., Chen, G., Chen, Y., & Huang, C. (2006). Effect of deposition potential and bath temperature on the electrodeposition of SnS film. *Optical Materials*, 29(4), 439–444.

[149] Wu, C. C., Cheng, K. W., Chang, W. S., & Lee, T. C. (2009). Preparation and characterizations of visible light-responsive (Ag–In–Zn) S thin-film electrode by chemical bath deposition. *Journal of the Taiwan Institute of Chemical Engineers*, 40(2), 180–187.

[150] Vangari, M., Pryor, T., & Jiang, L. (2013). Supercapacitors: Review of materials and fabrication methods. *Journal of energy engineering*, 139(2), 72–79.

[151] McPeak, K. M., Opasanont, B., Shibata, T., Ko, D. K., Becker, M. A., Chattopadhyay, S., ... & Baxter, J. B. (2013). Microreactor chemical bath deposition of laterally graded Cd1–x Zn x S thin films: A route to high-throughput optimization for photovoltaic buffer layers. *Chemistry of Materials*, 25(3), 297–306.

[152] Pourshaban, E., Abdizadeh, H., & Golobostanfard, M. R. (2015). ZnO nanorods array synthesized by chemical bath deposition: Effect of seed layer sol concentration. *Procedia Materials Science*, 11, 352–358.

[153] Liu, J., Wei, A., & Zhao, Y. (2014). Effect of different complexing agents on the properties of chemical-bath-deposited ZnS thin films. *Journal of Alloys and Compounds*, 588, 228–234.

[154] Sankaran, S., Glaser, K., Gärtner, S., Rödlmeier, T., Sudau, K., Hernandez-Sosa, G., & Colsmann, A. (2016). Fabrication of polymer solar cells from organic nanoparticle dispersions by doctor blading or ink-jet printing. *Organic Electronics*, 28, 118–122.

[155] Lukong, V. T., Ukoba, K., & Jen, T. C. (2022). Review of self-cleaning TiO_2 thin films deposited with spin coating. *The International Journal of Advanced Manufacturing Technology*, 122(9–10), 3525–3546.

12 Smart and Sustainable Energy Systems

Tamer Guclu, Pinar Mert Cuce, and Erdem Cuce

12.1 INTRODUCTION

Energy plays a critical role for humanity, which has duties like combating climate disasters, ensuring sustainable growth, coping with large energy crises in the world, protecting the environment, and protecting the diversity of living things. Today, the use of fossil fuels still has a more important share in the supply of energy needed (Zamfirescu et al. 2012). It is known that environmental problems caused by these sources pose a significant threat to our world (Cuce et al. 2020). On the other hand, the increase in energy requirements around the world and decreasing reserves of fossil fuels necessitate new approaches and solutions (Silva, Khan, and Han 2018). Therefore, in recent years, the energy policies of countries have been changing to stimulate the use of clean, renewable and sustainable energy sources. In fact, more than 190 countries around the world stated that they aim to keep the increase in average temperature below 2°C with the global climate agreement they accepted at the Paris Climate Conference. The main goal among European Union countries is to minimize greenhouse gas emissions and to ensure energy security by accelerating the transition to renewable energy sources in energy production (Bačeković and Østergaard 2018).

The conversion from fossil-fuelled energy sources to clean energy sources in the energy sector, which constitutes a huge amount of carbon emissions on a global scale, is a very important issue in tackling environmental and climate problems. It is important to define the energy transition design and sustainable system operations in this energy transformation. With the energy transition design, which aims to control storage, generation, and energy transmission, demand and supply changes can be controlled for certain time scales (Zhao et al. 2023). Today, various electronic devices are widely used in energy system modelling and analysis processes to control energy transition. Since it is a very common and diverse device network, the right modelling and design method must be selected specifically for the system to be applied. Equilibrium tools, simulation and optimization are widely used in energy system modelling, although they are very common (Lund et al. 2021). In addition, terms such as smart and sustainable energy systems have entered the literature due to the need to use these environmentally friendly and renewable energy sources, which have begun to replace fossil fuel energy sources, together, more efficiently and in accordance with automation. Smart energy systems, which are formed by integrating various energy sectors, are considered as a promising technology, especially in the near future, as they promise an economical and accessible sustainable energy system (Shakeri et al. 2017).

DOI: 10.1201/9781032651958-12

12.2 SMART AND SUSTAINABLE ENERGY SYSTEMS

The definition of smart energy systems was defined and used with an understanding representing the transition from single systems to an integrated system understanding, after traditional energy systems could not meet the multidisciplinary requirements of the 21st century (Lund et al. 2016). Smart energy systems enable active energy control and thus energy savings by integrating different energy sectors (Xu et al. 2020). Smart energy systems also ensure the scientific base for the transition from a single-sector approach to a multi-sector approach and the development of the most appropriate strategies in terms of cost reduction (Lund 2018). Sustainable energy systems are energy systems represented by renewable energy sources like wind and solar energy in the most general form. Interest in these systems is rising day after day. Unlike conventional energy system structures, sustainable energy systems can be defined as the interaction of supply. In addition, they have features such as intermittentness, volatility, and randomness (Liu et al. 2021). The smart energy system can be seen as a flexible transformation between energy systems. Since these flexible forms of transformation are achieved using new infrastructure and technologies, they are simple, systems consisting of energy sources integrated into each other rather than end-use from a single source. Another expected effect of smart energy systems is to offer flexible usage patterns that enable the efficiency provided by fossil fuels to be obtained from renewable energy sources (Connolly et al. 2013).

Transformation to smart energy systems, contains some important objectives like sustainability and minimizing harmful effects on the environment. In line with these purposes, designs, analyses, and product developments are being put forward. Smart energy systems offer an important approach to responding to the increasing global energy need efficiently, especially without harming the environment. On the other hand, in order for a system to be defined as a smart energy system, it must meet expectations in many aspects, such as efficiency, effectiveness, and sustainability. Dincer and Acar explained the potential effects expected from smart energy systems under various titles as bellow:

- Exergetically sound,
- Energetically secure,
- Environmentally benign,
- Economically feasible,
- Commercially viable,
- Socially acceptable,
- Integrable,
- Reliable (Dincer and Acar 2017).

They defined integration as the combination of existing energy systems or energy sources to improve the system efficiency, reduce costs, reduce the use of resources, and reduce detrimental effects on the environment. Smart energy systems are more efficient, safer, and interactive than conventional energy systems (Yuan et al. 2014).

When we first look at the source energy forms in classifying smart energy systems, we can classify them into three main groups: thermal energy, electrical energy, and chemical energy. It is possible to divide these three main categories into various sub-categories within themselves (Su 2020). With the developments in technology, the share of renewable and sustainable energy sources in energy systems is increasing, and as a result, it has become more economically efficient (Klass 2003). Smart energy systems used to convert these source energy forms are divided into three groups: smart thermal systems, smart electrical systems, and smart gas systems. Thanks to these smart energy systems, various energy sources such as solar, wind, nuclear, water, and oil are used as inputs and converted into products such as electricity, heating, cooling power, and fuel energy (Rong and Su 2017). The primary purpose of smart energy systems is to ensure secondary and even tertiary benefits in addition to the primary objective, such as using the waste heat discharged from the chimneys in industrial facilities in heating applications and obtaining regional cooling power from heat pumps.

12.2.1 Smart Thermal Energy Systems

Smart thermal energy systems are systems that enable the use of intermittent, distributed thermal energy sources more efficiently, as well as fulfil the functions of classical thermal energy systems, and provide the needed energy where it is needed through intelligent management. While traditional energy systems provide thermal energy from the producer to the consumer, smart thermal energy systems allow the consumer to be a producer (Yang et al. 2017). This two-way energy transfer possibility is the major feature that distinguishes smart thermal energy systems from traditional energy systems. Smart thermal energy systems convert renewable energy sources like wind, geothermal or solar and waste heat sources of industrial facilities into efficient energy (Stănişteanu 2017). For example, the electrical energy required for a cooling system can be supported by a microgrid PV unit. A solar hot water generator can be integrated into a central thermal energy system. Or, to provide additional benefits, a heat pump can be used for cooling and heating. Fischer and Madani (2017) investigated the effectiveness of heat pumps on smart energy systems. The flexibility, application areas, and control mechanisms provided by heat pumps have been examined based on the literature. Especially in recent years, heat pump applications integrated with renewable energy sources have attracted attention as smart thermal energy systems applications, thanks to their secondary benefits. According to Hennessy et al. (2019), energy systems should become more flexible day by day. One of the ways to provide this flexibility is to link storage and usage areas, which is possible by converting thermal energy systems with smart thermal energy systems. Thanks to smart thermal energy systems, the use of a thermal energy store obtained from renewable energy sources reduces heating energy costs from the primary source. Smart thermal energy systems also allow consumers to be included in the production system. At the same time, by integrating renewable and sustainable energy sources into this production process, the environmental damage caused by fossil fuel-based energy sources can be prevented. In this respect, district heating systems that allow a certain region, building, neighbourhood or city to be heated/cooled from a central facility or several heat generators are very important systems with regard to energy efficiency. With regard to sustainable development and energy supply, smart thermal energy systems are created by integrating renewable energy sources like wind, solar, geothermal, and heat pumps into these different heating systems or by recycling waste heat. These systems play an essential role in minimizing the harmful effects of traditional energy sources on the environment and meeting the increasing energy demand (Lund et al. 2014).

12.2.2 Smart Electrical Energy Systems

Electricity generation by steam power, which is obtained by using fossil fuels in very large central electric power plants, is losing its popularity day by day. Instead, the transition to small-scale, decentralized electricity generation in a location close to the usage area continues rapidly. In this transition, negative environmental effects caused by fossil fuels and energy losses in the distribution line play an important role (Colson and Nehrir 2009). The most effective method of obtaining this distributed energy generation is to produce using renewable energy sources at a location close to the point of use. Increases in energy costs and demand intensity around the world also encourage governments to give importance to this distributed energy production (Ustun, Ozansoy, and Zayegh 2011). Smart electrical energy systems, also called smart grids, are systems that integrate renewable energy sources like wind, sun, and water into an electrical energy source (Currie et al. 2004). For example, some of the electrical energy needed in a chalet is supplied by the photovoltaic system connected to the grid. In fact, if more electricity is produced from photovoltaic panels than needed, the excess electricity can be sold to the grid, allowing the consumer to become a producer. In addition, it is possible to obtain domestic hot water and heating of space by utilizing waste heat generated in the panels. Traditional power generation systems are limited to central power generation sources and provide a one-way energy flow to the customer. For this reason, the only point at which users are active in this process is paying bills. In addition, these systems are not flexible enough to add an alternative energy source from any point or to include the consumer in production. On the other hand, smart electrical energy systems save energy as they allow consumers to be involved in production. In addition, it prevents energy losses as it encourages production close to the usage location instead of long energy transmission lines. As it increases the share of renewable energy sources in energy production, it contributes to minimizing environmental damage caused by fossil fuels (Gharavi and Ghafurian 2011). A concept in which smart electrical energy systems are widely used is smart buildings. In the smart building concept, some of the electrical energy needed is provided by these panels, thanks to photovoltaic panels or wind turbines placed on the roof. In case of excess electricity generation, the excess energy is sent back to the grid. In addition, in rooms where people are not present, energy consumption can be reduced by automatically turning the Heating Ventilation and Air Conditioning (HVAC) systems off and on. Unnecessary lights can be turned off with a simple computerized control system, and devices that need to be charged can be activated during low usage hours (Molderink et al. 2010). Another concept in which smart electrical systems are used is electric or hybrid vehicles. Users can charge their vehicles with the energy they obtain by using renewable energy sources in their own areas, thus minimizing fuel costs (Abdallah and El-Shennawy 2013). Smart electrical energy systems provide modern monitoring and control for any voltage value. On the other hand, energy production fluctuations due to the instability of renewable energy sources create the need for additional control (Strasser et al. 2015).

12.2.3 SMART GAS ENERGY SYSTEMS

Renewable energy sources utilization, particularly for distributed energy generation, make significant contributions to the creation of sustainable energy systems. Based on these sources, electricity generating systems, combined heat power plants and high-efficiency heat pumps are produced. In this context, smart fuel energy systems, which contain fossil fuels, biofuels, and syngas in various proportions, are gradually replacing traditional fossil fuels that are used as a stand-alone energy source. These smart gas energy systems provide energy efficiency, especially with the effective completion of the supply chain. In addition, smart fuel systems provide the opportunity to respond quickly and effectively to variable production demands. It also allows energy to be stored in the network or specially prepared warehouses (Crisostomi et al. 2013). Smart fuel energy systems improve the efficiency of energy production, transmission, and use with the flexibility they provide in traditional energy production management (Mathiesen et al. 2015).

12.2.4 COST OF SMART AND SUSTAINABLE ENERGY SYSTEMS

Replacing existing traditional energy systems with smart and sustainable energy systems creates a serious initial cost. However, with this transformation, the efficient usage of energy resources will pay for itself economically at the end of certain periods and will provide economic benefits in case of long-term use (Hamidi, Smith, and Wilson 2010). Moreover, since the usage of smart and sustainable energy systems will prevent environmental disasters caused by fossil fuels, it will also contribute to the reduction of the budgets and environmental taxes allocated by governments to combat environmental problems. Cost and efficiency calculations of smart energy systems vary depending on the installed system and application area. In these calculations, electricity unit prices in the region, government tax policies, incentives, and capacities are taken into account. Therefore, a smart energy system that is economical for one region may not be advantageous for another region.

For this reason, it is not appropriate to carry out the economic analysis of smart energy costs with a standard method (Moretti et al. 2017). This situation stands out as one of the difficulties encountered in making a clear definition of smart energy systems.

12.3   CONCLUSION

Energy systems play a key role in terms of environmental, economic, social, and vitality in the globalizing and industrializing world day by day. For this reason, careful use of existing energy systems, protection of resources, and increasing their efficiency are extremely important issues for the continuation of vitality, protection and safety of future generations. Smart energy systems play an essential role in establishing a sustainable world order and preventing environmental disasters caused by fossil fuels, thanks to making renewable energy sources a part of production. It also enables distributed production in a location close to the usage area instead of a single large energy source and enables consumers to join the production. Although it brings new simulation and control tools, it is preferred because of the environmental and economic benefits it provides. From an economic point of view, the transmission of traditional energy systems to smart energy systems requires an initial cost investment. However, considering that the installed systems have been used more efficiently for many years, it would be wise to cover these initial investment costs.

REFERENCES

Abdallah, Lamiaa, and Tarek El-Shennawy. 2013. "Reducing Carbon Dioxide Emissions from Electricity Sector Using Smart Electric Grid Applications." *Journal of Engineering (United Kingdom)* 2013.

Bačeković, Ivan, and Poul Alberg Østergaard. 2018. "Local Smart Energy Systems and Cross-System Integration." *Energy* 151(2018): 812–25.

Colson, C. M., and M. H. Nehrir. 2009. "A Review of Challenges to Real-Time Power Management of Microgrids." *2009 IEEE Power and Energy Society General Meeting, PES '09.*

Connolly, D. et al. 2013. "Smart Energy Systems." *It – Information Technology* 55(2): 43–44. www.degruyter.com/document/doi/10.1524/itit.2013.9002/html

Crisostomi, Emanuele, Marco Raugi, Alessandro Franco, and Giuseppe Giunta. 2013. "The Smart Gas Grid: State of the Art and Perspectives." In *IEEE PES ISGT Europe 2013*, IEEE, 1–5.

Cuce, Erdem, Pinar Mert Cuce, Tamer Guclu, and Ahmet Burhaneddin Besir. 2020. "On the Use of Nanofluids in Solar Energy Applications." *Journal of Thermal Science* 29(3): 513–34.

Currie, R. A. F., Ault, G. W., Foote, C. E. T., Burt, G. M., and McDonald, J. R. 2004, September. Fundamental research challenges for active management of distribution networks with high levels of renewable generation. In *39th International Universities Power Engineering Conference*. UPEC 2004. (Vol. 3, pp. 1024–1028). IEEE.

Dincer, Ibrahim, and Canan Acar. 2017. "Smart Energy Systems for a Sustainable Future." *Applied Energy* 194: 225–35. http://dx.doi.org/10.1016/j.apenergy.2016.12.058

Fischer, David, and Hatef Madani. 2017. "On Heat Pumps in Smart Grids: A Review." *Renewable and Sustainable Energy Reviews* 70(May 2016): 342–57.

Gharavi, H, and R Ghafurian. 2011. "Smart grid: The electric energy system of the future [scanning the issue]." *Proceedings of the IEEE* 99(6): 917–21.

Hamidi, V., K. S. Smith, and R. C. Wilson. 2010. "Smart Grid Technology Review within the Transmission and Distribution Sector." In *2010 IEEE PES Innovative Smart Grid Technologies Conference Europe (ISGT Europe)*, IEEE, 1–8. http://ieeexplore.ieee.org/document/5638950/

Hennessy, Jay, Hailong Li, Fredrik Wallin, and Eva Thorin. 2019. "Flexibility in Thermal Grids: A Review of Short-Term Storage in District Heating Distribution Networks." *Energy Procedia* 158: 2430–34. https://doi.org/10.1016/j.egypro.2019.01.302

Klass, Donald L. 2003. "A Critical Assessment of Renewable Energy Usage in the USA." *Energy Policy* 31(4): 353–67.

Liu, Qiang, Songlin Sun, Bo Rong, and Michel Kadoch. 2021. "Intelligent Reflective Surface Based 6G Communications for Sustainable Energy Infrastructure." *IEEE Wireless Communications* 28(6): 49–55.

Lund, Henrik et al. 2014. "4th Generation District Heating (4GDH). Integrating Smart Thermal Grids into Future Sustainable Energy Systems." *Energy* 68: 1–11. http://dx.doi.org/10.1016/j.energy.2014.02.089

Lund, Henrik. 2018. "Renewable Heating Strategies and Their Consequences for Storage and Grid Infrastructures Comparing a Smart Grid to a Smart Energy Systems Approach." *Energy* 151: 94–102. https://doi.org/10.1016/j.energy.2018.03.010

Lund, Henrik et al. 2021. "EnergyPLAN – Advanced Analysis of Smart Energy Systems." *Smart Energy* 1: 100007. https://doi.org/10.1016/j.segy.2021.100007

Lund, Henrik, Neven Duic, Poul Alberg Østergaard, and Brian Vad Mathiesen. 2016. "Smart Energy Systems and 4th Generation District Heating." *Energy* 110(2016): 1–4.

Mathiesen, B. V. et al. 2015. "Smart Energy Systems for Coherent 100% Renewable Energy and Transport Solutions." *Applied Energy* 145: 139–54.

Molderink, Albert et al. 2010. "Management and Control of Domestic Smart Grid Technology." *IEEE Transactions on Smart Grid* 1(2): 109–19.

Moretti, M. et al. 2017. "A Systematic Review of Environmental and Economic Impacts of Smart Grids." *Renewable and Sustainable Energy Reviews* 68: 888–98. http://dx.doi.org/10.1016/j.rser.2016.03.039

Rong, Aiying, and Yan Su. 2017. "Polygeneration Systems in Buildings: A Survey on Optimization Approaches." *Energy and Buildings* 151: 439–54. http://dx.doi.org/10.1016/j.enbuild.2017.06.077

Shakeri, Mohammad et al. 2017. "An Intelligent System Architecture in Home Energy Management Systems (HEMS) for Efficient Demand Response in Smart Grid." *Energy and Buildings* 138: 154–64. http://dx.doi.org/10.1016/j.enbuild.2016.12.026

Silva, Bhagya Nathali, Murad Khan, and Kijun Han. 2018. "Towards Sustainable Smart Cities: A Review of Trends, Architectures, Components, and Open Challenges in Smart Cities." *Sustainable Cities and Society* 38(February): 697–713. https://doi.org/10.1016/j.scs.2018.01.053

Stănişteanu, Cristina. 2017. "Smart Thermal Grids – A Review." *The Scientific Bulletin of Electrical Engineering Faculty* 1(36). DOI: 10.1515/SBEEF-2016-0030

Strasser, Thomas et al. 2015. "A Review of Architectures and Concepts for Intelligence in Future Electric Energy Systems." *IEEE Transactions on Industrial Electronics* 62(4): 2424–38.

Su, Yan. 2020. "Smart Energy for Smart Built Environment: A Review for Combined Objectives of Affordable Sustainable Green." *Sustainable Cities and Society* 53(April 2019): 101954. https://doi.org/10.1016/j.scs.2019.101954

Ustun, Taha Selim, Cagil Ozansoy, and Aladin Zayegh. 2011. "Recent Developments in Microgrids and Example Cases around the World – A Review." *Renewable and Sustainable Energy Reviews* 15(8): 4030–41. http://dx.doi.org/10.1016/j.rser.2011.07.033

Xu, Yizhe et al. 2020. "Smart Energy Systems: A Critical Review on Design and Operation Optimization." *Sustainable Cities and Society* 62(200): 102369. https://doi.org/10.1016/j.scs.2020.102369

Yang, Libing, Evgueniy Entchev, Antonio Rosato, and Sergio Sibilio. 2017. "Smart Thermal Grid with Integration of Distributed and Centralized Solar Energy Systems." *Energy* 122: 471–81. http://dx.doi.org/10.1016/j.energy.2017.01.114

Yuan, Jiahai et al. 2014. "Smart Grids in China." *Renewable and Sustainable Energy Reviews* 37: 896–906. http://dx.doi.org/10.1016/j.rser.2014.05.051

Zamfirescu, C., I. Dincer, M. Stern, and W. R. Wagar. 2012. "Exergetic, Environmental and Economic Analyses of Small-Capacity Concentrated Solar-Driven Heat Engines for Power and Heat Cogeneration." *International Journal of Energy Research* 36(3): 397–408. https://onlinelibrary.wiley.com/doi/10.1002/er.1811

Zhao, Ning et al. 2023. "Emerging Information and Communication Technologies for Smart Energy Systems and Renewable Transition." *Advances in Applied Energy* 9(October 2022): 100125. https://doi.org/10.1016/j.adapen.2023.100125

13 **Minimizing High Free Fatty Acid (FFA) in Rubber Seed Oil (RSO) Using Ethanol as an Alcohol**

Optimization by Taguchi Method

Olusegun David Samuel, Jephtha Akporhuarho, Ogaga Ovuedhere, Akinrodoye Ibidapo, Ivrogbo D. Eseoghene, and Christopher C. Enweremadu

13.1 INTRODUCTION

The designation of biodiesel as a replacement for fossil diesel has been contingent on its compliance with international norms for automobile adoption. Currently, non-edible oil feedstocks (NEOFs), such as jatropha, castor, rubber, tobacco, etc., are used to produce biodiesel that is not competitive with food. Due to their underutilization and lack of competition with food intake, these non-edible oils have gained widespread acceptance. However, the critical challenge confronting researchers is the NEOF's high free fatty acid (FFA) concentration [1]. Knothe and Razon [2] emphasized that high FFA levels in NEOFs are notably associated with low conversion of (m) ethyl esters of non-edible oils, increased downstream costs, undesired soap production, and catalyst diminution. Hence, the esterification process can be a major approach for improving and rendering NEOFs suitable for transesterification [3].

To reduce high FFA component in NEOFs higher alcohols such as ethanol are included in the reaction mixture to make green diesel environmentally friendly and biobased and to enhance the cetane number. Methanol with sulfuric acid is mostly employed to esterify and pre-treat the oil. The adoption of methanol as an alcohol does not make the biodiesel entirely biobased [4].

The succinct summary of inedible oil esterification is shown in Table 13.1. As can be seen, methanol was used as an alcohol to pre-treat rubber seed oil (RSO) [3], neem-castor seed oil [5], jatropha crude oil (JCO) [6], neem seed oil [7], palm kernel oil [8], a mixture of rubber-palm mix [9]. Ethanol was employed to esterify waste cooking oil [10], oleic acid [11], and Butia Yatay coconut oil (BYCO) [12]. Additionally, Zhou et al. [11] and Zanuttini et al. [12] employed one-variable-at-a-time (OVAT) to determine the esterification conditions for oleic acid and BYCO, respectively, while Guo et al. [10] explored Response Surface Methodology (RSM) to predict the esterification condition for Waste Cooking Oil (WCO). Therefore, OVAT cannot be effectively used to assess several response characteristics of the experimental design. This is a result of OVAT's inability to simultaneously display the correlation between the input and response variables [13].

DOI: 10.1201/9781032651958-13

TABLE 13.1

A Concise Review of Optimization of Esterification of High FFA Oil

Oily Source	Types of Alcohols	Operation Esterification Conditions/AV (mg KOH/g)	Optimization Tools	Countries or Regions	Remarks	Refs.
Rubber seed oil	Methanol	ME:O = 44.21:1, Ti = 3.4 h, $Fe_2(SO_4)_3$ = 16.97%/ FFA = 0.56%	RSM-ANFIS (Response Surface Methodology -Adaptive neuro-fuzzy inference system)	Tropics	The capability of the RSM-ANFIS model for the heterogeneous acid esterification of RSO's pre-treatment showcased	Jisieike et al. [3]
Nee-castor seed oil	Methanol	ME:O = 4.5:1, T_1 = 2 h, N/CSO ratio = 20, C_a = 2 wt.%/ AV = 2	RSM-ANFIS technique	Tropics	RSM-ANFIS proved to predict esterification conditions for hybrid oils	Samuel et al. [5]
Waste cooking oil	Ethanol	E:O = 12:1, T_1 = 3 h,Ca = 4 wt%,	RSM	China	orthogonal design of the RSM predicted esterification condition	Guo et al. [10]
Jatropha crude oil (JCO)	Methanol	ME:O = 12.3:1, T_i = 149.8 min, H_2SO_4/ Ca = 0.23 vol.%	RSM	Malaysia	The optimum condition for esterifying JCO by RSM established cation reaction.	Farouk et al. [6]
Neem seed oil	Methanol	ME:O = 18.51, T_1 = 62.8 min, ferric sulfate dosage = 6 wt%,/FFA = 0.52	RSM-ANN-GA	Tropics	The capability of esterification of ANN and GA is superior to that of RSM in NSO's pre-treatment	Okpalaeke et al. [7]
Palm kernel oil	Methanol	ME:O = 3.4:1, T_1 = 24.06 min, Ca = 0.39 vol.%	RSM-ANN-ANFIS	Tropics	Various superiority of the ternary tools highlighted	Betiku et al. [8]
Oleic acid	Ethanol	ME:O = 7:1, T_i = 2 h,T_e = 90°C	OVAT	China	Oleic oil pre-treated under the esterification condition	Zhou et al. [11]
Butia Yatay coconut oil (BYCO)	Ethanol	Two stages of esterification	OVAT	Argentina	Suitability of two-step esterification effective in pre-treating BYCO	Zanuttini et al. [12]
A mixture of rubber-palm oils	Methanol	ME:O = 15:1, T_i = 3 h, H_2SO_4/ Ca = 0.5 wt%, T_e = 65°C	Taguchi	Malaysia	The capability of the Taguchi technique in pre-treating hybrid oils established	Khan et al. [9]

13.1.1 Enthusiasm, Objective, and Novelty of the Study

A literature review revealed that the Taguchi approach has been used to detect and optimize bio-diesel production (BPR) and engine characteristics of IC driven by various types of fuels. To the best of the authors' knowledge, little is known about Taguchi-based optimization of ethylic esterification of NEOFs with high FFA. Relatively limited research about the use of the Taguchi approach in the methylic esterification of oils has been documented. Only a mixture of rubber-palm oil has been used in the Taguchi technique [9]. It has been difficult for the rubber industry to implement RSM, Artificial Neural Network (ANN), and other metaheuristics algorithms for pre-treating high FFA for potential transesterification because of the complexity and laborious process of using the existing tools.

The study aimed to conduct parametric analyzes for the optimal esterification of rubber seed oil with high FFA using ethanol as the alcohol. The study's objectives were to produce ethyl ester of rubber seed oil (RSOEE) under conventional conditions, evaluate the financial costs of RSOEE, and determine if the biodiesel produced would be commercially viable.

13.2 MATERIALS AND PROCEDURE

13.2.1 Material and RSORSO Assessment

For the synthesis of biodiesel, rubber seed oil (RSO or RUSO), ethanol, potassium hydroxide, sulfuric acid (99.0%, Merck), and potassium hydroxide (Merck) were the main chemicals employed. These chemicals were all of analytical grades. The RSO was characterized in accordance with American Society for Testing and Materials (ASTM) etiquettes. The acid value (AV) of RSO was detected by ASTM D974. Eqs. (13.1) and (13.2) were adopted in estimating the FFA and acid value of RSO

$$FFA = \frac{28.2 \; x \; N_o \; x \; t_v}{m} \tag{13.1}$$

$$AV = \frac{56.1 \; x \; N_o \; x \; t_v}{m} \tag{13.2}$$

where t_v, N_o, and mw_{RSO} are the titre value, normality, and mass of RSO adopted while the molecular weight of RSO is 867.9 g/mol.

13.2.2 Ethylic-Based Esterification of RSO Via TM

Due to the high FFA content of the RSO (15.99 mm KOH/g) pre-treatment of the oil was conducted using a schematic setup depicted in Figure 13.1a. The reactor was initially filled with 150 ml of RSO and preheated to a temperature of 65°C. A freshly prepared blend of ethanol and sulfuric acid (H_2SO_4) was homogeneously mixed and heated at 65°C. The combination was then added to the pre-heated oil in the reactor. Ethanol-esterified rubber seed oil was then separated from the excess ethanol (see Figure 13.1b).

The esterification of RSO is influenced by various key variables in the process. Therefore, it is necessary to investigate these factors within specific constraints to determine the optimal settings that result in minimal reduction. An L4 matrix with three esterification variables is defined at two levels, as shown in Tables 13.2 and 13.3. The number of experiments (N) required depends on the number of parameters being optimized (P) and the total number of levels for each parameter (L) in Eq. (13.3). To evaluate RSO's AV, the Taguchi technique criterion adopted is

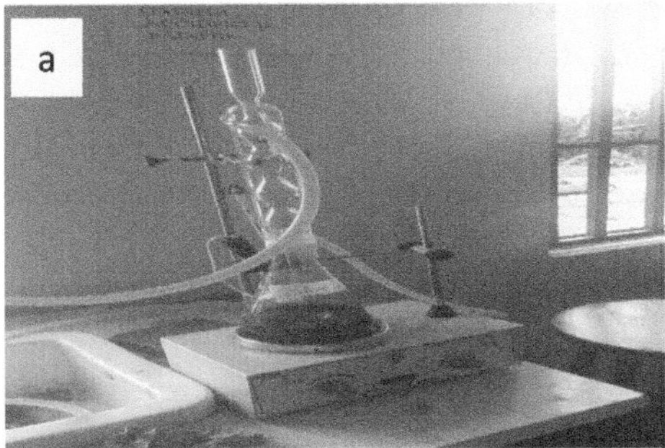

FIGURE 13.1   Esterification process for RSO: (a) esterification of RSO with ethanol, (b) separation of esterified RSO and impurities with excess FFA component.

TABLE 13.2
Matrix Develop for RSO Esterification

| | Levels | |
Factors	1	2
A [1]E/CRSO MR	6:1	9:1
B [2]DoC (wt.%)	0.5	1
C [3]RT (min)	45	60

1, ethanol/CRSO molar ratio; 2, dosage of catalyst; 3, reaction time.

TABLE 13.3
L4 Orthogonal Array for RSO Pre-Treatment

| | Parameters and Their Structures | | |
Experiment No.	E/CRSO MR	DoC (wt%)	RT (mins)
1	1	1	1
2	1	2	2
3	2	1	2
4	2	2	1

"smaller-the-better" (LBT), where a lower AV indicates a better RSO's FFA value (%) and acid value (mg KOH/g).

$$N = (L-1)P + 1 \tag{13.4}$$

13.2.3   Statistical Analysis of Variance (ANOVA)

The TM is capable of estimating the signal-to-noise ratio (SNR) from the experimental databases. The ratio of the mean of investigational response: AV to the standard deviation is equivalent to the SNR obtained. This methodology can be very expedient in explaining the influence of the esterification constraints.

The valuation of the statistical parameters using analysis of variance (ANOVA) aids in determining the distinctive effect of individual parameters on the ultimate result acquired, which can be estimated based on their contribution factors (AV in mg KOH/g). This was estimated from the aggregation of squares of each factor (SS_f) divided by the summation of squares of the entire model (SS_T). The percentage contribution factor can be appraised using Eq. (13.5).

$$\% \text{ contribution factor } \frac{SS_f}{SS_T} \times 100 \tag{13.5}$$

13.2.4 ETHYLIC PROTOCOL FOR THE ESTERIFIED RSO

The ethylic route for conversion of esterified RSO was conducted in a 2 L reactor containing a condenser and a heating device with a magnetic agitator. About 500 g RSO was transferred into the reactor and stationed on the heating device. Potassium ethoxide was produced by dissolving 5 wt.% KOH in 31.85 g of ethanol. The potassium ethoxide was transferred into the reactor, which had the pre-heated RSO, and the entire reactor content was agitated at a rate of 650 rpm and heated for 1 h. The details of RSOEE production from the pre-treated RSO are depicted in Figure 13.2. Significant properties of RSOEE were checked in accordance with recognized international standards.

13.2.5 COST ANALYSIS

For evaluating the expenses of BPR from RSO, the approaches stated by Marousek et al. [14], Rajak et al. [15], Barman and Jash [16], and Shrivastava et al. [17] are adopted. As can be seen, the cost

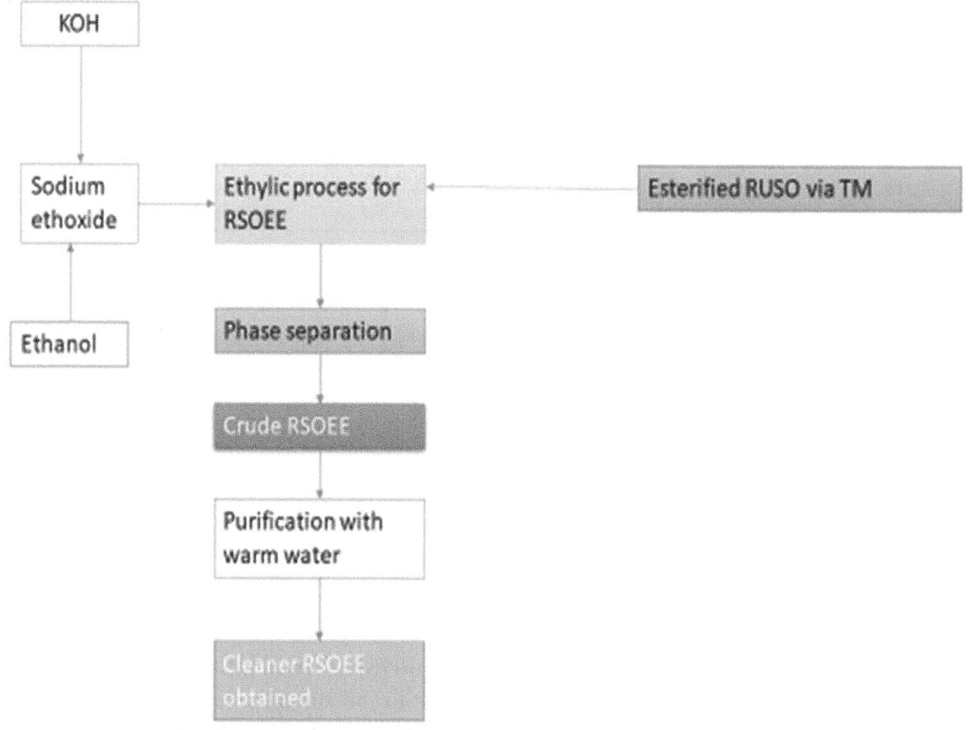

FIGURE 13.2 Protocol for RSOEE.

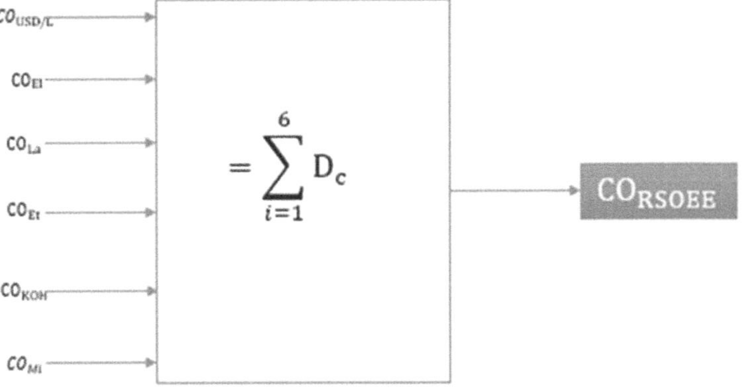

FIGURE 13.3   Schematic for the cost of biodiesel production.

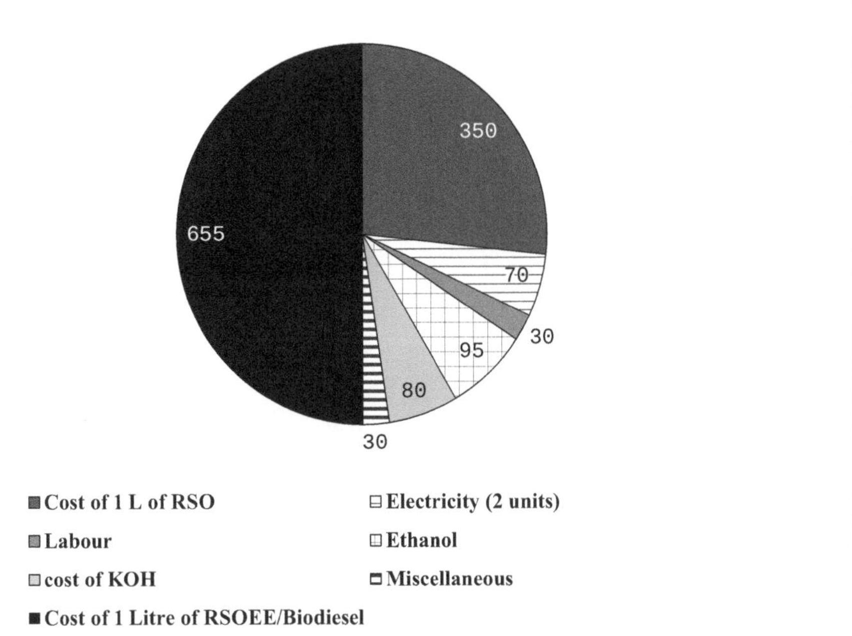

FIGURE 13.4   Cost assessment for RSO biodiesel production (Naira).

includes the expenditures related to the BP from a litre of RSO, ethanol, KOH, labour, electricity, catalyst preparation, and assorted. Figure 13.3 shows the schematic for the mathematical computation of the BPR from RSO, while Figure 13.4 exhibits the cost comparison of the component of BD from RSO.

13.3   RESULTS AND DISCUSSION

13.3.1   PROPERTIES OF THE RSO

The basic properties and fatty acid content of RSO were analyzed to assess its suitability for biodiesel production. Figures 13.5–and 13.6 show the fatty acid content and important properties

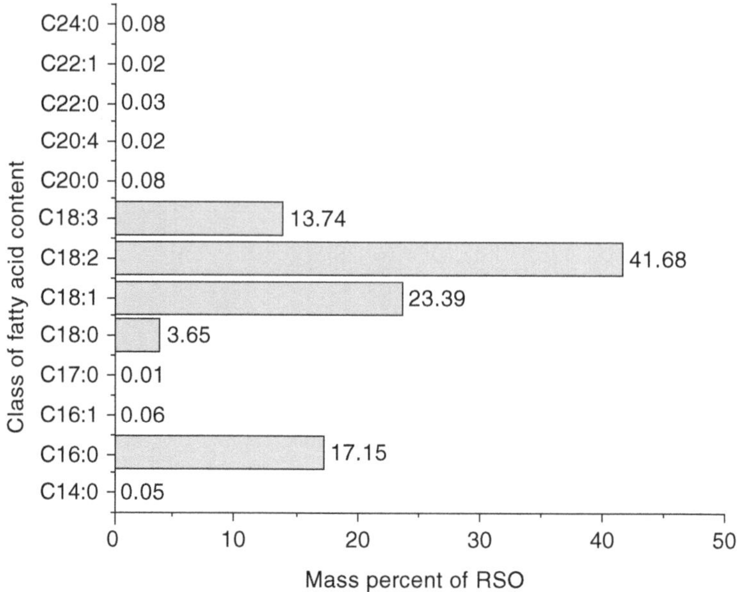

FIGURE 13.5 Fatty acid structure of RSO.

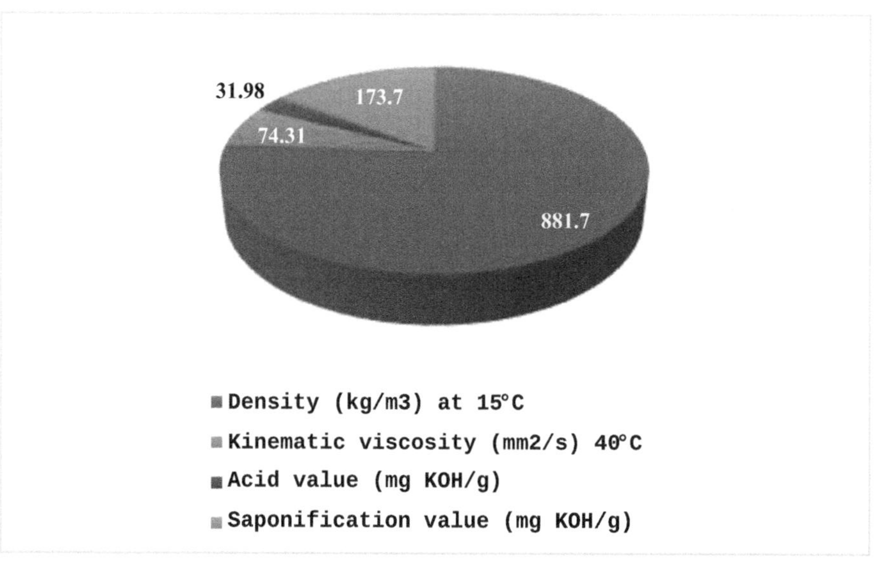

FIGURE 13.6 Basic properties of RSO.

of RSO, respectively. As shown in Figure 13.1, the unsaturation level of RSO is higher (55.6%) compared to the saturation level (44.4%). Niculescu et al. [18] suggested that RSO with high unsaturation can improve engine performance. Furthermore, the high kinematic viscosity and AV indicate the crude RSO can undergo an ethylic process without pre-treatment.

13.3.2 Determination of Optimal Experimental Conditions by Taguchi Method

The AV of esterified RSO under the considered established experiments, their SNRs, and complete mean SNR are presented in Table 13.4. The current study focused on AV minimization as

TABLE 13.4
AV of Pre-Treated RSO and Its SNR

Exp. No.	E/CRSO MR	DoC (wt%)	RT (mins)	Acid value	SNR
1	6:1	0.5	45	3.12	9.8831
2	6:1	1.0	60	2.10	6.44439
3	9:1	0.5	60	3.19	10.0758
4	9:1	1.0	45	5.10	14.1514

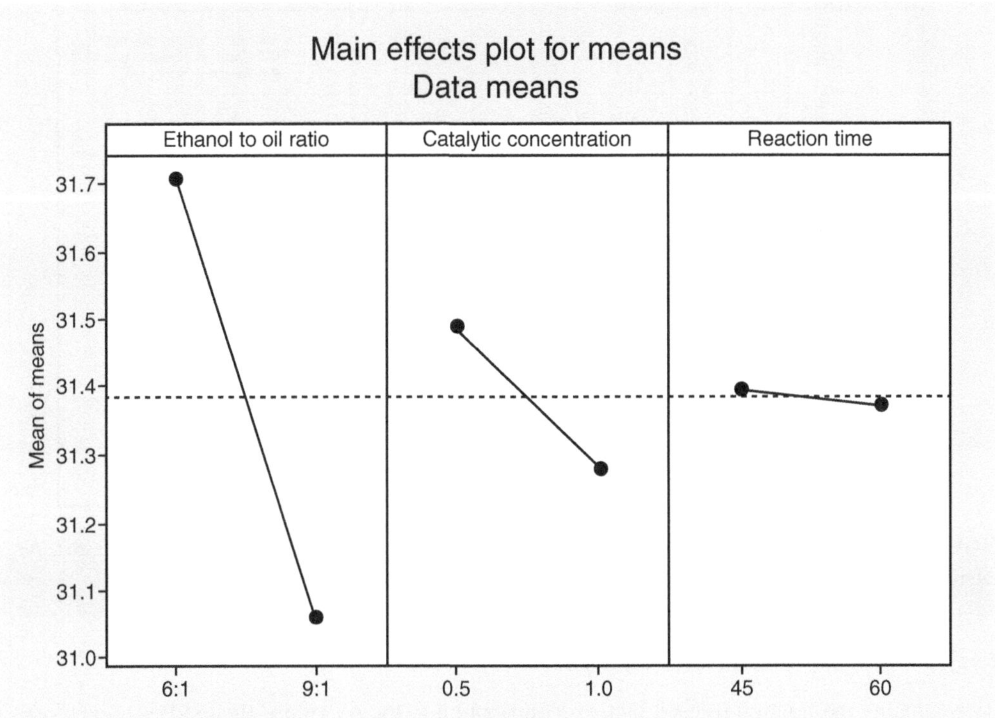

FIGURE 13.7 SNRL of each parameter at different levels.

the targeted function, so the smaller, the better (STB) model was implemented for SNR. As can be seen from the experiment outcomes, the minimum AV of the esterified RSO was achieved at 3.19 mg KOH/g in the final phase of the experiments, while the minimum AV was detected as 2.10 mg KOH/g with the second run of the experiments. However, this may not be the optimum combination of parameters. Figure 13.7 depicts the impact of each operational variable executed in the esterification process at assorted levels on the AV of the esterified RSO in relation to the signal-to-noise ratio level (SNRL). A higher SNRL indicates a greater influence of a specific parameter at that rank.

Furthermore, the supreme figures of SNRL in each portrayal represent the optimal operating conditions for those fixed parameters on pre-treated RSO. Consequently, the optimized effective variables to minimize the esterification of RSO having high FFA were predicted to be A (E/CRSO MR) at 6:1, B (DoC) at 0.5 wt.%, and C (RT) at 60 min.

TABLE 13.5
Percentage Contribution of Process
Parameters for Ethylic Esterification of RSO

Variables	SS_{fy}	% Contribution
E/CRSO MR	0.3856	30.20
DoC (wt%)	0.9210	59.30
RT (min)	0.0513	10.50

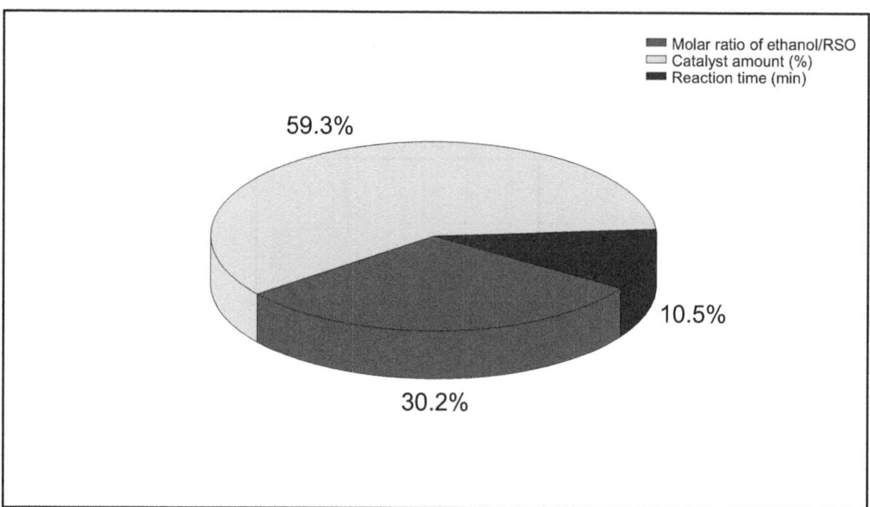

FIGURE 13.8 The proportion involvement of the operative esterification constraints on the AV of esterified RSO.

13.3.3 Analysis of Variance (ANOVA)

By evaluating the fraction of each factor's contribution to the AV of pre-treated RSO, the most significant operating variable was chosen. Table 13.5 summarizes the evaluated SS_f and percentage contribution, while the pictorial view of the contribution is portrayed in Figure 13.8. As seen, catalyst dosage was the most remarkable parameter with a 59.3% contribution to the AV of esterified RSO followed by the molar ratio of ethanol/RSO with 30.20%. The reaction time was the least compelling variable with 10.50%.

13.3.4 Fatty Acid Components & Key Fuel-Related Properties of RSOEE

Table 13.6 and Figure 13.9 present the basic properties and fatty acid components of RSOEE. As seen in Table 13.6, the fundamental properties of RSOEE align with the literature. To determine whether the synthesized cleaner fuel is suitable for automotive applications, it must meet the requirements outlined in endorsed standards. These standards ensure that the fuel possesses essential properties for effective operation in IC diesel engines. The key properties of RSOEE, as well as those of its counterparts and high-speed fossil fuel (HSFF), are shown in Table 13.6. The key properties of RSOEE are comparable to those of different species of RSO, such as heterogeneous- RSOME[19] and Malaysia spp. RSOME [20]. The slightly higher viscosity compared to HSFF indicates that diesel engine hardware will not need to be modified [1].

TABLE 13.6
Cogent Properties of RSOEE

Fuel Properties	RSOEE	Heterogeneous-RSOME Roschat et al [19]	Malaysia spp RSOME Ahmad et al. [20]	ASTM D6751	EN 41214	HSFF
Viscosity (mm²/s)	3.87	4.84	3.89	1.9–6.0	3.5–5.0	2.91
Density @ 15°C (kg/m³)	854	880	885	Nil	850–900	852
Flash point (°C)	154	184	152	Min 130	Min 120	76
Acid value (mgKOH/g)	0.25	0.35	0.42	0.50 max	0.50 max	0.13
Pour point (°C)	−4	–	–			ND

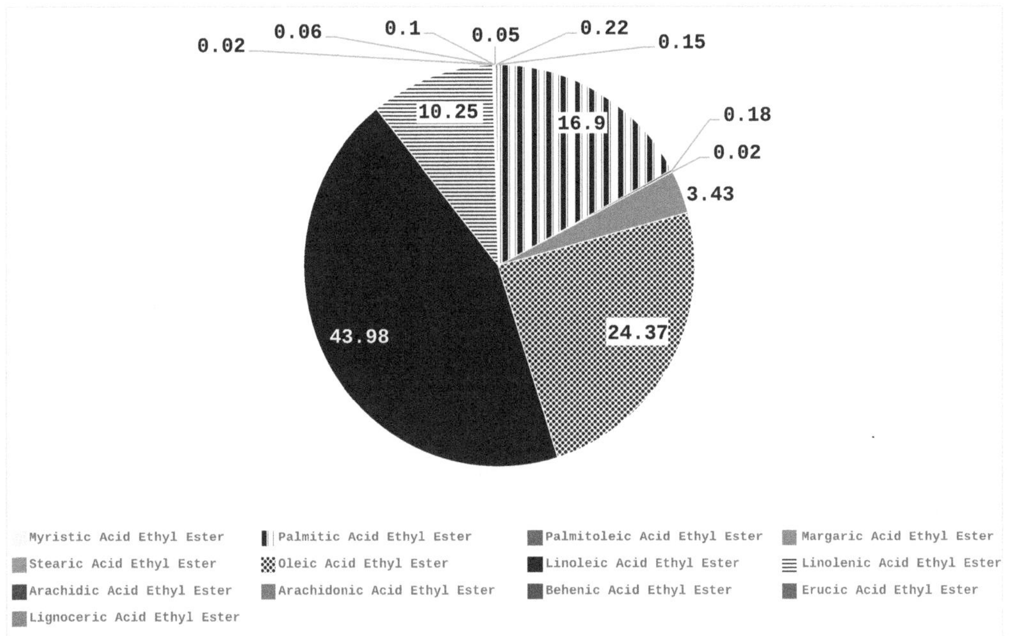

FIGURE 13.9 Fatty acid component of RSOEE.

As seen in Figure 13.9, the overall saturated and unsaturated (UN) fatty acids in the RSOEE were 45.12% and 54.46% by weight, respectively. Benjumea et al. [21] and Milano et al. [22] remarked that a high level of UN fatty acids in biodiesel leads to an increase in the emission of NOx, reducing thermal emissions but improving the cetane number.

13.3.5 Cost of RSOEE

The cost of producing RSOEE is highlighted in Table 13.7. It was found that the projected productioncharges of RSOEE ($1.24/litre) compared favourably with those in the literature. Cost variationscan be attributed to the ethylic and methylic production of oily feedstocks, as well as the nature ofthe catalyst. If mass-produced, the cost of RSOEE could be competitive with that of fossil diesel.However, given the rising diesel prices and the reliability of biodiesel feedstock sources, as wellas the environmental benefits, relevance to the local community, socioeconomic and topological stability, and energy security, the pricing of RSOEE can be justified [23].

TABLE 13.7
Cost of RSOEE with Biodiesels from the Different Non-Edible Feedstock

Generational Feedstock	Effective Parameters on the Aggregate Cost	Package	Production Cost (USD/ $)	Refs.
[a]MSW	Recovery ratio	Software	1.47/liter	Gaeta-Bernardi and Parente [24]
WCO	Cleaned oil	[b]AS	0.23–0.66/liter	Khan et al. [25]
Sludge	Removal and transesterification from a slush	Superpro Designer	0.67–1.07/kg	Chen et al. [26]
Karanja oil	Feedstocks expenditure, worth and assorted charge	Economic simulation structure	1.23/kg	Kumar et al. [27]
Castor	Plant scope and availability, oil content, labour accountability, and harvesting	AS	0.28/liter	Rahimi and Shafiei [28]
RSOEE		Excel package	1.42 per liter	Present study
Diesel	Nigeria	...	2.17/ liter	Price @ March. 2023

[a] *MSW = Municipal solid waste;*
[b] *AS = Aspen.*

13.4 CONCLUSION

In this experimental exploration, the production, characterization, and optimization of critical process parameters influencing the esterification process of a complete bio-based biodiesel synthesized from RSO have been considered and reported. Ethanol instead of methanol has been utilized with H2SO4 for the first time to pre-treat RSO, which has high FFA. Ethanol to CRSO molar ratio (ECM), temperature, and dosage of catalyst (H_2SO_4) were the three influencing variables considered for the acid-esterification of RSO exploring the Taguchi method (TAM). RSOEE was produced from esterified RSO under classical conditions. The fuel properties of RSOEE were developed, and the cost implications were studied. The following conclusion can be deduced:

- The optimal conditions for the AV of 2.10 mg KOH/g (Potassium Hydroxide per gram) were ECM (6:1), temperature (45 min), and H_2SO_4 (0.5 wt.%). Catalyst dosage was found to have the most influence on the reduction of FFA, followed by molar ratio, while temperature had minimal influence.
- Ethanol could be considered as a potential replacement for methanol to address the case of partial bio-based and to ensure 100% renewability during esterification and transesterification protocols.
- TAM proved appropriate in the pre-treating and model esterification process of RSO's high for complete bio-based RSOEE. The fundamental fuel attributes f of the RSOEE conformed with the global green fuel standards.
- The cost estimation for RSOEE (1.42 USD dollar) detected was more economically compared to fossil diesel (2.17 USD dollar). Also, the mass of RSOEE can be further reduced if mass-produced.

REFERENCES

[1] I. Khan, N. Prasad, A. Pal, and A. Yadav, "Efficient production of biodiesel from Cannabis sativa oil using intensified transesterification (hydrodynamic cavitation) method," *Energy Sources, Part A: Recovery, Utilization, and Environmental Effects*, vol. 42, no. 20, pp. 2461–2470, 2020.

[2] G. Knothe and L. Razon, "Biodiesel fuels.," *Progress in Energy and Combustion Science*, vol. 58, pp. 36–59, 2017.

[3] C. Jisieike, N. Ishola, L. Latinwo, and E. Betiku, "Crude rubber seed oil esterification using a solid catalyst: Optimization by hybrid adaptive neuro-fuzzy inference system and response surface methodology," *Energy*, vol. 263, p. 125734, 2023.

[4] O. Samuel and M. Okwu, "Comparison of Response Surface Methodology (RSM) and Artificial Neural Network (ANN) in modelling of waste coconut oil ethyl esters production," *Energy Sources, Part A: Recovery, Utilization, and Environmental Effects*, vol. 41, no. 9, pp. 1049–1061, 2019.

[5] O. Samuel, J. Emajuwa, M. Kaveh, E. Emagbetere, F. Abam, P. Elumalai, C. Enweremadu, P. Reddy, I. Eseoghene, and A. Mustafa, "Neem-castor seed oil esterification modelling: Comparison of RSM and ANFIS," Materials Today: Proceedings, 2023.

[6] H. Farouk, S. Zahraee, A. Atabani, M. Mohd Jaafar, and F. Alhassan, "Optimization of the esterification process of crude jatropha oil (CJO) containing high levels of free fatty acids: a Malaysian case study," *Biofuels*, vol. 11, no. 6, pp. 655–662, 2020.

[7] K. Okpalaeke, T. Ibrahim, L. Latinwo, and E. Betiku, "Mathematical modeling and optimization studies by Artificial neural network, genetic algorithm and response surface methodology: a case of ferric sulfate–catalyzed esterification of Neem (Azadirachta indica) seed oil," *Frontiers in Energy Research*, vol. 8, p. 614621, 2020.

[8] E. Betiku, V. Odude, N. Ishola, A. Bamimore, A. Osunleke and A. Okeleye, "Predictive capability evaluation of RSM, ANFIS and ANN: a case of reduction of high free fatty acid of palm kernel oil via esterification process," *Energy Conversion and Management*, vol. 124, pp. 219–230, 2016.

[9] M. Khan, S. Yusup, and M. Ahmad, "Acid esterification of a high free fatty acid crude palm oil and crude rubber seed oil blend: Optimization and parametric analysis," *Biomass and Bioenergy*, vol. 34, no. 12, pp. 1751–1756, 2010.

[10] M. Guo, W. Jiang, C. Chen, S. Qu, J. Lu, W. Yi, and J. Ding, "Process optimization of biodiesel production from waste cooking oil by esterification of free fatty acids using La^{3+}/ZnO-TiO_2 photocatalyst," *Energy Conversion and Management*, vol. 229, p. 113745, 2021.

[11] Y. Zhou, S. Niu and J. Li, "Activity of the carbon-based heterogeneous acid catalyst derived from bamboo in esterification of oleic acid with ethanol," *Energy Conversion and Management*, vol. 114, pp. 188–196, 2016.

[12] M. Zanuttini, M. Pisarello, and C. Querini, "Butia Yatay coconut oil: Process development for biodiesel production and kinetics of esterification with ethanol," *Energy Conversion and Management*, vol. 85, pp. 407–416, 2014.

[13] O. Samuel, M. Kaveh, T. Verma, A. Okewale, S. Oyedepo, F. Abam, C. Nwaokocha, M. Abbas, C. Enweremadu, E. Khalife, and M. Szymane, "Grey wolf optimizer for enhancing *Nicotiana tabacum* L. oil methyl ester and prediction model for calorific values." *Case Studies in Thermal Engineering*, vol. 35, p. 102095, 2022.

[14] J. Maroušek, S. Hašková, A. Maroušková, K. Myšková, R. Vaníčková, J. Váchal, M. Vochozka, R. Zeman, and J. Žák, "Financial and biotechnological assessment of new oil extraction technology," *Energy Sources, Part A: Recovery, Utilization, and Environmental Effects*, vol. 37, no. 16, pp. 1723–1728, 2015.

[15] U. Rajak, P. Chaurasiya, P. Nashine, M. Verma, T. Kota, and T. Verma, "Financial assessment, performance and emission analysis of *Moringa oleifera* and *Jatropha curcas* methyl ester fuel blends in a single-cylinder diesel engine," *Energy Conversion and Management*, vol. 224, p. 113362, 2020.

[16] S. Barman and T. Jash, "Study on optimization of process parameters for biodiesel production from waste cooking oil and its cost of production," *International Journal of Engineering Research & Technology*, vol. 4, no. 11, pp. 326–332, 2015.

[17] P. Shrivastava, T. Verma, O. Samuel, and A. Pugazhendhi, "An experimental investigation on engine characteristics, cost and energy analysis of CI engine fuelled with Roselle, Karanja biodiesel and its blends," *Fuel,* vol. 275, p. 117891, 2020.

[18] R. Niculescu, A. Clenci, and V. Iorga-Siman, "Review on the use of diesel–biodiesel–alcohol blends in compression ignition engines," *Energies*, vol. 12, no. 7, p. 1194, 2019.

[19] W. Roschat, T. Siritanon, B. Yoosuk, T. Sudyoadsuk, and V. Promarak, "Rubber seed oil as potential non-edible feedstock for biodiesel production using heterogeneous catalyst in Thailand," *Renewable Energy*, vol. 101, pp. 937–944, 2017.

[20] J. Ahmad, S. Yusup, A. Bokhari, and R. Kamil, "Study of fuel properties of rubber seed oil based biodiesel," *Energy Conversion and Management*, vol. 78, pp. 266–275, 2014.

[21] P. Benjumea, J. Agudelo, and A. Agudelo, "Effect of the degree of unsaturation of biodiesel fuels on engine performance, combustion characteristics, and emissions," *Energy & Fuels*, vol. 25, no. 1, pp. 77–85, 2011.

[22] J. Milano, H. Ong, H. Masjuki, A. Silitonga, F. Kusumo, S. Dharma, A. Sebayang, M. Cheah, and C. Wang, "Physicochemical property enhancement of biodiesel synthesis from hybrid feedstocks of waste cooking vegetable oil and Beauty leaf oil through optimized alkaline-catalysed transesterification," *Waste Management*, vol. 80, pp. 435–449, 2018.

[23] V. Perumal and M. Ilangkumaran, "Experimental analysis of operating characteristics of a direct injection diesel engine fuelled with *Cleome viscosa* biodiesel," *Fuel*, vol. 224, pp. 379–387, 2018.

[24] A. Gaeta-Bernardi and V. Parente, "Organic municipal solid waste (MSW) as feedstock for biodiesel production: A financial feasibility analysis," *Renewable Energy*, vol. 86, pp. 1422–1432, 2016.

[25] H. Khan, C. Ali, T. Iqbal, S. Yasin, M. Sulaiman, H. Mahmood, M. Raashid, M. Pasha, and B. Mu, "Current scenario and potential of biodiesel production from waste cooking oil in Pakistan: An overview," *Chinese Journal of Chemical Engineering*, vol. 27, no. 10, pp. 2238–2250, 2019.

[26] J. Chen, R. D. Tyagi, J. Li, X. Zhang, P. Drogui, and F. Sun, "Economic assessment of biodiesel production from wastewater sludge," *Bioresource Technology*, vol. 253, pp. 41–48, 2018.

[27] D. Kumar, B. Singh, A. Banerjee, and S. Chatterjee, "Cement wastes as transesterification catalysts for the production of biodiesel from Karanja oil," *Journal of Cleaner Production*, vol. 183, pp. 26–34, 2018.

[28] V. Rahimi and M. Shafiei, "Techno-economic assessment of a biorefinery based on low-impact energy crops: A step towards commercial production of biodiesel, biogas, and heat," *Energy Conversion and Management*, vol. 183, pp. 698–707, 2019.

14 Energy Alternatives and Efficiency Options for Sustainable Development in Nigeria

Daniel R.E. Ewim, Samuel Abiodun Olatubosun, and Sogo Mayokun Abolarin

14.1 INTRODUCTION

The need for energy for almost every human activity and its requirement to drive nearly all systems, equipment, and machinery for production cannot be over-emphasized. These needs have made every issue associated with energy attract serious attention, as man's existence is very dependent on one form of energy or the other. In developing a nation like Nigeria, where there are urgent and great needs for rapid development in all sectors of the economy, the need for sufficient, effective, and efficient sources of energy is a prerequisite and thus mandatory. It has been argued by various researchers that electricity is an initiator and propeller of rapid economic growth and development [1–3]. Therefore, a developing nation like Nigeria requires a regular supply of electricity to boost social and economic development locally, regionally, and globally. The nation needs to come out of the lingering energy crisis by embracing a mix of energy sources, especially renewable and nuclear energy sources, which have proven to be environmentally friendly and contributing to the achieving Sustainable Development Goal 7 (SDG 7) of the United Nations. The potential of the country to harness the various energy alternatives such as hydro, solar, wind, nuclear, geothermal, natural gas, and biogas, among others, that could provide immediate remedies to the abysmal energy supply in the country was examined in this research.

In a bid to achieve the United Nations SDG 7 (Affordable and Clean Energy) and SDG 13 (Climate Actions), many countries have committed themselves to achieving sustainable development by gradually transitioning from the consumption of conventional energy sources to clean energy sources. This is because the traditional sources are not friendly to the environment and pose a significant threat to life and living. Clean energy sources, on the other hand, are renewable sources that are environmentally friendly, abundantly available in almost every country, and are now being promoted, adopted, and utilized in many countries across the world to generate a significant portion of the energy required to achieve economic growth and development.

Nwulu and Agboola [4] stated that most developed nations are increasingly adopting renewable energy (RE) to satisfy huge energy requirements. In the study, Brazil was identified to be deploying ethanol as fuel to meet its energy needs for transportation. Considering the state of the energy industry (generation, distribution, and consumption) in Nigeria, the modes and capacity of power being generated, there is a need to improve the status of the nation in these respects by adopting all possible means of boosting power generation. Revamping, modernizing, and increasing the capacity

DOI: 10.1201/9781032651958-14

of the existing hydropower plants, upscaling the existing off- and mini-grid solar power plants, and introducing large-scale grid-tied solar photovoltaic and wind power stations as well as nuclear energy into the energy mix equation would contribute in boosting the energy prosperity of Nigeria. They also noted that Nigeria could effectively meet its energy requirements and also export electric power to the neighboring states due to the enormous renewable energy resources available in different parts of the country. Cogent ideas on solving the nation's lingering energy crises were also suggested for consideration.

Furthermore, to reduce her unemployment rate and its associated challenges to the nation (such as poverty, increasing waves of crimes, low productivity, etc.), emergency steps need to be taken to rapidly harness the nation's potential to generate energy from renewable and nuclear sources. In addition, climate change and its attendant consequences may not be mitigated given the nation's current dependence on fossil fuels. Investing in renewable energy can help to reduce Nigeria's dependence on fossil fuels and improve energy security. This is because renewable energy can help to increase access to energy, reduce carbon emissions, and create employment opportunities [5–12].

Nigeria has significant potential for renewable energy, particularly in solar and wind, and a shift toward renewable energy would align with the country's commitment to reducing its carbon footprint. Transitioning to renewable energy resources as well as deployment to meet a larger percentage of the nation's energy needs is the strategy toward reducing global warming and unemployment. As of today, the two most promoted carbon emission reduction measures are energy efficiency (EE) and renewable energy (RE) utilization. While there are abundant renewable energy resources, which when sustainably deployed, will largely contribute to reducing emissions, prioritizing energy efficiency ahead of renewable would not only save the environment and infrastructure but also the investment, operating and maintenance costs of the energy supply components as well. Energy efficiency ought to be first implemented so as to reduce the size of the renewable energy infrastructure required to achieve the nation's energy needs [13].

The recognition that renewable energy technologies represent the future of energy, coupled with the belief that traditional sources of energy, such as oil and gas (which Nigeria heavily relies on), are exhaustible and will ultimately cease to dominate the energy sector, has spurred research into renewable energy in Africa. Nigeria's economy is powered by crude oil exports, which contribute slightly over 30% to its gross domestic product (GDP) and account for more than 70% of the government's total earnings [14–17]. Irrespective of this huge earning from petroleum, the nation is still experiencing inadequacies in power production and supply related problems. With these, the level of power generation and distribution in Nigeria is not commensurate with the vast energy potential and energy demand of the country.

After recognizing the challenges and obstacles faced by Nigeria in generating, transmitting, and distributing sufficient power, a number of solutions can be implemented to address these issues. Such solutions include the proper maintenance and overhaul of existing power facilities, the reliable supply of gas, adequate financial support and backing from the government, the establishment of efficient billing and collection systems, the implementation of measures to ensure the security of power facilities across the country, the utilization of gas flared in oil and gas fields for the efficient generation of electricity, and so on.

Given the wealth of energy resources Nigeria possesses, particularly its significant renewable energy potential such as large and small hydroelectric power, solar energy, biomass, wind, hydrogen utilization possibilities, and the development of geothermal and ocean energy, the nation could overcome its energy crisis and achieve energy security within a few years. The abundant energy resources available in Nigeria are summarized in the National Energy Master Plan, approved in April 2022, as shown in Table 14.1.

TABLE 14.1
Renewable Energy Resources in Nigeria [18]

Sources of Energy	Capacity
Hydropower (Large Scale)	24,000 MW
Hydropower (Small Scale)	35,00 MW
Fuelwood	11 million hectares
Animal Waste	245 million tonnes/year (2001)
Crop Residue (waste)	72 million tonnes of agric land
Municipal waste	30 million tonnes/year
Solar radiation	3.5–7.0 kWh/m^2/day
Wind	2–4 m/s (10 m height)

14.1.1 CHAPTER OUTLINE

Based on the foregoing, this book chapter provides an overview of energy generation and consumption in Nigeria, highlighting the trends and challenges in the energy sector. It also discusses alternative sources of energy and their potential in Nigeria, along with the pros and cons of each source. The chapter further delves into Nigeria's nuclear power program, including the journey so far, the factors driving the country's nuclear energy mix, and the potential benefits and drawbacks. Additionally, the chapter examines the importance of energy efficiency and conservation measures and strategies, including the role of research, development, and training in promoting energy efficiency and conservation. It highlights the challenges of implementing these measures in Nigeria's energy industry and recommends ways to improve the level of energy efficiency and conservation. The chapter concludes with a summary of its key points and a call to action for sustainable energy development in Nigeria.

14.2 ENERGY GENERATION AND CONSUMPTION (TRENDS)

Energy is a critical infrastructure for the economic growth and development of any nation. It has become an important factor that distinguishes the developed countries from the underdeveloped ones. The extent of energy utilization and demand in a national economy are, on a large scale, indicative of its level of economic, social, and political development [19,20]. Enhanced food production, sufficient healthcare, effective transportation, appropriate housing, and increased employment opportunities—all indicators of an improved standard of living—rely on the availability of energy. This demonstrates that for any nation to rise among the ranks of developed countries globally, a dependable, affordable, and sustainable energy supply aligned with the United Nations SDG 7 is essential. Countries capable of providing affordable and reliable electricity tend to experience economic growth, marked by increased GDP, which in turn lays the foundation for national development [21–24]. In the pursuit of sustainable energy, numerous sources have been developed. However, due to the surge in global population, relying on a single energy source alone cannot satisfy the growing energy demands. To meet the increasing demand for energy, many countries have embraced the approach of combining various energy sources as the most viable strategy.

Nigeria just like any sub-Saharan African country has had its own share of energy deficit, and with little that has been done in solving this energy deficit, economic growth, sustainable development, and improvement of citizens' welfare have been on the decline [25]. As with other developing nations, if Nigeria is to achieve its goal of higher economic growth, it is imperative to have access to plentiful and cost-effective electricity via a diverse mix of power generation, as noted in reference [26].

Nigeria's journey in electric power generation has progressed through several stages. The initial phase began with the installation of a diesel-fired plant in Ijora, Lagos, in 1896 [27]. Subsequently, numerous small-scale diesel-powered plants were established in various locations, including Ibadan, Kano, Warri, and Port-Harcourt [28]. Following independence in the early 1960s, hydro and thermal power generation started in Kainji, Jebba, Shiroro, Afam, Delta (Ughelli), Sapele (Ogorode), and Egbin (Lagos) to supplement the diesel-powered plants [29,30]. Although electricity has been generated for over a century, Nigeria's current electricity demand far exceeds the supply, adversely impacting the nation's socio-economic and technological development. Only about 51–70% of Nigeria's population is connected to the national electricity grid [31,32], and the connected population often experiences power issues 60% of the time [33]. The load-shedding incidents have led individuals and manufacturing companies to install self-electrical generating units (commonly known as generators) [34]. Due to the use of generators, energy costs account for 40% of production costs in Nigeria [35].

A power grid transmission system has been established to connect all the state capitals to the large power stations in Nigeria. The country's abundance of natural fuels, such as coal, crude oil, natural gas (estimated at 180 trillion cubic feet), and rivers, has led to a focus on steam and gas turbines as well as hydropower plant generation. As a result, numerous power stations have been installed across the country [36]. By 2009, Nigeria's installed capacity for electricity generation was 5000 MW, but only 2900 MW was generated as of November 2009 [37]. Power generation in Nigeria has faced significant challenges due to societal, governmental, natural, and political factors [29,38]. Recently, renewable energy, primarily in the form of solar energy, has been added to the generation mix, with new projects being developed [39,40].

Over the years, Nigeria has relied on its petroleum reserves and aging hydro-power plants for electricity generation to supply 40% of its total population connected to the national grid. Aliyu et al. [27] conducted an extensive literature survey on the prevailing energy issues in Nigeria. The study analyzes electricity generation and the government's future expansion plans over a 20-year period. The plan includes introducing new electricity generation technologies that were not in use in the base year (2010).

14.3 ALTERNATIVE SOURCES OF ENERGY

According to the literature [27], clean energy technologies have been found to have lower CO_2 emissions. Nigeria is fortunate to have abundant natural energy resources, including crude oil, natural gas, coal and lignite, wind, solar radiation, biomass, and nuclear. Nigeria boasts the largest natural gas reserves in Africa and the seventh largest in the world, with an estimated 37 billion barrels of untapped oil reserves [41]. However, Nigeria's natural gas reserves have been underdeveloped due to limited infrastructure, leading to the flaring of associated gas with oil production. To accelerate growth in the sector, policies to ban gas flaring, the expansion of the Liquified Natural Gas (LNG) infrastructure, and the development of regional pipelines are expected to be implemented for both export and domestic use in electricity generation. Investment policies and regulatory frameworks have been uncertain, causing delays in the development of oil and gas exploration projects, including LNG projects. Although Nigeria has coal reserves estimated at 40 billion tons, they have not been fully utilized for electricity generation, with the majority of focus being on oil, gas, and hydropower. Hydro, as a renewable source, has played a significant role in Nigeria's power sector. Around the world, the proportion of renewable energy has been rising, with hydropower and wind power expected to contribute the largest share of the anticipated growth in total renewable generation [42]. Nigeria's heavy reliance on petroleum has impeded the development of alternative fuels, making it essential to achieve a diverse mix of energy supplies for improved energy security. The energy crisis in Nigeria has highlighted the need to explore alternative energy sources to satisfy the nation's growing energy demands. Various energy sources, such as hydro, solar, wind, nuclear, geothermal, natural gas, and biogas, are available and could be harnessed to provide immediate solutions to Nigeria's energy crisis. In this section, we will examine these energy alternatives and their potential to fulfill Nigeria's energy needs.

14.3.1 HYDROPOWER

Hydropower is a renewable source of energy that harnesses the power of water to generate electricity. Nigeria has the potential to generate up to 13,000 MW of electricity from its numerous rivers and dams. However, only a fraction of this potential has been harnessed, with hydropower accounting for only 15% of Nigeria's total electricity generation capacity. The potential for hydropower in Nigeria remains largely untapped, and there is a need for the government to invest in the development of this source of energy.

14.3.2 SOLAR POWER

Solar power is another renewable source of energy that harnesses the power of the sun to generate electricity. Nigeria has abundant sunshine, which makes it an ideal location for the deployment of solar power. The country has the potential to generate up to 427,000 MW of electricity from solar power. However, only a small fraction of this potential has been harnessed, with solar power accounting for less than 1% of Nigeria's total electricity generation capacity. The potential for solar power in Nigeria is enormous, and the government needs to invest in the development of this source of energy.

14.3.3 WIND POWER

Wind power is a renewable source of energy that harnesses the power of the wind to generate electricity. Nigeria has the potential to generate up to 11,000 MW of electricity from wind power. However, only a fraction of this potential has been harnessed, with wind power accounting for less than 1% of Nigeria's total electricity generation capacity. The potential for wind power in Nigeria is significant, and there is a need for the government to invest in the development of this source of energy.

14.3.4 NUCLEAR POWER

Nuclear power is an alternative source of energy that harnesses the power of nuclear reactions to generate electricity. Nigeria has recently signed a deal with Russia's state-owned Nuclear Corporation (ROSATOM) to collaborate on the design, construction, and operation of 4000 MW capacity nuclear power plants by 2035. This would be achieved by installing four nuclear power plants in two certified locations in Nigeria. The potential for nuclear power in Nigeria is significant, and the government needs to invest in the development of this source of energy.

14.3.5 GEOTHERMAL POWER

Geothermal power is a renewable source of energy that harnesses the heat generated from the earth's core to generate electricity. Nigeria has a potential for geothermal power, but there are a few locations where it could be harnessed. The potential for geothermal power in Nigeria is limited, and the government needs to invest in the exploration and development of this source of energy.

14.3.6 NATURAL AND BIOGAS

Natural gas and biogas are non-renewable and renewable sources of energy, respectively, that are harnessed to generate electricity. Nigeria has the seventh-largest natural gas reserve in the world and the largest in Africa. The country has the potential to generate up to 12,500 MW of electricity from natural gas. Biogas, on the other hand, is produced from the decomposition of organic matter and can be harnessed from waste products such as animal manure and sewage. The potential for biogas in Nigeria is significant, and there is a need for the government to invest in the development of this source of energy.

To achieve a sustainable energy future, Nigeria needs to develop a comprehensive energy policy that promotes the development of these alternative energy sources while minimizing the negative environmental impacts associated with energy production and consumption.

14.4 INTRODUCING NUCLEAR ENERGY INTO NIGERIA'S ENERGY MIX

14.4.1 INTRODUCTION

According to sources [43,44], Nigeria has decided to include nuclear power in its energy mix, as evidenced by its updated energy policy and the ongoing nuclear power program. This section intends to highlight the main factors motivating Nigeria to pursue its nuclear power program. These driving forces primarily stem from the insufficiency and unsustainability of the existing hydropower resources, the expected depletion of gas reserves by the early 2030s, and the forecasted growth in energy demand.

14.4.2 NUCLEAR POWER PROGRAM IN NIGERIA

14.4.2.1 The Journey So Far

As pointed out by Lowbeer-Lewis [45], Nigeria's nuclear program began in the late 1970s, initially established primarily in response to South Africa's acquisition of nuclear weapons and India's testing of a nuclear device. However, over the years, the country has shifted its focus toward using nuclear power for peaceful purposes. This led to the establishment of the Nigeria Atomic Energy Commission (NAEC) in 1976. NAEC was designed to serve as the central national agency responsible for developing a framework and master plan to explore, exploit, and harness atomic energy for peaceful applications in all aspects to support Nigeria's socio-economic development [46]. To expedite the development and deployment of nuclear technologies in the country, two other nuclear energy centers, the Center for Energy Research and Training (CERT) in Zaria and the Center for Energy Research and Development (CERD) in Ife, were also established [47].

Other nuclear-based centers in Nigeria include the Centre for Nuclear Energy Studies and Training (CNEST) at the University of Port Harcourt, the Radiological Protection and Nuclear Safety Centre (RPNSC) at the University of Ibadan, the Centre for Nuclear Energy Research and Training at the University of Maiduguri, the Nuclear Technology Centre (NTC) at the Sheda Science and Technology Complex (SHESTCO) in Abuja, the FGN-IAEA Marine Contamination Coastal Field Monitoring Station in Koluama II (MCCFMS), and the Nuclear Safety and Radiation Protection Directorate (NSRPD) at the Nigerian Nuclear Regulatory Authority (NNRA) in Abuja.To align with international standards, the NNRA was established by the Nuclear Safety and Radiation Protection Act in 1994 and tasked with regulating all nuclear activities in the country, including enforcing all nuclear laws and regulations. The NNRA commenced operations in 2001 and has since begun implementing regulations [46,48]. In 2004, the poor state of electricity generation in Nigeria prompted the federal government to create an Inter-Ministerial Committee on Energy Resources to assess the country's energy resources, availability, and estimated derivable electricity, among other factors. The committee's report identified nuclear energy as a significant potential source, leading to the inauguration of an Inter-Ministerial Committee and Technical Committee on Nuclear Power Project (NPP) in 2005. The NPP committee assessed the feasibility of deploying the NPP, and in 2006, the Nigeria Atomic Energy Commission (NAEC) became fully operational [49,50]. Under the supervision of the Federal Ministry of Power, the Nigerian government signed a technical roadmap for the country's nuclear energy program a year after NAEC's inauguration. The roadmap aimed to provide guidance for reactor design and certification, regulatory procedures, and infrastructure development for the nuclear power project. In 2009, the government adopted a strategic plan to proceed with the projects, and a cooperation agreement with Russia followed. The agreement involves the exploration and mining of uranium in Nigeria, as well as the construction of a Russian research reactor [48,51]. In 2010, NAEC identified four potential sites for nuclear power stations in Nigeria, strategically located across the country's four geopolitical regions. These sites were Geregu/Ajaokuta in Kogi state for the North-central zone, Itu in Akwa Ibom state for the southeast, Agbaje, Okitipupa in Ondo state for the Southwest zone, and Lau in Taraba state for the North-east zone. An environmental and siting team from NAEC conducted further evaluations and assessments of the sites, ultimately selecting Geregu and Itu as the preferred locations for nuclear power plant installations in 2014. In 2015, Nigeria signed an agreement with ROSATOM to collaborate on the design, construction, and operation of nuclear power plants with a total capacity of 4000 MW by 2035 at the two certified locations, each planned to house two plants. The first plant, with a capacity of 1000 MW, is expected to begin operation by 2025 [46,48,52]. To achieve their objectives, the Federal Government reconstituted the Russian-Nigerian Joint Coordination Committee (JCC) of NAEC in July 2021 after three years of inactivity. This move aimed to revive the relationship between the two countries concerning the use of nuclear energy. The JCC was established to facilitate effective negotiations between Nigeria and ROSATOM regarding the implementation of their various agreements [53].

14.4.3 FACTORS DRIVING NIGERIA NUCLEAR ENERGY MIX

Several factors are driving Nigeria's push toward nuclear energy. These factors include Nigeria's long-term vision for industrial development, rising demand for electricity, a rapidly growing population, and limited availability of hydro and gas resources. Additionally, other considerations such as energy diversification, climate change mitigation, and industrialization are also playing a role in Nigeria's decision to pursue nuclear energy as an option for economic diversification and transformation.

14.4.3.1 Visions of Nigeria's Development Plans

Nigeria holds the title of the most populous African nation and ranks seventh globally, with a population of approximately 206 million people. With an annual average growth rate of 2% and a land area of around 1 million km^2, Nigeria also has the largest economy on the continent. Its economy, which grows at an annual rate of 6%, relies heavily on the export of oil products.

Figure 14.1 shows Nigeria's annual GDP per capita since 1960, which was 2,097 Dollars ($) in the year 2020. As illustrated in Figure 14.2, it can be affirmed in terms of economic development Nigeria is still very behind compared with developed nations of the world. The country's electricity production capacity is about 7000 MW but as a result of weak infrastructure, gas supply problems, and water shortage only about 4000 MW gets to the national grid [55].

Developed countries have greatly benefited from long-term development strategies, which have resulted in abundant electricity supplies. Nations such as Japan, Sweden, South Korea, the United States, and Australia have implemented electricity supply plans spanning 30 years or more

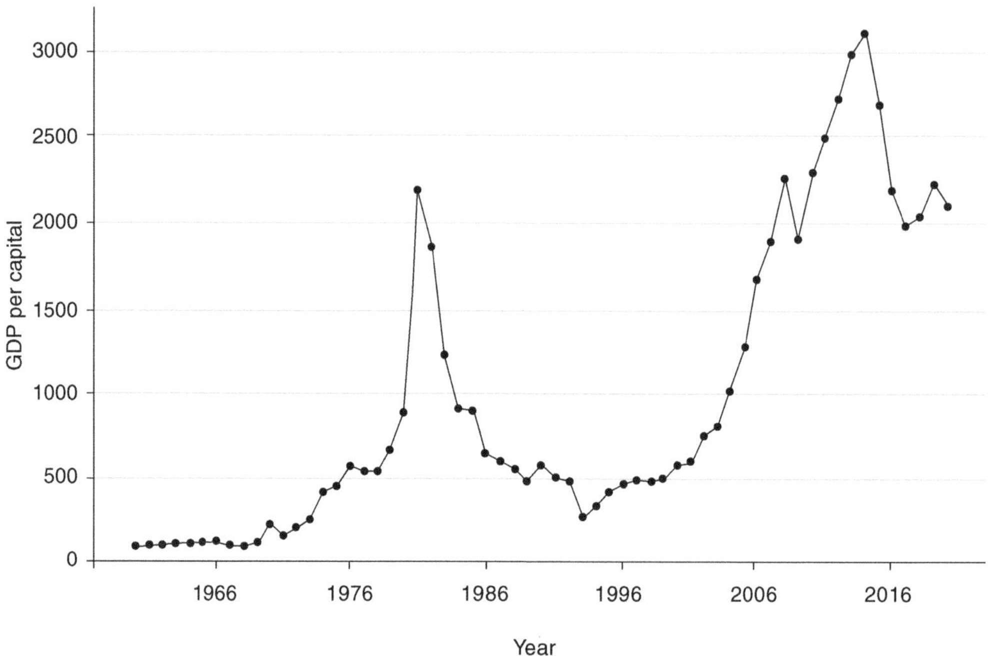

FIGURE 14.1   Nigerian GDP per capital from 1960 to 2020 [54].

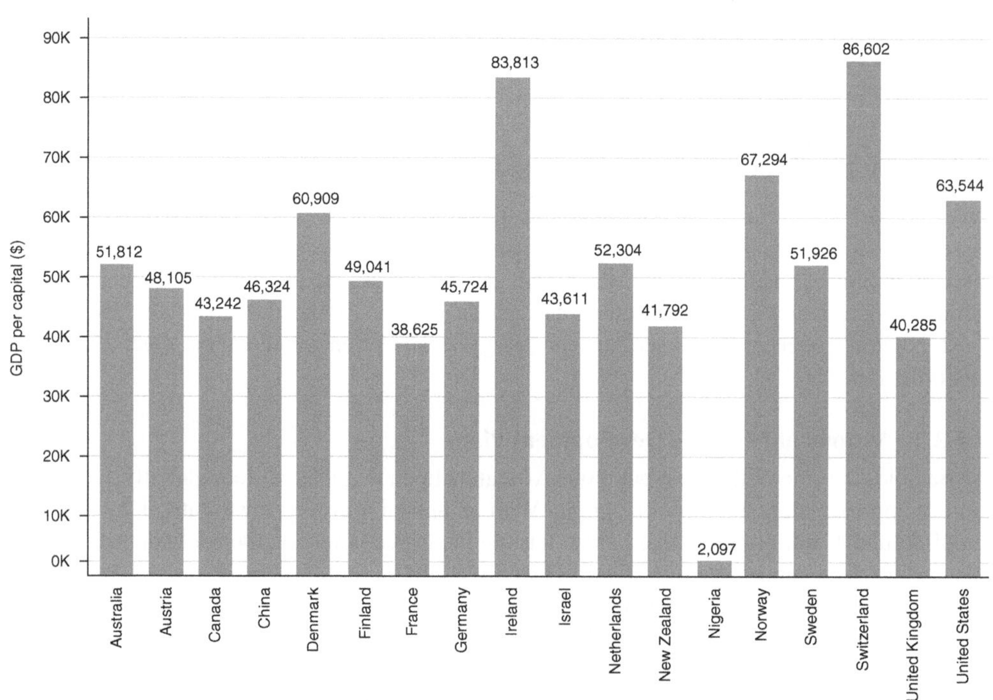

FIGURE 14.2   Comparing Nigeria GDP with other developed nations of the world [54].

[56–58]. One key factor that has advanced these countries' development beyond Nigeria is their increasing focus on long-term energy planning for periods exceeding 30 years [59,60]. Successive Nigerian governments have introduced various development plans since gaining independence, but challenges such as corruption, plan indiscipline, lack of continuity in plans, projects, and policies, and inadequate synergy between plans and budgets have prevented them from achieving sustainable development [61–63].

Several statutory planning bodies have been established by the Nigerian government, including the Joint Planning Board (JPB), National Council on Development Planning (NCDP), National Economic Council (NEC), and Federal Executive Council (FEC), among others. These bodies are currently implementing the Medium-Term National Development Plan (MTNDP) for 2021–2025 and are also working on Nigeria Agenda 2050, which will succeed the Economic Recovery and Growth Plan (ERGP) for 2017–2020 and Nigerian Vision 20:20. These plans are being prepared through a highly participatory, consultative, and inclusive process, with active involvement from the private sector and sub-national entities. The MTNDP serves as a guide for annual budget preparation and is expected to build on government policies, initiatives, programs, and projects. It will also guide economic management and budgeting processes in the short to medium term and provide clear policies and plans for potential domestic and foreign investors [64]. A draft of the MTNDP for 2021–2025 was published in March 2021 [65].

The Nigerian government is working on a long-term plan called Nigeria Agenda 2050, which aims to guide the nation's development and position Nigeria among the top ten global economies, with a projected GDP per capita of around $15,960. The plan envisions an export-based, industrialized, and diversified economy with high electricity demand. If successfully executed, the plan will transform Nigeria into a leading global economy [64,66,67].

To achieve the country's long-term aspirations for high-income levels and long-term growth, which are expected to drive economic infrastructure expansion, a large amount of baseload electricity will be required. Research shows that energy released in a nuclear fission reaction is ten million times greater than that released when burning fossil fuels [68], making nuclear energy an ideal candidate to provide the energy density needed for a strong baseload [69,70]. Introducing nuclear energy as part of Nigeria's energy mix has the potential to create jobs in various areas, leading to a diversified economy.

14.4.3.2 Increasing Electricity Demand

The United Nations project that the overall population of Nigeria will reach about 401.31 million by the end of the year 2050 [71], which is expected to give rise to an expanded economy resulting in social amenities and infrastructure expansion. As projected by the Energy Commission of Nigeria (ECN), the country's electricity demand is expected to increase, as shown in Figures 14.3 and 14.4 in the year 2050 [72]. Hence, the current 7000 MW electricity generation capacity has to undergo a serious upgrade if the country is ever going to meet up with the forecast or continue on the path of a socio-economic downturn [73].

As shown in Figure 14.3, there is a need for the country to first reconsider its energy consumption approach either via energy efficiency (technology being used to consume the energy) or conservation (the behaviour of the energy consumers). This is very important because procurement and use of energy-inefficient appliances will absolutely continue to lead to a rise in energy consumption [8,74]. When it has been identified that more power stations are needed to achieve economic growth and acquire more electricity-producing plants to meet the ever-growing demand [8,74,75].

Figure 14.4 makes it lucid that in order to meet the supply target sustainably, other energy sources will play a significant role [73]. However, the country's power generation is still largely from fossil fuels, which include coal, oil and natural gas. The combustion of fossil fuels releases greenhouse gases into the atmosphere, which contributes to global warming resulting in climate change [13,76]. Hence, the contribution of nuclear power to meet the increasing energy demand and most

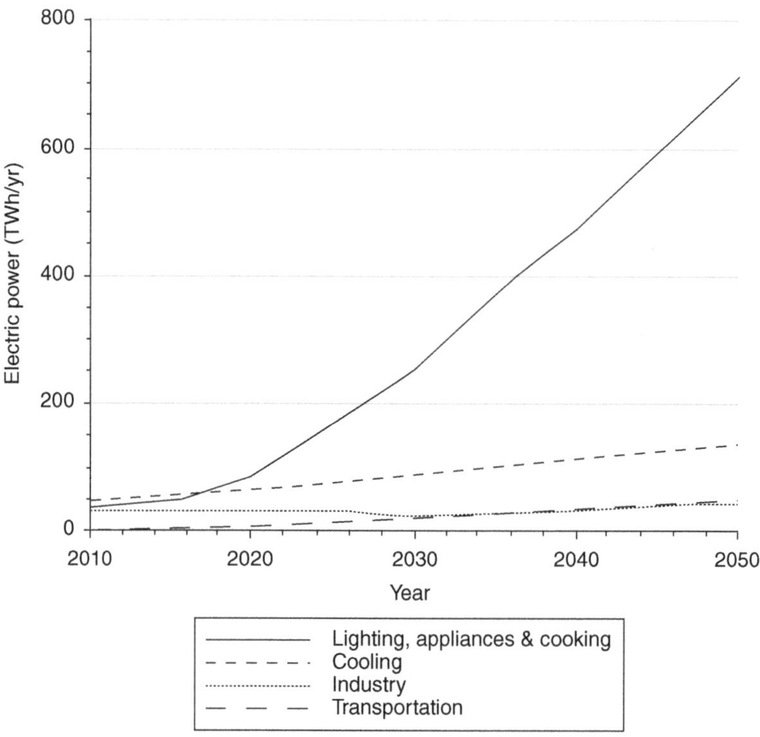

FIGURE 14.3 Nigerian Energy demand by sectors in 2050 [72].

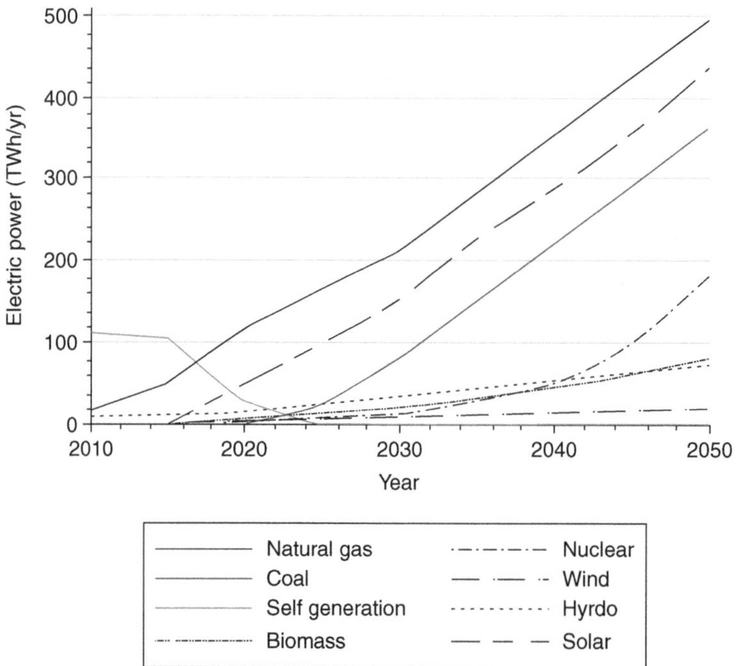

FIGURE 14.4 Nigeria energy 2050 electrical power generation projection and their sources [72].

importantly providing clean, buoyant, affordable and abundant energy makes it an important source of the generation mix for Nigeria [77–79].

14.5 ENERGY EFFICIENCY AND CONSERVATION MEASURES AND STRATEGIES

Energy efficiency is widely recognized as a fundamental technology in the push for clean energy. It involves producing the same level of energy services while using less energy input and without negatively affecting comfort. This means reducing energy use per energy service without compromising their quality or level, including services such as lighting, cooling, heating, manufacturing, cooking, transport, and entertainment [80]. A thorough energy audit is critical for good energy management, just as a financial audit is essential for good financial management [81,82]. The energy audit is conducted to determine how, where, and when energy is being used or wasted, identify opportunities for low-, medium-, and high-cost energy management, reduce energy usage, and formulate prioritized recommendations for energy performance improvements in all energy-consuming sectors. Energy conservation refers to behaviors that contribute to a reduction in energy consumption [74]. One example of an energy-efficient measure is the procurement of energy-consuming appliances that do not consume a lot of energy to provide the intended service. For example, when lighting is needed to improve illumination in a particular space, the installation of lighting technology such as Light Emitting Diode (LED) over incandescent light bulbs would increase lighting efficiency. Similarly, installing air conditioning systems with an Energy Efficiency Ratio (EER) that complies with the minimum energy performance standard contributes to improved energy efficiency compared to those with an EER less than 2.8. To promote energy conservation, applying human knowledge or artificial intelligence to control how energy-consuming appliances are used can significantly reduce energy consumption in all energy-consuming sectors. This includes turning off these appliances when not in use, regulating the temperature of cooling or heating devices to provide indoor-outdoor temperature differences suitable for human comfort, maintaining them to reduce heat loss, and regularly storing, profiling, and evaluating energy consumption data to identify possible saving opportunities. Additionally, locating workplaces closer to public transport or living areas can significantly reduce energy usage in transporting workers to and from their workplaces [83].

Energy efficiency and conservation, therefore, involve adjusting and optimizing energy-using (active) systems and procedures to reduce energy requirements per unit of output while holding constant or reducing the total cost of providing output or maximizing the usage of energy, whether through behavior (practices), improved management, or the introduction of new technology.

Effective energy utilization, efficiency, and conservation aim to maximize the usage of the nation's non-renewable energy resources, enhance energy security and self-reliance, reduce the cost of production of manufactured goods and services, and minimize the adverse effects of energy utilization on the environment and populace. In Nigeria, energy utilization is still far from an efficient level, and inefficiencies result in losses in energy, excessive investment in infrastructure (equipment and resources) more than required, increased environmental issues, and increased cost of production/ goods [80]. Opportunities for effective energy savings and utilization in the Nigerian economy are substantial, especially in the three main energy demand sectors: households, industries, and transportation. For example, significant energy loss in homes results from leaving corridors and interior lighting on perpetually, not switching off electrical appliances/devices when not in use, and using traditional three-stone stoves in rural settings. In the manufacturing sector, using energy-efficient equipment can significantly conserve energy. Energy audit studies indicate that simple housekeeping measures and practices can save over 25% of residential and industrial energy [13,84,85]. In the transportation sector, promoting energy-efficient transportation through the use of electric and hybrid vehicles can reduce energy consumption and greenhouse gas emissions. Nigeria can encourage the adoption of electric and hybrid vehicles by providing incentives and developing charging infrastructure [86].

In general, effective energy utilization, efficiency, and conservation are geared toward achieving the following objectives:

i. maximizing the usage of the nation's non-renewable energy resources.
ii. enhancing energy security and self-reliance.
iii. reduction of the cost of production of manufactured goods and services.
iv. minimizing the adverse effects of energy utilization on the environment and populace.

14.5.1 Strategies for Efficient Energy Utilization and Conservation

The strategies being put in place as stipulated by the nation's energy policies [87] to ensure efficient utilization and conservation of energy are:

1. The first strategy involves ensuring strict adherence to regulations set by the energy industry and relevant agencies. This includes regulations on energy resource exploitation and environmental conservation to prevent the wastage of resources and negative impacts on the environment.
2. The second strategy focuses on the provision and enforcement of policies and institutional arrangements that promote energy conservation and the use of energy-efficient methods. These policies may include incentives for individuals and organizations that adopt energy-efficient practices, regulations on energy consumption and emissions, and promotion of renewable energy sources.
3. The third strategy emphasizes the development of building and construction regulations to reduce energy consumption. This may involve designing buildings to take advantage of natural climatic conditions to reduce the need for artificial cooling and heating.
4. The fourth strategy involves the importation of energy-efficient equipment and machinery to replace less efficient ones, thereby reducing energy consumption.
5. The fifth strategy aims to reduce energy consumption by improving and expanding mass transportation and communication systems across the country. This may include the promotion of public transportation systems that run on renewable energy sources, such as electric buses and trains.
6. The sixth strategy promotes research and development activities for energy conservation and efficiency, including the development and manufacture of energy-efficient equipment and devices locally.
7. The seventh strategy encourages the production and use of improved and more energy-efficient electrical appliances and cooking devices at home. This involves technologies that help to conserve or better use energy, also known as end-use efficiency.
8. The eighth strategy focuses on taking appropriate measures to reduce energy storage, transmission, and distribution losses, which account for significant losses at present, especially during transmission.
9. The ninth strategy focuses on sensitizing the public about the benefits of improved energy efficiency and conservation, which can be done by raising awareness, providing information, highlighting savings, engaging the community, encouraging behavior change, showcasing success stories, and offering incentives. A multifaceted approach that combines education, engagement, and incentives is needed to promote these practices and create a more sustainable future.

14.5.2 ROLE OF RESEARCH, DEVELOPMENT, AND TRAINING IN PROMOTING ENERGY EFFICIENCY AND CONSERVATION

For the Nigerian energy sector, especially power generation, transmission, and distribution sub-sectors, research, development, and training should be given adequate attention on a self-sustainable basis. The interests should be in key issues such as harnessing energy resources, utilizing techno-logical innovations on conventional sources (introduction of new technologies), and introducing efficient sources of energy into the energy mix of the nation. Research, training, and development can help to achieve a lot through strengthening, upskilling, and promoting local capabilities in the design and fabrication of energy equipment/machinery, demonstration and dissemination of renew-able energy devices and technology, initiating and encouraging result-oriented research in the energy sector, and organizing training for building a workforce for the energy industry. Energy conservation education is an important strategy for promoting energy efficiency and conservation. The Nigerian government can promote energy conservation education in schools and through public campaigns to raise awareness about the importance of energy conservation and provide practical tips on how to conserve energy [88]. Energy conservation education could lead to improved energy efficiency and conservation behaviors in Nigeria.

14.5.3 PROMOTING ENERGY EFFICIENCY AND CONSERVATION THROUGH STANDARDS AND LABELLING

The Nigerian government can promote energy efficiency standards and labeling for appliances and equipment as a means of improving energy efficiency. This is because energy labeling and standards have been successful in increasing the adoption of energy-efficient appliances and equipment [89]. Establishing and enforcing building codes and standards that promote energy efficiency is a neces-sary strategy for improving energy efficiency in Nigeria. This recommendation is supported by studies that have shown that building codes and standards can significantly reduce energy con-sumption and greenhouse gas emissions [90–92]. Nigeria can adopt and enforce building codes and standards that promote energy efficiency in new and existing buildings to reduce energy waste and improve energy efficiency. Overall, here are the relevant steps that must be taken:

1. **Develop standards**: Nigeria can develop energy efficiency standards for appliances and equipment, such as air conditioners, refrigerators, and lighting. These standards would set minimum efficiency levels that products must meet to be sold in the Nigerian market.
2. **Implement labeling**: Nigeria can also implement energy efficiency labeling for appliances and equipment. These labels would display the energy efficiency rating of the product, making it easier for consumers to make informed purchasing decisions.
3. **Enforce compliance**: To ensure that products sold in Nigeria meet energy efficiency standards and labeling requirements, the government can enforce compliance through testing and certification.
4. **Provide incentives**: Nigeria can provide incentives for manufacturers and retailers to produce and sell energy-efficient products, such as tax breaks or subsidies.
5. **Raise awareness**: To encourage consumers to choose energy-efficient products, Nigeria can raise awareness about the benefits of energy efficiency and conservation through public edu-cation campaigns, advertisements, and social media.
6. **Collaborate with international organizations**: Nigeria can collaborate with international organizations, such as the International Energy Agency and the United Nations Development Programme, to develop and implement energy efficiency standards and labeling programs.

Implementing energy efficiency standards and labeling programs can help Nigeria reduce energy consumption, save money, and reduce greenhouse gas emissions. By promoting energy efficiency and conservation, Nigeria can improve energy security and increase access to electricity in rural areas.

14.5.4 Benefits of Energy Efficiency and Conservation in Nigeria

Effective energy efficiency and conservation measures and strategies are crucial for achieving various goals, including those suggested by Uyigue [84]. One such goal is the minimization of capital expenditure on new power stations, which can free up capital for other investments in the future. By implementing energy efficiency measures, the demand for electricity can be reduced, which in turn reduces the need for additional power plants.

Another goal is to reduce consumption on the part of consumers, which can result in lower electricity bills. Energy efficiency measures such as the use of energy-efficient appliances, LED lighting, and building insulation can help reduce energy consumption and save money for consumers. Energy efficiency measures can also lead to more energy availability to other parts of the population and/ or purposes. This is because energy efficiency reduces overall energy consumption, allowing more energy to be distributed to areas that may have previously faced shortages. Diversification of the energy sector is another benefit of energy efficiency and conservation measures. By reducing the reliance on oil and gas, energy efficiency measures can encourage the adoption of renewable energy sources such as solar, wind, and hydro. This diversification can lead to greater energy security and resilience against energy price shocks. Energy efficiency measures can also better position industries for competition internationally with minimum local and global benchmarks, resulting in a reduction in the cost of production per unit output. By reducing energy consumption, industries can reduce their operating costs, allowing them to be more competitive in the global market. The negative effects of current energy production methods on the environment and populace can also be reduced through energy efficiency measures. For example, the use of energy-efficient appliances and LED lighting can reduce greenhouse gas emissions, which contribute to climate change. Additionally, reducing energy consumption can help reduce air pollution, which can have negative effects on both the environment and human health.

Finally, energy efficiency measures can also increase employment opportunities. Interventions such as the promotion of small and medium-scale enterprises, as well as investments in the housing and transport sector, can create jobs and stimulate economic growth. In conclusion, implementing effective energy efficiency and conservation measures and strategies can have numerous benefits, including reducing capital expenditure on new power stations, lowering energy bills, increasing energy availability, diversifying the energy sector, improving industrial competitiveness, reducing negative environmental and societal impacts, and creating employment opportunities. These benefits make energy efficiency and conservation an essential component of any sustainable development strategy.

14.5.5 Challenges of Energy Efficiency and Conservation in Nigeria

The successful implementation of energy efficiency and conservation (EEC) measures in Nigeria is hindered by several factors. One major challenge is the lack of political drive to establish the necessary legislative and policy frameworks for EEC implementation. Without the proper legal and regulatory structures in place, it becomes difficult to enforce energy-saving standards and incentives and to promote the production and procurement of energy-efficient products in the energy industry. Another hindrance to EEC implementation is the inadequacy of incentives and packages that encourage local production, procurement, and penetration of energy-efficient products in the industry. Incentives such as tax breaks, rebates, and grants can go a long way in promoting EEC practices and products. Furthermore, there is a shortage of human capital and expertise needed for the implementation of EEC measures, programs, and operations. This shortage of skilled personnel can lead to delays and inefficiencies in the implementation process. There are also insufficient national sensitization and awareness programs on the adoption and benefits associated with energy efficiency and conservation. Many people in Nigeria are still unaware of the advantages of EEC practices, and this lack of awareness poses a significant barrier to adopt EEC measures. In addition, energy-saving barriers experienced in household settings include non-awareness or inadequate sensitization of residents, poor maintenance of appliances, non-challant attitude to energy facilities, and a poor system of billing and payment. These barriers can be addressed by improving awareness and education about EEC practices, ensuring regular maintenance of appliances, and implementing a more efficient billing and payment system. Inappropriate energy pricing and cost subsidies also pose a significant barrier to the implementation of EEC measures. The current pricing structure for energy in Nigeria does not reflect the actual cost of production, and subsidies provided by the government can lead to wasteful energy consumption and a lack of incentive to adopt EEC practices. Finally, the proliferation of inefficient equipment and the desire to minimize initial costs can lead to a lack of interest in adopting energy-efficient technologies. This can be addressed by promoting the long-term cost savings associated with EEC practices and products and providing incentives for their adoption. In conclusion, the successful implementation of EEC measures in Nigeria requires the establishment of the necessary legislative and policy frameworks, incentives for local production and procurement of energy-efficient products, the availability of skilled human capital, and the promotion of national sensitization and awareness programs. Addressing the barriers to energy-saving practices and products at the household level, improving energy pricing and subsidies, and promoting the long-term cost savings of EEC measures and products are also crucial for successful implementation.

14.5.6 RECOMMENDATIONS FOR IMPROVING THE LEVEL OF ENERGY EFFICIENCY AND CONSERVATION IN NIGERIA ENERGY INDUSTRY

Some possible ways to improve energy efficiency and conservation are highlighted below, parts of which were enumerated by Oyedepo [93].

Improving energy efficiency and conservation is crucial for Nigeria to meet its energy requirements and minimize the negative impacts of energy production on the environment and populace. One of the ways to achieve this is by increasing nationwide sensitization and awareness programs on the benefits of energy efficiency and conservation. This will help to create a better understanding of the importance of energy conservation and motivate individuals and industries to adopt energy-efficient practices. Another strategy is the development and adoption of energy-efficient technologies. This involves the introduction of energy-efficient appliances, devices, and systems, which can help reduce energy consumption and promote conservation. The establishment and enforcement of an enabling institutional and legislative framework is also essential for implementing energy efficiency and conservation programs. This can help to create an environment that supports and encourages the adoption of energy-efficient practices and technologies.

To achieve cost-reflective energy tariffs and effective systems for billing and payment, full and transparent deregulation of the energy sector is necessary. This will create a competitive environment where energy providers are incentivized to improve energy efficiency and minimize energy waste. Human capital building is crucial for the implementation of energy efficiency and conservation measures, programs, and operations. This includes training programs for energy auditors, energy managers, measurement and verification professionals, and related professionals to ensure that the necessary skills and knowledge are available to implement energy-efficient practices. Effective collaboration of all stakeholders in energy efficiency and conservation is also necessary. This includes the government, private sector, civil society organizations, and energy consumers. By working together, stakeholders can identify and implement energy-saving measures that are mutually beneficial. The establishment of energy efficiency and conservation and renewable energy equipment/device manufacturing plants in the country can also promote energy efficiency and conservation. This will encourage local production and procurement of energy-efficient products, which can reduce the reliance on imported products and help to create employment opportunities. Demand-side management is another strategy for improving energy efficiency and conservation. This involves the implementation of policies and practices that encourage users to use energy in efficient ways or to shift their energy use away from peak demands. Load shifting can delay the immediate need to construct new generating stations/units and ensure the effective deployment of generating capacity. Energy management can also contribute to energy efficiency. Effective and reliable energy management practices include ensuring that waste spare heat is not vented away but recovered, activating lighting only when there is a need for it, and conducting energy audits to optimize all aspects of energy management. Dispersed generation is another strategy for improving energy efficiency and conservation. By generating energy locally and feeding it directly into the distribution systems, energy losses due to grid transmission over long distances can be minimized. This approach is particularly relevant to Nigeria, where the current supply output is below the stipulated minimum energy requirement of between 12,000 MW and 15,000 MW [81].

In summary, Nigeria needs to adopt various energy efficiency and conservation strategies in line with the adoption of renewable energy mix technologies to satisfy the nation's energy requirement and promote energy security. By implementing these strategies, Nigeria can minimize the negative impacts of energy production on the environment and populace, reduce energy consumption and costs, promote economic growth, and create employment opportunities.

14.6 CONCLUSION

In conclusion, it is evident that the energy crisis in Nigeria has become increasingly complex and requires a multifaceted approach to address it. This article has identified and analyzed several energy sources that have the potential to contribute to solving Nigeria's energy challenges. It is clear that no single energy source can provide a comprehensive solution to the country's energy problems, but rather a combination of different sources and strategies. While traditional sources such as oil, gas, and coal remain essential to Nigeria's energy mix, there is a growing need to diversify and invest in cleaner and renewable energy sources, such as solar, wind, hydro, and biomass. These sources have immense potential to provide reliable, affordable, and sustainable energy and promote energy security in the long run. Additionally, energy efficiency and conservation measures such as smart grid technology, energy audits, and building codes can also play a vital role in reducing energy waste and enhancing energy security in Nigeria. By adopting such measures, the country can reduce its carbon footprint, mitigate the negative impact of climate change, and save costs associated with energy consumption. Moreover, the energy sector is crucial for Nigeria's economic growth, and the adoption of a diverse energy mix that emphasizes clean, renewable energy sources is essential for achieving sustainable development. Therefore, the Nigerian government and other stakeholders must work together to implement policies and strategies that prioritize a sustainable energy future for the country. This would require adequate funding, human capital development, regulatory frameworks, and incentives that encourage innovation and investment in clean energy technologies. In conclusion, addressing Nigeria's energy challenges will require a concerted effort from all stakeholders, including the government, private sector, civil society, and international partners. The adoption of a diverse and sustainable energy mix, coupled with energy efficiency and conservation measures, is crucial to achieving energy security and sustainable development in the country.

REFERENCES

[1] A. I. Lawal, I. Ozturk, I. O. Olanipekun, and A. J. Asaleye, "Examining the linkages between electricity consumption and economic growth in African economies," *Energy*, vol. 208, p. 118363, 2020. https://doi.org/10.1016/j.energy.2020.118363

[2] R. Nepal and N. Paija, "Energy security, electricity, population and economic growth: The case of a developing South Asian resource-rich economy," *Energy Policy*, vol. 132, pp. 771–781, 2019, September. https://doi.org/10.1016/j.enpol.2019.05.054

[3] J. Wang and H. Li, "The impact of electricity price on power-generation structure: Evidence from China," (in English), *Frontiers in Environmental Science*, vol. 9, p. 933809, 2021, September. https://doi.org/10.3389/fenvs.2021.733809

[4] N. I. Nwulu and O. P. Agboola, "Utilizing renewable energy resources to solve Nigeria's electricity generation problem," *International Journal of Thermal & Environmental Engineering*, vol. 3, no. 1, pp. 15–20, 2011. [Online]. Available: https://iasks.org/articles/ijtee-v03-i1-pp-15-20.pdf

[5] B. Ugwoke, S. Sulemanu, S. P. Corgnati, P. Leone, and J. M. Pearce, "Demonstration of the integrated rural energy planning framework for sustainable energy development in low-income countries: Case studies of rural communities in Nigeria," *Renewable and Sustainable Energy Reviews*, vol. 144, p. 110983, 2021, July. https://doi.org/10.1016/j.rser.2021.110983

[6] T. R. Ayodele, A. S. O. Ogunjuyigbe, O. D. Ajayi, A. A. Yusuff, and T. C. Mosetlhe, "Willingness to pay for green electricity derived from renewable energy sources in Nigeria," *Renewable and Sustainable Energy Reviews*, vol. 148, p. 111279, 2021, September. https://doi.org/10.1016/j.rser.2021.111279

[7] M. O. Dioha, A. Kumar, D. R. Ewim, and N. V. Emodi, "Alternative scenarios for low-carbon transport in Nigeria: a long-range energy alternatives planning system model application," *Economic Effects of Natural Disasters*, 2021, pp. 511–527.

[8] D. R. E. Ewim, S. S. Oyewobi, and S. M. Abolarin, "COVID-19 – Environment, Economy, And Energy: Note from South Africa." *Journal of Critical Reviews*, vol. 08, no. 03, pp. 67–80, 2021.

[9] D. R. Ewim, S. S. Oyewobi, M. O. Dioha, C. E. Daraojimba, S. O. Oyakhire, and Z. Huan, "Exploring the perception of Nigerians towards nuclear power generation," *African Journal of Science, Technology, Innovation and Development*, vol. 14, no. 4, pp. 1059–1070, 2022.

[10] J. M. Egieya, R. M. Ayo-Imoru, D. R. Ewim, and E. C. Agedah, "Human resource development and needs analysis for nuclear power plant deployment in Nigeria," *Nuclear Engineering and Technology*, vol. 54, no. 2, pp. 749–763, 2022.

[11] M. Ntuli, M. Dioha, D. Ewim, and A. Eloka-Eboka, "Review of energy modelling, energy efficiency models improvement and carbon dioxide emissions mitigation options for the cement industry in South Africa," *Materials Today: Proceedings*, vol. 65, pp. 2260–2268, 2022.

[12] D. R. E. Ewim, S. M. Abolarin, T. O. Scott, and C. S. Anyanwu, "A survey on the understanding and viewpoints of renewable energy among south african school students," *The Journal of Engineering and Exact Sciences*, vol. 9, no. 2, pp. 15375-01e, 2023.

[13] S. M. Abolarin et al., "A collective approach to reducing carbon dioxide emission: A case study of four University of Lagos Halls of residence," *Energy and Buildings*, vol. 61, pp. 318–322, 2013. https://doi.org/10.1016/j.enbuild.2013.02.041

[14] I. A. Ologunde, F. M. Kapingura, and K. Sibanda, "Sustainable development and crude oil revenue: A case of selected crude oil-producing african countries," (in eng), *International Journal of Environmental Research and Public Health*, vol. 17, no. 18, 2020, September. https://doi.org/10.3390/ijerph17186799

[15] K. Musa, R. Maijama'a, H. Shaibu, and A. Muhammad, "Crude oil price and exchange rate on economic growth: ARDL approach," *Open Access Library Journal*, vol. 6, pp. 1–5, 2019. https://doi.org/10.4236/oalib.1105930

[16] N. J. Ikue, L. I. Amabuike, J. O. Denwi, A. U. Mohammed, and A. U. Musa, "Economic growth and crude oil revenue in Nigeria: A control for industrial shocks," *International Journal of Research in Business and Social Science* (2147–4478), vol. 10, no. 8, pp. 218–227, 2022, January. https://doi.org/10.20525/ijrbs.v10i8.1500

[17] O. E. Ogbonna, I. A. Mobosi, and O. W. Ugwuoke, "Economic growth in an oil-dominant economy of Nigeria: The role of financial system development," *Cogent Economics & Finance*, vol. 8, no. 1, p. 1810390, 2020. https://doi.org/10.1080/23322039.2020.1810390

[18] ECN, *National energy master plan*. Abuja, 2022, pp. 1–242.

[19] A. Castellano, A. Kendall, M. Nikomarov, and T. Swemmer, "Brighter Africa: The growth potential of the sub-Saharan electricity sector," 2015, pp. 1–58.

[20] *Energy Commision of Nigeria*, National Energy Policy, 2003.

[21] *United Nations, Word economic situation and prospects 2020*. 2020. www.un.org/development/desa/dpad/wp-content/uploads/sites/45/WESP2020_Annex.pdf. [Online] Available: www.un.org/development/desa/dpad/wp-content/uploads/sites/45/WESP2020_Annex.pdf

[22] R. Bryce, *Power Hungry: The Myths of "Green" Energy and the Real Fuels of the Future*. PublicAffairs, 2011.

[23] D. Ahuja and M. Tatsutani, "Sustainable energy for developing countries," *Surveys and Perspectives Integrating Environment and Society*, vol. 2, ed 1, pp. 1–16, 2009.

[24] C. Brandoni and B. Bošnjaković, "HOMER analysis of the water and renewable energy nexus for water-stressed urban areas in Sub-Saharan Africa," *Journal of Cleaner Production*, vol. 155, pp. 105–118, 2017.

[25] E. Salakhetdinov and O. Agyeno, "Achieving energy security in Africa: prospects of nuclear energy development in South Africa and Nigeria," *African Journal of Science, Technology, Innovation and Development*, pp. 1–9, 2020.

[26] C. Uzoma, C. Nnaji, and M. Nnaji, "The role of energy mix in sustainable development of Nigeria," *Continental Journal of Social Sciences*, vol. 5, no. 1, p. 21, 2012.

[27] A. S. Aliyu, A. T. Ramli, and M. A. Saleh, "Nigeria electricity crisis: Power generation capacity expansion and environmental ramifications," *Energy*, vol. 61, pp. 354–367, 2013. https://doi.org/10.1016/j.energy.2013.09.011

[28] S. Uwaifo, *Electric Power Distribution Planning and Development*. Malthouse Press, 1994.

[29] A. Sule, "Major factors affecting electricity generation, transmission and distribution in Nigeria," *International Journal of Engineering and Mathematical Intelligence*, vol. 1, no. 1, pp. 159–164, 2010.

[30] S. Nwani, E. Ozegbe, and Y. Olunlade, "*An Overview of Nigerian Energy Sector: Prospects and Challenges*," ed, 2020.

[31] E. N. Vincent and S. D. Yusuf, "Integrating renewable energy and smart grid technology into the Nigerian electricity grid system," *Smart Grid and Renewable Energy*, vol. 05, no. 09, pp. 220–238, 2014, Art no. 49592. https://doi.org/10.4236/sgre.2014.59021

[32] S. O. Babalola, M. O. Daramola, and S. A. Iwarere, "Socio-economic impacts of energy access through off-grid systems in rural communities: A case study of southwest Nigeria," (in eng), *Philosophical Transactions Series A, Mathematical Physical and Engineering Science*, vol. 380, no. 2221, p. 20210140, 2022, April. https://doi.org/10.1098/rsta.2021.0140

[33] E. N. C. Okafor and C. K. A. Joe-Uzuegbu, "Challenges to development of renewable energy for electric power sector in Nigeria," *International Journal of Academic Research*, vol. 2, no. 211, 2010.

[34] U. K. Elinwa, J. E. Ogbeba, and O. P. Agboola, "Cleaner energy in Nigeria residential housing," *Results in Engineering*, vol. 9, p. 100103, 2021, March. https://doi.org/10.1016/j.rineng.2020.100103

[35] O. Olatunji et al., "Electric power crisis in Nigeria: A strategic call for change of focus to renewable sources," *IOP Conference Series: Materials Science and Engineering*, vol. 413, no. 1, p. 012053, 2018. https://doi.org/10.1088/1757-899X/413/1/012053

[36] P. Amotsuka, "Sustainable agriculture in the Niger Delta region: a case study of Finima community of rivers state," *African Journal of General Agriculture*, vol. 6, no. 3, pp. 115–121, 2010.

[37] R. Babalola, "6000Mw power target report card," *Daily Trust Newspaper*, vol. 22, no. 94, p. 1, 2009.

[38] C. Onyekwena, J. Ishaku, and C. Akanonu, "Electrification in Nigeria: Challenges and Way Forward," *Report of Centre for the Study of the Economies of Africa (CSEA) Abuja, Nigeria*, pp. 1–32, 2017.

[39] O. Bamisile, M. Dagbasi, A. Babatunde, and O. Ayodele, "A review of renewable energy potential in Nigeria; solar power development over the years," *Engineering and Applied Science Research*, vol. 44, no. 4, pp. 242–248, 2017.

[40] J. Oji, N. Idusuyi, T. Aliu, M. Petinrin, O. Odejobi, and A. Adetunji, "Utilization of solar energy for power generation in Nigeria," *International Journal of Energy Engineering*, vol. 2, no. 2, pp. 54–59, 2012.

[41] O. Nnodim, "Nigeria's oil reserves hit 37.046 billion barrels, gas, 208.62TCF–FG," in *Punch Newspaper*, ed. Lagos, Nigeria: Punch Newspaper, 2022.

[42] J. Vujić, D. P. Antić, and Z. Vukmirović, "Environmental impact and cost analysis of coal versus nuclear power: The US case," *Energy (Oxford)*, vol. 45, no. 1, pp. 31–42, 2012. https://doi.org/101016/jenergy201202011

[43] O. Kehinde, K. Babaremu, K. Akpanyung, E. Remilekun, S. Oyedele, and J. Oluwafemi, "Renewable energy in Nigeria – A review," *International Journal of Mechanical Engineering and Technology*, vol. 9, no. 10, pp. 1085–1094, 2018.

[44] L. Bello, "Review of preparedness as Nigeria match towards nuclear power plant as part of its energy mix," in *Scientific Initiative of Foreign Students and Graduate Students: Collection of Reports of the 1st International Scientific and Practical Conference, Tomsk, April 27–29, 2021 T. 2. – Tomsk, 2021*, 2021, vol. 2, pp. 377–381.

[45] N. Lowbeer-Lewis, *"Nigeria and Nuclear Energy: Plans and Prospects," Nuclear Energy Futures Paper No. 11, January 2010.* Waterloo, Ontario, Canada: The Centre for International Governance Innovation, 2010. Retrieved from www.cigionline.org/sites/default/files/nuclear_energy_wp11-web.pdf

[46] A. R. Ejiogu, "A nuclear Nigeria: How feasible is it?," *Energy Strategy Reviews*, vol. 1, no. 4, pp. 261–265, 2013.

[47] L. Bello, "Review of preparedness as Nigeria match towards nuclear power plant as part of its energy mix," in *Научная инициатива иностранных студентов и аспирантов: сборник докладов1 Международной научно-практической конференции, Томск, 27-29 апреля 2021 г. Т. 2.—Томск, 2021*, 2021, vol. 2: Томский политехнический университет, pp. 377–381.

[48] F. A. Ishola, O. O. Olatunji, O. O. Ayo, S. A. Akinlabi, P. A. Adedeji, and A. Inegbenebor, "Sustainable nuclear energy exploration in Nigeria–A SWOT analysis," *Procedia Manufacturing*, vol. 35, pp. 1165–1171, 2019.

[49] F. E. Osaisai, "Nuclear power introduction in Nigeria: organization and way forward," Presented at the *Presentation at the Second Regional Conference on Energy and Nuclear Power in Africa, Cape Town*, May 30–31, 2011.

[50] H. Muhammed, "Safeguards of nuclear materials: the Nigerian experience," *Presented at the Poster Presentation at the International Atomic Energy Agency Symposium on International Safeguards: Preparing for Future Verification Challenges, Vienna, Australia*, 1–5 November 2010.

[51] E. C. Praxede, "Nuclear energy and global security risk: The implication of Nigeria's Quest as a source of its energy," *International Journal of Socical Science Humanities Review*, vol. 6, no. 2, pp. 71–76, 2016.

[52] A. S. Aliyu, A. T. Ramli, and M. A. Saleh, "First nuclear power in Nigeria: an attempt to address the energy crisis?" *International Journal of Nuclear Governance, Economy and Ecology*, vol. 4, no. 1, pp. 1–10, 2013.

[53] C. Muanya, "Advancing nuclear energy in Nigeria," *The Guardian.* www.guardian.ng/features/advancing-nuclear-energy-in-nigeria/amps, 2021.

[54] World Bank Group. *GDP Per Capita (Current US$), World Bank National Accounts Data, and OECD National Accounts Data Files.* https://data.worldbank.org/indicator/NY.GDP.PCAP.CD?end=2020&name_desc=true&start=1960&view=chart, 2020.

[55] I. Gerretsen, "To power its growing population, Nigeria is turning to renewables," *World Economic Forum.* www.weforum.org/agenda/2018/05/oil-rich-nigeria-turns-to-renewable-energy-as-population-booms/, 2018.

[56] M. Asif and T. Muneer, "Energy supply, its demand and security issues for developed and emerging economies," *Renewable and Sustainable Energy Reviews*, vol. 11, no. 7, pp. 1388–1413, 2007.

[57] K. Byman, "Electricity production in Sweden: IVA's electricity crossroads project," *Royal Swedish Academy of Engineering Sciences*, 2016.

[58] M. Gustavsson, E. Särnholm, P. Stigson, and L. Zetterberg, "Energy scenario for Sweden 2050," *Swedish Environmental Research Institute, Gothenburg*, 2011.

[59] G. Morrison et al., *"Long-Term Energy Planning in California: Insights and Future Modeling Needs,"* Citeseer, 2014.

[60] F. Riva, A. Tognollo, F. Gardumi, and E. Colombo, "Long-term energy planning and demand forecast in remote areas of developing countries: Classification of case studies and insights from a modelling perspective," *Energy Strategy Reviews*, vol. 20, pp. 71–89, 2018.

[61] A. Oladapo, "Achieving Nigeria's development goals," *Thisday, June*, vol. 21, 2004.

[62] K. Ejumudo, "The problematic of development planning in Nigeria: A critical discourse," *Developing Country Studies*, vol. 3, no. 4, pp. 67–80, 2013.

[63] D. K. Ologbenla, "Leadership, governance and corruption in Nigeria," *Journal of Sustainable Development in Africa*, vol. 9, no. 3, pp. 97–118, 2007.

[64] *Sustainable path to economic growth & development: Imperatives of development planning*, 2021.

[65] *Nigeria Federal Government, Nigeria's Medium Term National Development Plan (MTNDP) – 2021–2025*, 2021.

[66] C. Agbabi, "Nigerian Government Launches Nigeria Agenda 2050 July 2021," *AllAfrica*, https://allafrica.com/stories/202010160773.html, 20202020101607, 2020.

[67] "Financial Nigeria International Limited, Nigeria to be among world's top 10 economies by 2050," www.financialnigeria.com/nigeria-to-be-among-world-s-top-10-economies-by-2050-news-457.html, 2016.

[68] C. Lane. "Nuclear energy pros and cons," www.solarreviews.com/blog/nuclear-energy-pros-and-cons (accessed 20 September 2023).

[69] I. A. E. Agency, *Non-Baseload Operations in Nuclear Power Plants: Load Following and Frequency Control Modes of Flexible Operation*. IAEA, 2018.

[70] B. W. Brook, A. Alonso, D. A. Meneley, J. Misak, T. Blees, and J. B. van Erp, "Why nuclear energy is sustainable and has to be part of the energy mix," *Sustainable Materials and Technologies*, vol. 1, pp. 8–16, 2014.

[71] *World Population Prospects 2019, Population Data, File: Total Population-Both Sexes, Estimates Tab*. United Nations Population Division, 2019. https://population.un.org/wpp/Download/Standard/Population/

[72] Energy Commission of Nigeria, Nigerian. Calculator Excel version, 2050. www.energy.gov.ng/index.php?option=com_docman&task=cat_view&gid=39&Itemid=49&limitstart=40

[73] C. Ogbonnaya, C. Abeykoon, U. Damo, and A. Turan, "The current and emerging renewable energy technologies for power generation in Nigeria: A review," *Thermal Science and Engineering Progress*, vol. 13, p. 100390, 2019.

[74] S. M. Abolarin, A. O. Gbadegesin, B. M. Shitta, and O. Adegbenro, "Energy (lighting) audit of four University Of Lagos halls of residence," *Journal of Engineering Research*, vol. 16, no. 2, pp. 1–10, 2011.

[75] S. M. Abolarin, B. M. Shitta, E. M. Aghogho, P. B. Nwosu, C. M. Aninyem, and L. Lagrange, "An impact of solar PV specifications on module peak power and number of modules: A case study of a five-bedroom residential duplex," *IOP Conference Series: Earth and Environmental Science*, vol. 983, no. 1, p. 012056, 2022. https://doi.org/10.1088/1755-1315/983/1/012056

[76] M. O. Dioha, N. V. Emodi, and E. C. Dioha, "Pathways for low carbon Nigeria in 2050 by using NECAL2050," *Renewable Energy Focus*, vol. 29, pp. 63–77, 2019.

[77] C. Muanya, "Advancing nuclear energy in Nigeria," *The Guardian*, https://guardian.ng/features/advancing-nuclear-energy-in-nigeria/#:~:text=Since%202004%2C%20Nigeria%20has%20a,Nuclear%20Power%20for%20Generation%20of, 2021.

[78] S. Ojolo, S. Abolarin, and O. Adegbenro, "Development of a laboratory scale updraft gasifier," *International Journal of Manufacturing Systems*, vol. 2, no. 2, pp. 21–42, 2012.

[79] S. J. Ojolo, J. I. Orisaleye, S. O. Ismail, and S. M. Abolarin, "Technical potential of biomass energy in Nigeria," *Ife Journal of Technology*, vol. 21, no. 2, 2012. [Online]. Available: http://ijt.oauife.edu.ng/index.php/ijt/article/view/110

[80] S. M. Abolarin et al., "An economic evaluation of energy management opportunities in a medium scale manufacturing industry in Lagos," *International Journal of Engineering Research in Africa*, vol. 14, pp. 97–106, 2015.

[81] G. N. Osaghae, H. Ahidjo, and A. M. Umar, "Energy efficiency and conservation: A positive step towards sustainable economic growth of Nigeria," *International Journal of Scientific & Engineering Research*, vol. 5, no. 3, pp. 187–194, 2014.

[82] S. M. Abolarin et al., "An approach to energy management: A case study of a medium scale printing press in Lagos, Nigeria," *International Journal of Energy and Power Engineering*, vol. 3, no. 1, pp. 7–14, 2014, https://doi.org/10.11648/j.ijepe.20140301.12

[83] J. Berg and J. Ihlström, "The importance of public transport for mobility and everyday activities among rural residents," *Social Sciences*, vol. 8, no. 2, pp. 2–13, 2019. [Online]. Available: www.mdpi.com/2076-0760/8/2/58

[84] E. Uyigue, "The concept of energy efficiency". A presentation on national dialogue to promote renewable energy and energy efficiency in Nigeria," Abuja, November 10–11, 2008.

[85] A. D. Mambo and C. M. F. Kebe, "Building energy audit in Nigeria: Some guides for energy efficiency building regulations," *ICST Institute for Computer Sciences, Social Informatics and Telecommunications Engineering*, vol. LNICST 249, pp. 41–51, 2018. https://doi.org/10.1007/978-3-319-98878-8_4

[86] N. Wang and G. Tang, "A review on environmental efficiency evaluation of new energy vehicles using life cycle analysis," *Sustainability*, vol. 14, no. 6, p. 3371, 2022. [Online]. Available: www.mdpi.com/2071-1050/14/6/3371

[87] ECN, *National energy policy*. Abuja, 2022, pp. 1–110.

[88] T. C. Ogbuanya and N. I. Nungse, "Effectiveness of energy conservation awareness package on energy conservation behaviors of off-campus students in Nigerian universities," *Energy Exploration & Exploitation*, vol. 39, no. 5, pp. 1415–1428, 2021. https://doi.org/10.1177/0144598720975133

[89] G. Si-dai, L. Cheng-Peng, L. Hang, and Z. Ning, "Influence mechanism of energy efficiency label on consumers' purchasing behavior of energy-saving household appliances," (in English), *Frontiers in Psychology*, Original Research vol. 12, pp. 1–15, 2021-October-18. https://doi.org/10.3389/fpsyg.2021.711854

[90] M. H. Elnabawi, "Evaluating the impact of energy efficiency building codes for residential buildings in the GCC," *Energies*, vol. 14, no. 23, p. 8088, 2021. [Online]. Available: www.mdpi.com/1996-1073/14/23/8088

[91] K. Ahmed Ali, M. I. Ahmad, and Y. Yusup, "Issues, impacts, and mitigations of carbon dioxide emissions in the building sector," *Sustainability*, vol. 12, no. 18, p. 7427, 2020. [Online]. Available: www.mdpi.com/2071-1050/12/18/7427

[92] X. Zhong et al., "Global greenhouse gas emissions from residential and commercial building materials and mitigation strategies to 2060," *Nature Communications*, vol. 12, no. 1, p. 6126, 2021, October. https://doi.org/10.1038/s41467-021-26212-z

[93] S. O. Oyedepo, "Energy efficiency and conservation measures: Tools for sustainable energy development in Nigeria," *International Journal of Energy Engineering*, vol. 2, no. 3, pp. 86–98, 2012.

15 Energy Improvement and Storage

Harun Sen, Pinar Mert Cuce, and Erdem Cuce

15.1 INTRODUCTION

The world we live in has hosted many populations for millions of years. It can be assumed that human life began long after the beginning of the world. Although this is the case, the increase in energy demand due to technological developments and the increasing human population in the last century causes humanity to consume the world rapidly. The fact that energy sources are predominantly fossil fuel continues to be dangerous for the sustainable environment and clean nature day by day. Although energy resources are mostly fossil fuels today, it is predicted that fossil fuel production will continue to increase in the next 30 years (Abas et al. 2015). This situation reveals the serious problems that will arise in terms of environmental pollution and global warming, which are currently reaching disturbing levels. The only way to eliminate the problems caused by using fossil fuels is to turn to renewable energy sources, which are clean energy sources (Cuce et al. 2022a). Renewable energy sources have the capacity to be an alternative to fossil fuels with their potential and use in wide geographies.

In addition to the resources used in energy production, the efficient use of energy also gains importance. The increasing human population, the increase in the density of people in living spaces, vertical architecture and technology, and the increasing use of energy have caused environmental problems that reach every person's close environment. The increase in the human population in indoor environments has triggered the increasing need for Heating, Ventilation, and Air Conditioning (HVAC). HVAC systems are among the largest energy consumers in buildings (Wang and Ma 2008). This consumption is tried to be minimized with innovative approaches. Building insulation techniques (Kumar et al. 2020), innovative approaches in windows (Akram et al. 2023), and the properties of building construction materials (Bribián et al. 2011) are extremely important in order to reduce heat losses from indoor environments in cold environments. This study aims to compile the strategies presented in recent years to improve energy performance in buildings.

15.2 ENERGY IMPROVEMENT AND STORAGE

With the increasing human population, the need for living space is increasing day by day all over the world. The transition from old traditional housing to innovative houses has already begun. Especially with the increasing technological opportunities, the increase in the comfort level of people has increased energy consumption. Buildings today have serious energy consumption. Today, buildings are responsible for approximately 30–40% of global energy consumption (Roy et al. 2022). This makes the application of innovative models in buildings and energy-efficient designs inevitable in

DOI: 10.1201/9781032651958-15

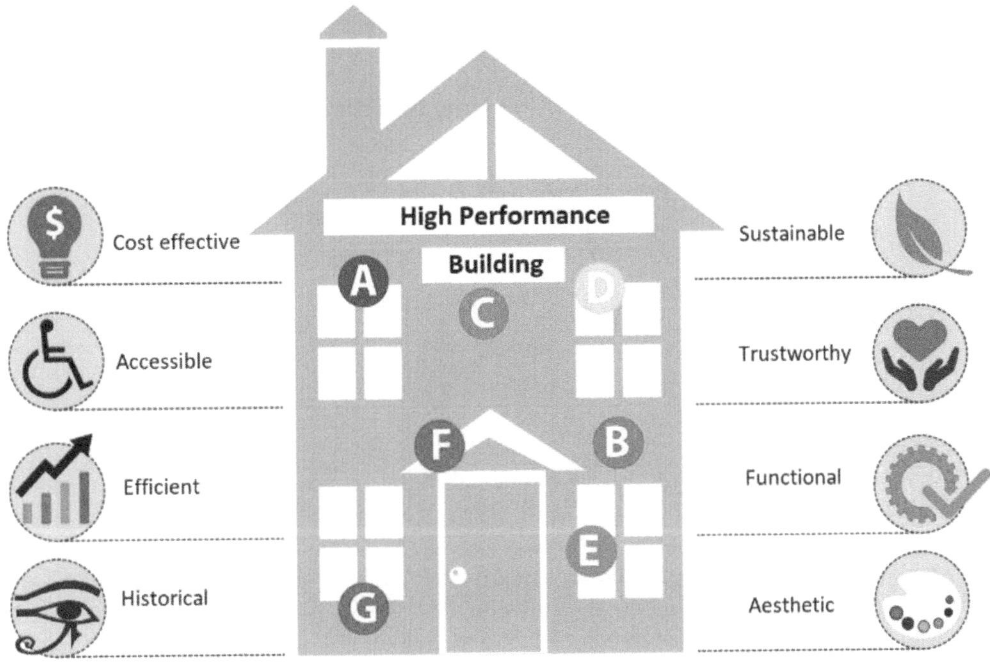

FIGURE 15.1   High-performance building parameters (Harputlugil 2016).

TABLE 15.1
Thermal Transmittance (W/m² K) Requirements
According to Norwegian Building Codes
(Gustavsen et al. 2007)

Building Section	Thermal Transmittance (W/m² K)
Outer wall	0.18
Window	1.2
Door	1.2
Roof	0.13
Floor	0.15

terms of energy consumption (Figure 15.1). Energy-efficient building designs stand out as high-performance buildings with their cost-effective, accessible, efficient, historical, sustainable, trustworthy, functional, and aesthetic properties (Harputlugil 2016).

In order to improve the energy performance in buildings, it is necessary to evaluate the heat losses caused by the basic building elements. Basically, buildings lose heat through exterior walls, windows, doors, roofs, and floors. The main reason for the heat loss is the high thermal permeability of these building elements. It is a legal requirement for building elements to have certain thermal permeability in order to minimize energy losses in building construction in countries. Table 15.1 gives legal thermal permeability values of building elements for a sample country. When the data are compared, it is seen that the thermal transmittance value is the highest for the window and door. Considering that there are many windows for buildings, it is clear that reducing heat losses from windows will contribute to a serious reduction in energy consumption.

TABLE 15.2
Heat Transmission Coefficients and Production Energies of Some Insulation Materials (Victoria University of Wellington 2023; Jelle 2011)

Material	Thermal conductivity (mW/mK)	Production energy (MJ/kg)
Wool	30–40	16.8–28
Cellulose	40–50	0.94–3.3
Cork	40–50	4
Polyurethane	20–30	72.1
Extruded polystyrene	30–40	117
Expanded polystyrene	30–40	88.6

Buildings have a great impact on environmental pollution and CO_2 emissions. Considering buildings are responsible for 60% of global energy consumption, the importance of energy-efficient building design and applications alone can be better understood. Studies have shown that wooden structures need less energy during their life cycles and have lower CO_2 emissions compared to other building elements (Gustavsson et al. 2006). Building materials may include options that do not have many alternatives due to the initial cost of installation, the difficulty of changing later, and procurement. Today, most buildings are not smart, as the integration of technical equipment such as heating, cooling, and lighting into the building, which is called "smart building" with the developing technical possibilities, is not designed during the construction phase of the buildings (Wigginton 2002).

Most of the energy consumption in buildings is due to thermal comfort load (Yang et al. 2014). For thermal comfort conditions, heating load in cold climates and cooling load in hot climates cause serious energy consumption. For this, it is important that buildings are designed with energy efficiency. Buildings not designed to be energy efficient during the initial installation phase may not be efficient even if refurbished afterward. In buildings under these conditions, insulation can be applied afterward to reduce the heat loss from the buildings. With the progress in insulation techniques and natural insulation materials, old buildings can be renewed, and lower emission levels can be reduced. With the application of insulation alone, greenhouse gas emissions can be reduced by 5.5% (Chitnis et al. 2012). With the advancement in insulation techniques, this rate may increase even more. Despite the low thermal resistance of the basic building materials, the materials produced with innovative methods have a very low thermal conductivity coefficient, which is promising for minimizing heat losses in buildings. In addition to reducing the energy consumption of insulation materials, production costs, and processes are also important. Table 15.2 gives heat transmission coefficients and production energy amounts of some insulation materials.

Another building element that will reduce heat loss from buildings, such as insulation, is windows. Windows are responsible for 60% of energy losses in buildings (Jelle et al. 2012a). In this case, reducing the heat loss from the windows can contribute to serious energy savings. Windows are structures placed in the spaces opened in the building walls to benefit from daylight in indoor environments. These building elements, which allow daylight to reach indoor environments, can cause heat losses while providing the advantage of lighting. For these reasons, there are numerous studies that will reduce heat transfer related to windows. Certain parameters are important in the work done for window glasses. These are solar heat gain coefficient (SHGC), apparent transmittance (VT), and heat transfer coefficient (U_w). The performance parameters of window glasses, which are widely used in the literature, are given in Table 15.3. It is seen that the window frames are effective in heat losses as well as the glass, and the U_w value is 0.59 W/m² K for "Double, low solar gain, low-e glass" with the aluminum frame type, while the U_w value is reduced by 56% to 0.26 W/m² K

TABLE 15.3
Solar Heat Gain Coefficient (SHGC), Visible Transmittance (VT) and Heat Transfer Coefficients (U_w) of Window Glasses in the Literature (Cuce and Riffat 2015a)

Glazing Type	U_w (W/m² K)	SHGC	VT
Single, clear	0.84	0.64	0.65
Double, clear	0.49	0.56	0.59
Double, high-performance tint	0.49	0.39	0.50
Double, high-solar gain, low-e	0.37	0.53	0.54
Double, low-solar gain, low-e	0.34	0.30	0.51
Triple, moderate solar gain, low-e	0.29	0.38	0.47

with the insulated fiberglass frame type (Cuce and Riffat 2015a). This shows the importance of the use of frames.

New developments to minimize heat losses through windows further reduce the U_w value. The U_w value can be reduced with the multilayer gas formed using air, krypton, and argon gases between the glasses. A U_w value of 0.49 W/m² K is obtained with 36 mm multilayer glass produced using 12 mm krypton gas between three pieces of 4 mm glass (Cuce and Riffat 2015a). The most important handicaps for multilayer glasses are the production costs and the decrease in the transmittance of visible light (Jelle et al. 2012a). It is an effective method to reduce the effects of transmission, convection, and radiation on the glass in order to reduce the heat loss from the windows. The vacuum glass method minimizes the transmission and convection effects by evacuating the gas between the glasses (Cuce et al. 2022b). Support pillars are used to prevent the glass from breaking in the vacuum environment created. The commercial product PILKINGTON produced by vacuum glass technique has a U_w value of 1.2 W/m² K, and the researchers claim that this value will be reduced to 0.67 W/m² K with the supports to be produced from aerogel (Cuce and Riffat 2015b). Researchers use low-e glass or low-e coating to reduce radiation effects in the vacuum glass technique (Fang et al. 2007). It is stated that Low-e coated vacuum glasses are commercially produced, and the U_w value is 40% lower than the uncoated condition (Jelle et al. 2012a). Low-e coatings prevent heat from entering indoor environments on hot days and reduce heat losses from windows in cold climate conditions. The biggest disadvantage of the use of coatings is that they make indoor environments dark due to their low light transmittance (Bülow-Hübe 1995). One of the biggest handicaps of windows is that they cannot vary for mixed climatic conditions. It is impossible to use glasses that will allow less sunlight to pass through when the sun is intense and show high permeability when it is not intense at the same time. This problem can be solved with smart windows. Smart glasses are available as products that can achieve the desired level of illumination and heating by adjusting their visibility and thermal permeability according to environmental conditions (Baetens et al. 2012). Electrochromic windows, which keep the daylight entering the indoor environment at the highest level in the transparent state and keep it at the lowest level in the colored state, are among the smart windows currently in production (Jelle et al. 2012a). Innovative windows try to minimize heat loss, and warm climate enclosures are successful in reducing energy consumption by blocking the sun's excess radiation from entering indoor environments. However, the need for energy-focused products is increasing day by day in order to reduce human consumption and energy consumption. This center enters photovoltaic (PV) systems. According to data transfers via PV systems to be downloaded into the building, most of the energy given can be met. Building heating loads can be reduced due to the heating of the PV modules in transitions that pass natural ventilation with PV systems surrounding the buildings and type PV module structures cages (Bloem 2008). Switching to conventional PV modules is possible in cabinets of transparent PV modules. In this way, transparent PV integrated structures, which have become commercially widespread, are promising with wide-area applications (Fath et al. 2002). In buildings, PV modules can be used not only in windows but also in wall and roof openings. Modules produced in different environments and systems with a power output of up to 144.2 W per meter have serious energy potential (Jelle et al. 2012b). External wall and roof covering methods have been used for many years to prevent heat losses in buildings. One of the materials used in this method is silica aerogel. These materials with pore diameters in the range of 10–100 mm are not new (Kistler and Caldwell 1934), and their 90% porous structure in nanometer size makes them an excellent insulating material for buildings (Caps and Fricke 1986). Since the permeability levels of aerogel materials are not yet at the desired level, they are still suitable for use on the roofs and facades of buildings. In addition, with the increase in the level of permeability, it will be possible to use transparent aerogel in window glasses. The biggest disadvantage of the material is its rapid deterioration with water and its low strength (Liu 2012). The increase in the heating and cooling load in buildings day by day with the increasing population causes passive heating and cooling methods to gain importance. In general, energy storage is a concept that can be used for all situations. One of the energy storage applications for exterior facades in buildings is phase change materials (PCM). PCMs change phase by absorbing the heat falling on them. Then, when the temperature drops below a certain level, the stored heat is released, and the PCM returns to its old phase. PCM can be used in regions with high-temperature fluctuations by taking advantage of this conversion. PCMs, which can be produced as translucent as they can be used on building exteriors, can also be used by integrating into window glasses for daylight illumination (Manz et al. 1997). It is shown that the U_w value of the window glass produced with PCM glazing industrially is less than 0.50 W/m^2 K (Cuce and Riffat 2015a). Researchers also propose innovative designs beyond the given ones for sustainable energy use and building concepts. Gas-filled glass, which is also widely used, is one of the most used products in the market. The low heat transmission coefficients of the gases between the double glazing reduce the heat loss. Noble gases have low transmission coefficients. For this reason, heat losses can be minimized by using double glazing. Self-cleaning glass products are also produced. These products, which have a U_w value of 1.20 W/m^2 K, have low daylight transmittance (Cuce and Riffat 2015a).

15.3 PASSIVE HEATING, PASSIVE COOLING, AND PASSIVE HEATING-COOLING CONCEPTS FOR BUILDINGS

Considering that today most of the energy consumption occurs in indoor environments, that is, buildings, the effects of improvements to be made in this direction on energy consumption will be extremely high. Currently, traditional methods are heating and cooling processes to be carried out in closed environments. In traditional methods, fossil fuels are used for heating, and electricity is used for cooling. Since electricity is predominantly produced from fossil fuels, cooling is indirectly based on fossil fuels. In this case, passive heating and passive cooling methods become important for buildings. In general, passive heating, passive cooling, and cooling techniques for buildings can be classified in Figure 15.2.

The first of the primary passive methods for heating buildings is to conserve existing heat. This is possible by reducing insulation and heat losses. Another alternative is to benefit more from the sun during the colder seasons. Windows can be used for solar heat gain. Energy transfer and lighting can be provided to indoor environments through windows, which are the only elements of buildings that can take solar energy directly into the interior. Building window systems need to be optimized (Lee et al. 2012). With these traditionally used windows, lighting, and heat losses can be adjusted by sizing. With the structures to be added to the buildings with innovative approaches, energy consumption can be reduced while serving different purposes. One of the systems used in this way is the application called Trombe wall. In this system, which is applied in the form of covering the building wall with transparent glass, the solar radiation passing through the glass is trapped and stored in the energy storage layer on the wall. In addition, the natural movement of the air is provided by

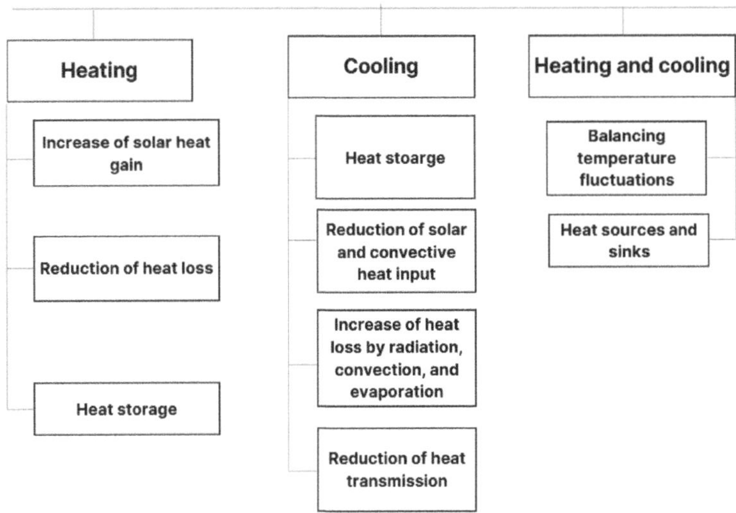

FIGURE 15.2 Some of the passive heating, passive cooling, and passive heating-cooling concepts for buildings.

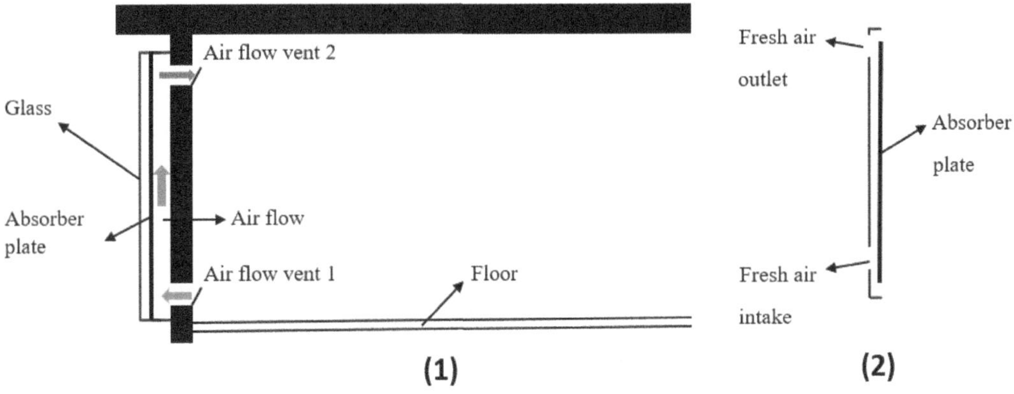

FIGURE 15.3 Schematic view of Trombe wall (Liu et al. 2013).

the greenhouse effect that occurs under the glass cover. The schematic view of the system is given in Figure 15.3 with no 1 (Liu et al. 2013). When the airflow vent 1 is open in the room, the solar radiation passes through the glass cover and is absorbed by the absorber plate, and the temperature increases on the surface. In the meantime, since the air entering the duct will come into contact with the hot surface and its temperature will rise, it moves upward and returns to the room from airflow 2. In this way, the air in the room circulates continuously. This applies to daytime conditions. Researchers claim that the system can be used by keeping the airflow vent 1 closed in the evening hours when there is no open sun during the daytime (Liu et al. 2013).

In the Trombe wall method, only the air in the room does not circulate continuously. Researchers can save energy by taking the fresh air from the outside into the room by warming it behind the glass cover so that the outdoor air can be taken indoors at a certain temperature in cold climate conditions. The visual of the design required for this is given in Figure 15.3 with no 2. Bevilacqua et al. (2022) implement a building energy simulation (BES) using the Computation Fluid Dynamic (CFD) method. They claim that by keeping the fresh air outlet valve open during the day in order to prevent overheating for the Trombe wall in summer, the energy consumption required for cooling will be reduced. In addition, they show that by keeping all the flaps closed during the winter months, when the sun is not present, heat transfer can be achieved from the wall, which is heated by the greenhouse effect, to the indoor environment by conduction. They state that in the building configuration in which this system is integrated, approximately 10% energy savings will be achieved in the summer and winter months. Some researchers emphasize that energy savings in the form of energy storage and diffusion can be achieved by placing PCM in the Trombe wall (Zalewski et al. 2012). Trombe wall may not be useful in regions where the winter climate is dominant and the average temperature is low. This disadvantage can be eliminated with PV-integrated Trombe wall applications (Ghazali et al. 2016). However, due to the additional heating that will occur in summer with the PV integrated, more cooling requirements may arise in indoor environments. In this case, it is necessary to calculate and analyze the additional loads required for heating in winter and cooling in summer (Bajc et al. 2015). Passive solar energy can be used directly to heat indoor environments. Especially on south-facing facades in the northern hemisphere, the sun directly penetrates the indoor environment and heats. Since this situation will cause additional cooling load in the summer months, the system can be idealized for summer and winter by placing additional structural elements on the roof (Chel and Kaushik 2018). A direct passive solar heating system called Sunspaces can be installed outside the building, and the heated air can be supplied to the indoor environment through a vent (Gainza-Barrencua et al. 2021). The direct passive solar heating scheme for the summer and winter months is given in Figure 15.4.

It is inevitable to use innovative approaches for a sustainable environment, as the increasing number of buildings brings the burden of heating, cooling, and ventilation in order to provide living space for the increasing human population. While direct sun can be used for heating by indirect methods, the use of electrical-based devices for cooling and ventilation is already very common. As mentioned before, the sun can penetrate indoors better in winter, and the energy consumption required for heating and cooling can be reduced by transferring solar energy to indoor environments in summer. Ventilation of closed spaces can be achieved without energy consumption with simple innovative approaches. Solar chimneys can be used for this situation. Based on simple physical laws, solar chimneys work with the upward movement of the heated air due to the density difference (Cuce et al. 2022c). With the chimney system integrated into the building wall or roof in indoor environments, ventilation can be provided without energy consumption by providing upward movement of the indoor air.

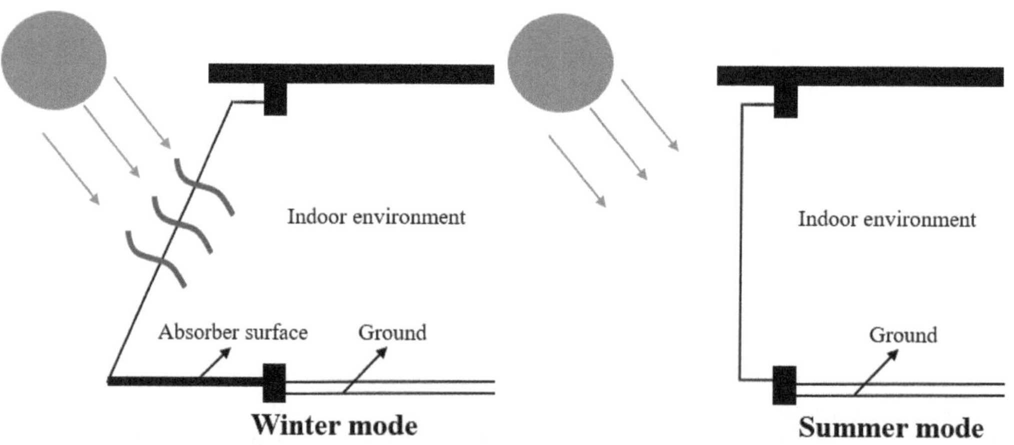

FIGURE 15.4 Direct passive solar heating for summer and winter (Chel and Kaushik, 2018).

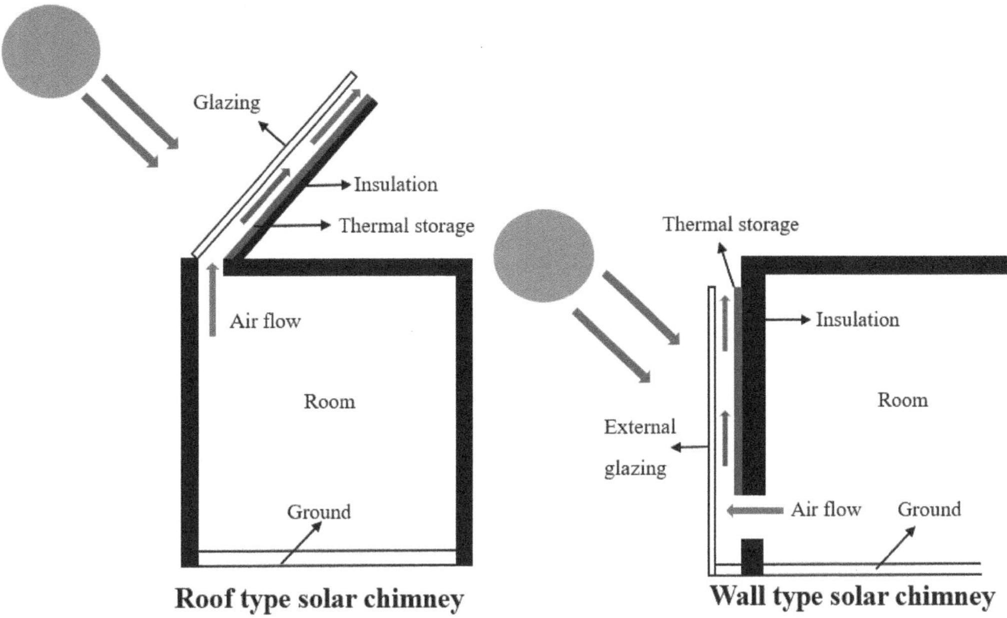

FIGURE 15.5 Roof-type and wall-type solar chimney concepts for passive ventilation (Harris and Helwig, 2007; Imran et al. 2015).

The solar radiation passing through the transparent glass layer is stored on the surface where it comes into contact with the thermal energy storage layer. As the stored energy causes the temperature to rise at the surface, the temperature of the air molecules in contact with the surface increases. In this way, the air molecules, whose temperature increases, begin to move upward. Natural circulation occurs as air movement provides continuity due to the conservation of mass. The solar chimney in the indoor environment continuously evacuates air from the upper part where the hot air is in the room; this type of solar chimney is called a roof-type solar chimney. Similarly, the same system can be applied with a vertical solar chimney to be placed parallel to the wall. Solar chimney diagrams with chimney and wall type are given in Figure 15.5. The energy consumed for cooling is expected to decrease as it evacuates the hot air in the upper part of the room, especially in regions with a summer climate (Khanal and Lei 2011). According to Khedari et al. (2000), the applicability of a solar chimney to reduce heat gain in a house is evaluated. They claim that using a solar chimney for a closed environment with a volume of 25 m^3 will be more useful than ventilation by opening a window or door, and the air exchange rate will be at the level of 8–15 with a solar chimney. As a result, they emphasize that the solar chimney is useful for reducing heat gain and thermal comfort. Doors and windows are not very useful for ventilation, as the wind effect is not important in very hot tropical climates. In addition, opening doors and windows can create unwanted situations such as dust and odor (Bassiouny and Koura 2008). In these conditions, the use of solar chimneys can be extremely effective. With the solar chimney to be integrated into a 27 m^3 room, 5–6 air changes per hour can be achieved (Mathur et al. 2006).

15.4 CONCLUSION

Energy consumption should be minimized to promote a sustainable environment and a clean nature. With the increasing human population and technological progress, this is only possible with the effective use of energy. Considering that the majority of energy consumption is from buildings, minimizing energy consumption in indoor environments is an effective way. The following conclusions can be drawn from the studies conducted for this purpose:

- Efficient use of energy is mandatory in closed environments.
- A smart building concept is essential for minimizing heat losses and low emissions in buildings.
- Studies should be carried out to reduce insulation and energy loss afterward for structures that are not smart buildings.
- Considering that the most heat loss is from the windows, innovative glass designs should be applied.
- Since vacuum and aerogel glass will provide the highest insulation in buildings with a U_w value of 0.30 W/m² K, its use will become widespread in the future.
- Building structural elements should be selected according to geography by considering climatic differences.
- Energy storage is very suitable for use in regions with high-temperature fluctuations.
- Natural ventilation can be done in indoor environments with methods such as Trombe wall and solar chimneys without energy consumption.
- With building concepts designed according to climatic conditions, the energy costs required for heating, cooling, and ventilation can be reduced to much lower levels.

REFERENCES

Abas, N., Kalair, A., Khan, N. (2015). Review of fossil fuels and future energy technologies. *Futures, 69*, 31–49.

Akram, M. W., Hasannuzaman, M., Cuce, E., Cuce, P. M. (2023). Global technological advancement and challenges of glazed window, facade system and vertical greenery-based energy savings in buildings: A comprehensive review. *Energy and Built Environment, 4*(2), 206–226.

Baetens, R., Jelle, B. P., Gustavsen, A. (2010). Properties, requirements and possibilities of smart windows for dynamic daylight and solar energy control in buildings: A state-of-the-art review. *Solar Energy Materials and Solar Cells, 94*(2), 87–105.

Bajc, T., Todorović, M. N., Svorcan, J. (2015). CFD analyses for passive house with Trombe wall and impact to energy demand. *Energy and Buildings, 98*, 39–44.

Bassiouny, R., Koura, N. S. (2008). An analytical and numerical study of solar chimney use for room natural ventilation. *Energy and Buildings, 40*(5), 865–873.

Bevilacqua, P., Bruno, R., Szyszka, J., Cirone, D., Rollo, A. (2022). Summer and winter performance of an innovative concept of Trombe wall for residential buildings. *Energy, 258*, 124798.

Bloem, J. J. (2008). Evaluation of a PV-integrated building application in a well-controlled outdoor test environment. *Building and Environment, 43*(2), 205–216.

Bribián, I. Z., Capilla, A. V., Usón, A. A. (2011). Life cycle assessment of building materials: Comparative analysis of energy and environmental impacts and evaluation of the eco-efficiency improvement potential. *Building and Environment, 46*(5), 1133–1140.

Bülow-Hübe, H. (1995). Subjective reactions to daylight in rooms: Effect of using low-emittance coatings on windows. *International Journal of Lighting Research and Technology, 27*(1), 37–44.

Caps, R., Fricke, J. (1986). Radiative heat transfer in silica aerogel. In *Aerogels: Proceedings of the First International Symposium, Würzburg, Fed. Rep. of Germany September 23–25, 1985* (pp. 110–115). Springer Berlin Heidelberg.

Chel, A., Kaushik, G. (2018). Renewable energy technologies for sustainable development of energy efficient building. *Alexandria Engineering Journal, 57*(2), 655–669.

Chitnis, M., Sorrell, S., Druckman, A., Firth, S., Jackson, T. (2012). *Estimating Direct and Indirect Rebound Effects for UK Households.* Sustainable Lifestyles Research Group: Working Paper, 01–12.

Cuce, E., Riffat, S. B. (2015a). A state-of-the-art review on innovative glazing technologies. *Renewable and Sustainable Energy Reviews, 41*, 695–714.

Cuce, E., & Riffat, S. B. (2015b). Aerogel-assisted support pillars for thermal performance enhancement of vacuum glazing: a CFD research for a commercial product. *Arabian Journal for Science and Engineering*, *40*, 2233–2238.

Cuce, E., Cuce, P. M., Carlucci, S., Sen, H., Sudhakar, K., Hasanuzzaman, M., Daneshazarian, R. (2022a). Solar chimney power plants: A review of the concepts, designs and performances. *Sustainability*, *14*(3), 1450.

Cuce, P. M., Cuce, E., Sen, H. (2022b). Impact of support pillars on the thermal insulation performance of vacuum glazing. In: Riffat, Su, ed., *Sustainable Energy Technologies: Proceedings of the 19th International Conference on Sustainable Energy Technologies, 16th–18th August 2022, Turkey.* University of Nottingham: Buildings, Energy & Environment Research Group. pp 239–246. Available from: nottingham-repository.worktribe.com/ [last access 30 March 2023].

Cuce, E., Saxena, A., Cuce, P. M., Sen, H., Eroglu, H., Selvanathan, S. P., Sudhakar, K., Hasanuzzaman, M. (2022c). Performance assessment of solar chimney power plants with natural thermal energy storage materials on ground: CFD analysis with experimental validation. *International Journal of Low-Carbon Technologies*, *17*, 752–759.

Fang, Y., Eames, P. C., Norton, B., Hyde, T. J., Zhao, J., Wang, J., Huang, Y. (2007). Low emittance coatings and the thermal performance of vacuum glazing. *Solar Energy*, *81*(1), 8–12.

Fath, P., Nussbaumer, H., Burkhardt, R. (2002). Industrial manufacturing of semitransparent crystalline silicon POWER solar cells. *Solar Energy Materials and Solar Cells*, *74*(1–4), 127–131.

Gainza-Barrencua, J., Odriozola-Maritorena, M., Hernandez_Minguillon, R., Gomez-Arriaran, I. (2021). Energy savings using sunspaces to preheat ventilation intake air: Experimental and simulation study. *Journal of Building Engineering*, *40*, 102343.

Ghazali, A., Salleh, E., Haw, L. C., Sopian, K., Mat, S. (2016). Research article photovoltaic Façade in Malaysia: The development and current issues. *Research Journal of Applied Sciences, Engineering and Technology*, *13*(8), 652–663.

Gustavsen, A., Jelle, B. P., Arasteh, D., Kohler, C. (2007). *State-of-the-Art Highly Insulating Window Frames – Research and Market Review*. Oslo.

Gustavsson, L., Pingoud, K., Sathre, R. (2006). Carbon dioxide balance of wood substitution: comparing concrete-and wood-framed buildings. *Mitigation and Adaptation Strategies for Global Change*, *11*, 667–691.

Harputlugil, G. U. (2016). Enerji Verimli Bina Tasarım Stratejileri. https://webdosya.csb.gov.tr/db/meslekihi zmetler/ustmenu/ustmenu845.pdf (access 25 March 2023).

Harris, D. J., Helwig, N. (2007). Solar chimney and building ventilation. *Applied Energy*, *84*(2), 135–146.

Imran, A. A., Jalil, J. M., Ahmed, S. T. (2015). Induced flow for ventilation and cooling by a solar chimney. *Renewable Energy*, *78*, 236–244.

Jelle, B. P. (2011). Traditional, state-of-the-art and future thermal building insulation materials and solutions – Properties, requirements and possibilities. *Energy and Buildings*, *43*(10), 2549–2563.

Jelle, B. P., Hynd, A., Gustavsen, A., Arasteh, D., Goudey, H., Hart, R. (2012a). Fenestration of today and tomorrow: A state-of-the-art review and future research opportunities. *Solar Energy Materials and Solar Cells*, *96*, 1–28.

Jelle, B. P., Breivik, C., Røkenes, H. D. (2012b). Building integrated photovoltaic products: A state-of-the-art review and future research opportunities. *Solar Energy Materials and Solar Cells*, *100*, 69–96.

Khanal, R., Lei, C. (2011). Solar chimney—A passive strategy for natural ventilation. *Energy and Buildings*, *43*(8), 1811–1819.

Khedari, J., Boonsri, B., Hirunlabh, J. (2000). Ventilation impact of a solar chimney on indoor temperature fluctuation and air change in a school building. *Energy and Buildings*, *32*(1), 89–93.

Kistler, S. S., Caldwell, A. G. (1934). Thermal conductivity of silica aerogel. *Industrial & Engineering Chemistry*, *26*(6), 658–662.

Kumar, D., Alam, M., Zou, P. X., Sanjayan, J. G., Memon, R. A. (2020). Comparative analysis of building insulation material properties and performance. *Renewable and Sustainable Energy Reviews*, *131*, 110038.

Lee, J. W., Jung, H. J., Park, J. Y., Lee, J. B., Yoon, Y. (2013). Optimization of building window system in Asian regions by analyzing solar heat gain and daylighting elements. *Renewable Energy*, *50*, 522–531.

Liu, H. (2012). *The Development of Novel Window Systems Towards Low Carbon Buildings* (Doctoral dissertation, University of Nottingham).

Liu, Y., Wang, D., Ma, C., Liu, J. (2013). A numerical and experimental analysis of the air vent management and heat storage characteristics of a Trombe Wall. *Solar Energy*, *91*, 1–10.

Manz, H., Egolf, P. W., Suter, P., Goetzberger, A. (1997). TIM–PCM external wall system for solar space heating and daylighting. *Solar Energy, 61*(6), 369–379.

Mathur, J., Bansal, N. K., Mathur, S., Jain, M. (2006). Experimental investigations on solar chimney for room ventilation. *Solar Energy, 80*(8), 927–935.

Roy, A., Ghosh, A., Mallick, T. K., Tahir, A. A. (2022). Smart glazing thermal comfort improvement through near-infrared shielding paraffin incorporated SnO_2-Al_2O_3 composite. *Construction and Building Materials, 331*, 127319.

Victoria University of Wellington. *Embodied Energy Coefficients.* www.victoria.ac.nz/cbpr/documents/pdfs/ee-coefficients.pdf (access 25 March 2023).

Wang, S., Ma, Z. (2008). Supervisory and optimal control of building HVAC systems: A review. *Hvac&R Research, 14*(1), 3–32.

Wigginton, M. (2002). *Intelligent Skins.* Butterworth-Heinemann Linacre House, Jordan Hill, Oxford.

Yang, L., Yan, H., Lam, J. C. (2014). Thermal comfort and building energy consumption implications – A review. *Applied Energy, 115*, 164–173.

Zalewski, L., Joulin, A., Lassue, S., Dutil, Y., Rousse, D. (2012). Experimental study of small-scale solar wall integrating phase change material. *Solar Energy, 86*(1), 208–219.

16 Unveiling Africa's R&I Landscape

Mapping Capacities, Horizontal Skills, and Actions

Chakib Seladji, Samir Brairi, Hakim Benyelles, Merwan Messaoudi, Amazigh Dib, Abdellatif Zerga, Emanuela Colombo, and Giacomo Crevani

16.1 INTRODUCTION

Africa as a continent confronts severe energy issues, but via a radical energy transformation, it has enormous potential for sustainable development. The need for energy in Africa is rising due to the continent's burgeoning economy and rapidly increasing population. However, many people still do not have access to dependable and clean energy sources. Due to the widespread dependence on fossil fuels and ineffective energy infrastructures, this energy imbalance not only impedes socio-economic development but also presents serious environmental and health risks.

Africa must immediately start on a thorough energy transformation path considering the pressing need to solve these issues. The construction of robust and sustainable energy systems, improved energy efficiency techniques, and a paradigm change in favour of renewable energy sources (RES) are all part of this transformation. African nations can satisfy their rising energy needs, promote economic development, and successfully combat the negative consequences of climate change by embracing these radical advances.

However, a successful energy shift requires more than just modern technology. To promote sustainable growth, it is necessary to provide a strong framework for capacity building and to foster horizontal skills. Initiatives to create capacity are essential for providing people, organizations, and communities with the information, tools, and resources they need to successfully implement and manage sustainable energy projects. These projects include a wide range of topics, including the creation of policies, project management, the use of technology, and funding sources.

Furthermore, it is essential for promoting sustainable growth in Africa's renewable energy industry to cultivate horizontal abilities that cut across disciplinary and sectoral barriers. The advancement of the energy transition depends on competences including competent project planning and administration, intelligent policy formation, the inventive design of renewable energy systems, application of energy efficiency practises, successful community participation, and expert financial modelling. People can successfully support the energy transition, spark innovation, and accelerate social and economic change by developing these adaptable talents.

Given the prevailing factors, this chapter is intended to explore the African energy landscape, underline the urgent need for an energy transition, highlight the significance of capacity-building programmes, and emphasize the importance of developing horizontal skills for sustainable development. Through a comprehensive examination of these interrelated aspects, valuable insights can be obtained into the possibilities and challenges faced by the continent. Furthermore, effective solutions can be suggested to accelerate the transition to a sustainable energy future in Africa.

DOI: 10.1201/9781032651958-16

African nations may provide a strong basis for sustainable development by promoting a thorough awareness of the energy environment, using capacity-building initiatives, and cultivating horizontal skills. This strategy will support equitable development and effective climate change mitigation in addition to promoting improved energy availability. The next parts of this book chapter will delve into these subjects in more depth, providing insightful analysis and practical suggestions for an effective and significant energy transition in Africa.

This study is funded by the European Union (EU) as part of the LEAPR-RE project, which aims to foster long-term partnerships and knowledge exchange between Europe and Africa in the field of renewable energy.

16.2   MAPPING THE MAIN R&I CAPACITY BUILDING ACTIVITIES AND NEEDS IN AFRICA

When considering RES, it is essential to address the exploitation of their available potential with adequate solutions and skills. These encompass a range of capacities beyond technical aspects, spanning from policy and regulatory design and management to project preparation, evaluation, development, implementation, and financing within government ministries, financing institutions, regulatory agencies, and utilities. The lack of comprehensive plans for knowledge, skill, and capacity development in the renewable energy sector poses a barrier to the widespread adoption of new renewable energy technologies (RETs) in developing countries [1]. This issue is particularly pronounced in the realm of research and innovation (R&I) activities. Previous literature has demonstrated a positive correlation between inadequate investments in R&I and the low rate of penetration of new RETs in local markets, subsequently diminishing the interest of external investors in these countries [2]. However, evaluating the overall impact achieved by such investments remains challenging [3].

Within this context, numerous capacity-building initiatives taking place in Africa have been collected. These actions are summarized in the subsequent sub-sections, leading to the focalization of a taxonomy in the final sections of this chapter.

16.2.1   ACTIONS IN PLACE IN AFRICA FOR RES

The research methodology employed a combination of online investigations and qualitative interviews conducted with different African institutions. The collected information was then compiled, shared, and disseminated among the partners.

The objective of this chapter is to present the findings and provide a comprehensive mapping of existing opportunities and ongoing actions in capacity building within the African RE sector. It is important to note that among e the mapping exercise covers a wide range of initiatives, it is not exhaustive due to the width and depth of the topic. Therefore, continuous monitoring and regular updates will be necessary to ensure accuracy and relevance.

Particular attention has been devoted to Sub-Saharan Africa (SSA), as the lack of information on capacity-building initiatives in the region has become apparent. By highlighting existing gaps and ongoing efforts, this study aims to shed light on areas that require further attention and investment to bridge the skills and knowledge gap in the African RE sector.

The comprehensive mapping of capacity-building initiatives presented in this chapter establishes a foundation for stakeholders and policymakers to assess the current landscape and identify opportunities for collaboration, knowledge sharing, and targeted interventions. By addressing the identified gaps and implementing effective capacity-building programmes, the development and adoption of RETs in Africa can be expedited, thereby contributing to sustainable energy access, job creation, and environmental stewardship.

The initial phase of the mapping exercise focuses on examining the distribution of skill and capacity training activities in the RE sector across the African continent. This encompasses academic programmes, such as higher education initiatives and curricular programmes, as well as practical

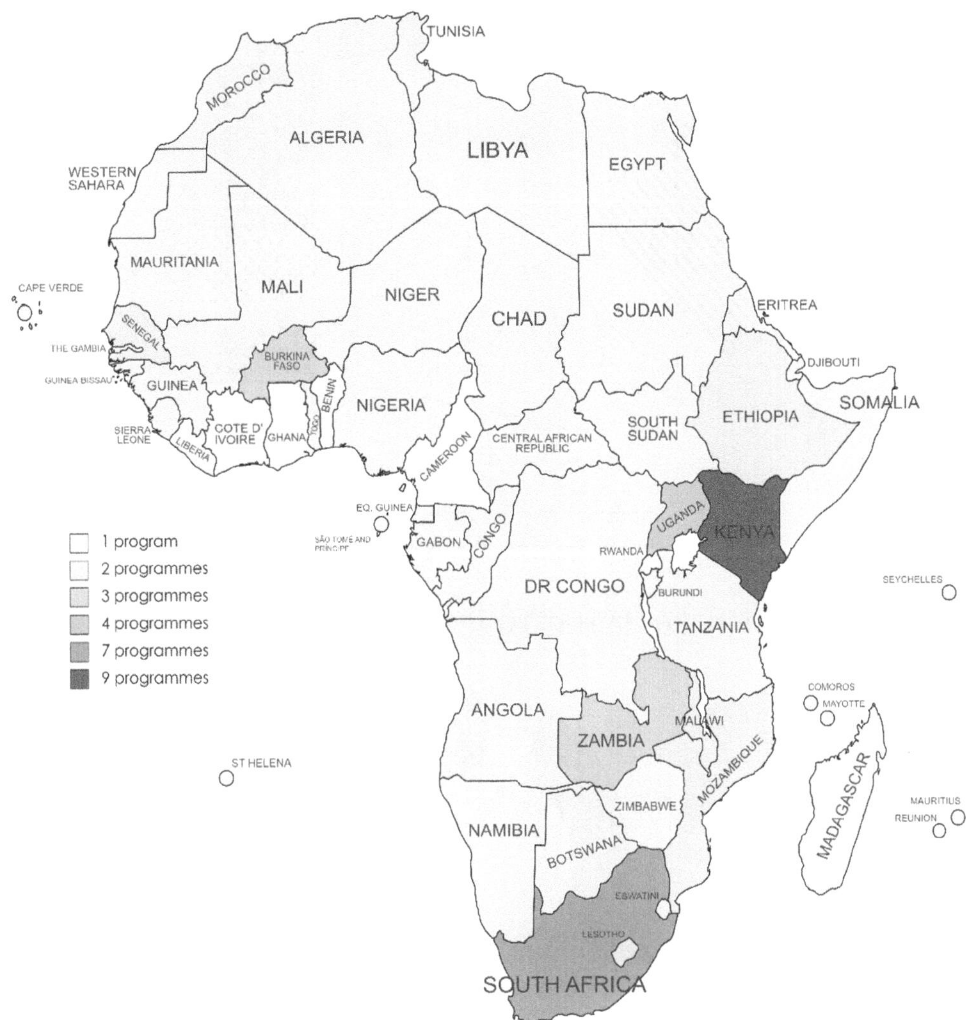

FIGURE 16.1 First mapping of RE programmes in the African continent.

training programmes, including vocational training initiatives and higher education vocational skills training. These activities encompass a diverse range of actions, such as pilot projects for innovative teaching, technical assistance training programmes, technical higher education programmes, and undergraduate or postgraduate courses. Additionally, programmes aimed at empowering disadvantaged women through technical training are explored.

The subsequent phase of the capacity-building mapping exercise involves compiling a selection of short-term courses that can be considered as Massive Open Online Courses (MOOCs) for technical training purposes. These courses target practitioners in the field and offer opportunities for continuous professional development and knowledge enhancement. By providing accessible and flexible learning options, these MOOCs contribute to building the necessary skills and expertise in the renewable energy sector.

16.2.1.1 Aggregated Results Analysis

To provide useful and updated information, all the activities considered are ongoing to date. The first aggregated map, in Figure 16.1, depicts the geographical distribution of these activities. Most of the

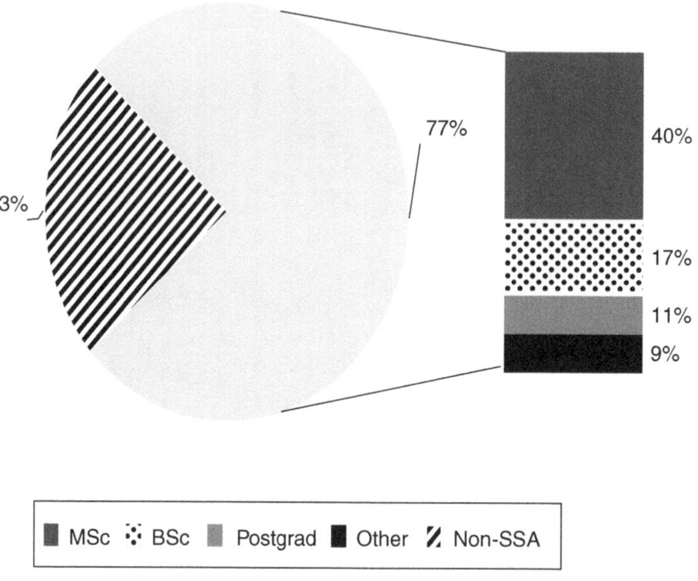

FIGURE 16.2   Mapping of academic RE programmes (total 86) in SSA and African non-SSA countries.

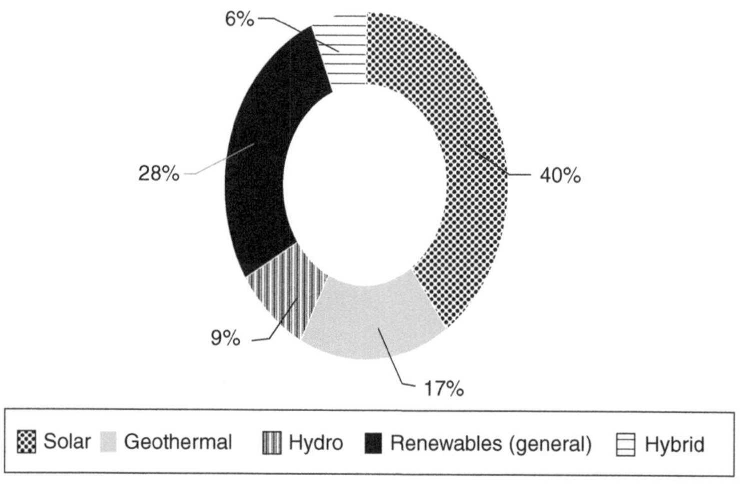

FIGURE 16.3   Mapping of practical training RE programmes (total 47) in Africa, by RES of interest.

tracked programmes, however, takes place in the south and south-east of the continent with South Africa and Kenya constituting more than the 60% of the sample. This shows much room for further programmes in the rest of SSA.

A more accurate mapping has then been conducted focusing on *academic programmes* on RE-related topics. The investigation concerned a total of 86 academic programmes, the majority of which are held in SSA. Among these, more than half of the mapped programmes are MSc courses, while very few concern PhD or other high-level programmes (in the category "Other"). The figures are depicted in Figure 16.2.

Finally, a third mapping was conducted focusing on a total of 47 *practical training programmes* on RE-related topics, such as academies, intensive technical courses, workshops, etc. These activities were then clustered by RES of interest, and the results are represented in Figure 16.3. A major interest can be noticed in the delivery of training programmes related to solar resources: all the countries investigated were provided with at least one course on the topic. A substantial number of programmes (28%) covered one or more RES in the same course, offering wider training to the participants. This results in a handful when dealing with RES since greater knowledge on their chances of exploitation expands the set of technological solutions to be adopted, as well. Hybrid technologies, such as third generation hybrid-mini grids, are crucial in this sense but still poorly covered by devoted training courses to date.

16.2.2 LIST OF ACTIONS IN PLACE IN AFRICA

Table 16.1 is based on the research conducted by mean of online investigation as well as qualitative interviews with institutions implicated in the field from both Europe and Africa.

16.3 REVIEWING HORIZONTAL SKILLS NEEDS IN AFRICA: LEAP-RE CASE STUDY

LEAP-RE, an EU-funded project under the Horizon 2020 programme [4], serves as a case study in this book chapter, particularly focusing on several projects within LEAP-RE that are targeted as case studies. The aim is to synthesize the horizontal skills required.

An ecosystem analysis was conducted to draft the Background Paper for Research and Capacity Building agendas. This analysis aimed to identify gaps, trends, and potential opportunities for renovating EU-AU collaboration in the renewable energy (RE) sector. The analysis relied on information provided by consortium members, relevant international literature, and existing studies with a similar synthesis task. As a result, an Initiatives & Network Matrix was constructed, comprising 89 selected initiatives that met specific criteria.

The ecosystem analysis underscores the vital importance of REs in addressing the global challenge of climate change and providing reliable energy access to millions of people worldwide. The following key recommendations for R&I in the field are summarized and schematically represented in Figure 16.4:

- *Technological development* needs to be advanced across the energy supply chain, including conversion technologies and end-user devices. Resource assessment remains crucial for certain sources, while R&I in distribution are essential for integrating renewables through smart hybrid mini-grids, whether in off-grid configurations or long-term integration into national grids. This presents an attractive research area where leveraging innovation with the ongoing digital revolution in Africa can enable source integration and additional storage opportunities.
- Technological development cannot exist in isolation. *A comprehensive methodological approach* is necessary, addressing distinct phases of the energy supply chain while considering societal needs, market evaluation, business models for long-term sustainability, solution deployment, and the long-term impact on society. Such an approach, as emphasized by the AU-EU High-Level Policy Dialogue (HLPD) roadmap on climate change and sustainable energies (CCSE), is crucial for ensuring the long-term social, economic, and environmental sustainability of technology.
- Renewed attention to *energy scenarios and policy* is crucial for understanding the contexts in which technologies and energy solutions will be developed, thereby minimizing unforeseen consequences. Support for further research and capacity building in energy scenario analysis, including modelling approaches and tools, is necessary to assist policy and decision-makers in creating long-term plans at the country and regional levels.

TABLE 16.1
Mapping of Actions in Place in Africa for Capacity Building for RES

Designation	Duration	Description	Link
		Training women to become Solar Engineers. More than 2500 engineers trained in villages across the globe	www.barefootcollege.org/solutions/
Household Solar Workforce Challenge		Solar training grants for third party training provides "for initiative to manage training staff, technicians, or sales agents for the off-grid energy sector"	www.usaid.gov/householdsolar/call-for-proposals
Short course on Solar PV	2 days	https://aceesd.ur.ac.rw/short-course-photo-voltaic-technology	www.strathmore.edu/open-africa-power-initiative/
	8 weeks	Technical and sales training through certified skills courses (solar design, installation, maintenance and servicing, safety, project management, solar water pumps, and solar water heaters. Also provides on the job coaching for "frontline staff and agents and employers for continuous development, sales and professional skills"	www.enlightinstitute.org
Solar irrigation training		The University of Development Studies (UDS) signed a funding agreement partnership with GIZ to provide "training to technicians, installers and agricultural extension agents"	www.faapa.info/en/2021/10/09/giz-uds-partner-to-train-actors-on-solar-powered-irrigation-system/
Solar cooling	3 days	This course is open but not limited to maintenance engineers, field engineers from the humanitarian or development sector, product managers, technical staff, and independent consultants	https://strathmore.edu/course/master-of-science-in-sustainable-energy-transitions/
Solar Water Pumping	5 days	Practical excersises in stand-alone and hybrid (grid or genset backup) configuration to learn the most important operational parameters. Proprietary manufacturer software will be used to design and calculate PV powered pumping and irrigation systems	https://serc.strathmore.edu/programmes/
T1 and T2	20 days	This course is the entry point to become a licensed Solar technician or Solar products vendor. Participants receive comprehensive know-how for photovoltaics, hands-on training with technical components, designing optimized PV stand-alone systems to apply for a T2 license from the Energy Regulatory Commission	https://serc.strathmore.edu/programmes/
T3 Grid Tie	20 days	Participants will learn the technical planning, design and cost calculations for larger PV systems. Positive course completion enables to apply for a T3 license at ERC	https://serc.strathmore.edu/programmes/

Certification of Solar PV Installers		For successful RE and EE installations and projects, there is a need for quality assurance at various levels. This includes quality assurance of equipment, e.g. through the establishment and enforcement of product standards. However, high-quality products will only provide the desired services if RE and EE systems are designed, installed and maintained by highly qualified individuals. Economic Community of West African States (ECOWAS) Centre for Renewable Energy & Energy Efficiency (ECREEE) and its partners, therefore, decided to support the development of the regional market for RE and EE services by establishing a scheme for certifying the skills of solar PV installers and other sustainable energy professionals. The objective is to introduce a quality mark for sustainable energy skills that is recognized by professionals and end users across borders in all 15 ECOWAS member states	www.ecreee.org/certification
L'Ente Nazionale per l'Energia Elettrica (ENEL) Green Power Photovoltaic Skills Training		Solar PV Rooftop installer course and PV Enterprise & Sales Development Course (free course funded by ENEL green power)	www.green-cape.co.za/assets/Uploads/Course-Brochure-2018.pdf
Solar PV Installer course	5 days	The SARETEC Solar PV Installer course is a 5-day short course and it emphasizes the physical practical aspects of mounting solar modules on the roof, wiring of the combiner box, the inverter and the distribution box	www.saretec.org.za/solar/saretec-solar-pv-course/
PV GreenCard	2 days	The PV GreenCard contains details of the installation such as, what sort of PV modules and PV inverters were used, as well a checklist of all of the necessary installation steps that were completed.	www.saretec.org.za/solar/pv-greencard-assessment/
The Basics of Solar PV Systems	2 days	Two-day interactive training course, hosted by SARETEC trainers and international Solar PV experts. Non-electricians permitted An attendance certificate will be provided	www.saretec.org.za/online-training/solar/
Solar Training		Our commitment does not end with the support. Through continuous training hundreds of participants in the pioneering field of solar energy can be trained year by year. Solar energy not only secures energy supply in resource-poor regions, but also creates sustainable jobs	https://sunfarming.de/en/business-areas/solar-training-center-en/south-africa
Capacity and human resource building for solar market development in Tunisia		The education and training courses available for the practice-oriented and needs-based qualification of PV specialists are improved. Training institutions network and cooperate closely.	www.giz.de/en/worldwide/58176.html
Basic and advanced courses	Ranges from 1 to 5 days per course	The GREEN Solar Academy PTY Ltd. is an independent training provider and a spin-off of the maxx-solar academy. Our academies follow the German DGS SolarSchool approach and all academies and courses are accredited by the German Solar Energy Society (DGS). GREEN stands for Global Renewable Energy & Efficiency Network as our goal is not only to provide training but to build up a network of PV installers all over Africa. GREEN is the link between manufacturers, wholesalers, associations and PV installers, our alumni	https://solar-training.org

(continued)

TABLE 16.1 (Continued)
Mapping of Actions in Place in Africa for Capacity Building for RES

Designation	Duration	Description	Link
		BRILHO is a five-year programme, 2019–2024, that will catalyse Mozambique's off-grid energy market in order to provide clean and affordable energy solutions to the country's off-grid population. BRILHO's overall goal is to improve and increase energy access for people and businesses, leading to money saving, better well-being and livelihood opportunities for the low-income population.	https://brilhomoz.com
Public Private Development Partnership for Renewable Energy Skills Training and Women's Economic Empowerment in Somalia		The project has two objectives: to economically empower women, especially women entrepreneurs in the MSME sector, and to develop skills to support an expanded electrical system with a greater proportion of renewable energy sourcing. The project will also establish a common curriculum for the design and installation of solar PV systems. At least 10% of the participants in each training are women	www.ilo.org/africa/countries-covered/somalia/ppdp-wee/lang--en/index.htm
Geothermal Exploration and Development of Geothermal Resources Series			www.grocentre.is/gtp/capacity-development-gtp/short-courses-and-workshops/workshops
Workshop for Decision Makers on Geothermal Projects and their Management		The objective of arranging Workshops for Decision Makers in various parts of the world is to increase the cooperation between specialists of neighbouring countries and to promote geothermal development by enlightening top level decision makers on geothermal development, regulations, workforce, equipment, financing, and other issues concerning geothermal utilization. The main goal being that they lay the foundaition for a continuous Short Course Series in support of the SDGs	
In development		GETRI was founded as a response to an existing need in the geothermal industry in developing capacity to exploit geothermal energy in the region. GeTRI is expected offer the first-ever comprehensive programme that brings together world-class experts to teach all elements of geothermal energy from resource discovery to utilization, including drilling, reservoir engineering, plant design, environmental impact, and applicable business principles. Its core service areas include postgraduate training, research, publication, seminars & conferences in close collaboration with the geothermal industry, local and international universities. Initially, the Institute will have programmes at MSc level and, later, Ph.D. level	https://getri.dkut.ac.ke
An introduction to geothermal science and technology	15 days	The main objective of the training is to strengthen and enhance the capacities and skills of young Africans in geothermal science and technology that came from 11 Eastern Africa countries. The trainees are expected to use the acquired knowledge and skill during this training in implementation of various geothermal projects in their respective countries	https://theargeo.org/AGCE/augustlectures.php

Name	Duration	Description	Reference
Geoscience and Reservoir Studies for Advanced 3D Models Training	4 days		http://theargeo.org/AGCE/files/ Leapfrog%20Training%20 Progress%20Report_Day%20 2%203%20%204_Report.pdf
Training on introduction to ggeothermal science and technology	11 days	The main objective of the training is to strengthen and enhance the capacities and skills of young Africans in geothermal science and technology that came from eleven Eastern Africa countries. The trainees are expected to use the acquired knowledge and skill during this training in implementation of various geothermal projects in their respective countries	https://theargeo.org/AGCE/lectures. php
Slimhole Drilling Webinar Series		Organized in collaboration with the New Zealand Geothermal Facility under the auspices of Interim Project Coordination Unit of the African Geothermal Center of Excellence (IPCU-AGCE). These webinars hope to build awareness and knowledge of slimhole drilling. Representatives from the ten ARGeo member countries: Comoros, Djibouti, Ethiopia, Kenya, Malawi, Mozambique, Rwanda, Uganda, Tanzania, and Zambia will attend the webinar series	https://theargeo.org/AGCE/NZ.html
Serveral	Ranges from 21 to 45 days	The Geothermal Training Centre currently offers courses in various disciplines ranging from scientific, engineering, statutory and other related areas	www.kengen.co.ke/index.php/ geothermal-center-of-excellence/ kengen-gtc.html
T3 Hybrid	20 days	Must have tertiary education to participate. Participants will learn all planning steps by designing and calculating their own PV hybrid systems.	https://serc.strathmore.edu/ programmes/
Nigerian Energy Support Programme II		Offers training courses as part of the broader programme	www.giz.de/en/worldwide/26374. html
		The Micro-grid Academy is the main programme of RES4Africa focused on providing the necessary skills and knowledge for mini-grid development. "It is a vocational capacity building programme aiming at creating a skilled and conscious workforce, in order to deploy decentralised renewable energy solutions and business skills." Local partners including the Institute of Energy Studies and Research (IESR, formerly the Kenya Power Training School), the University of Strathmore, St Kizito Vocational Training Institute, and Association of Volunteers in International Service (AVSI), a non-governmental organization	www.res4africa.org/ micro-grid-academy
		Voith will consult with the Ministry for Energy and Water of the Republic of Angola (MINEA) in 2020 to determine a suitable location for the training centre	https://voith.com/corp-en/2020-02-07-vh-voith-signs-memorandum-of-understanding-to-build-training-center-in-angola.html

(continued)

TABLE 16.1 (Continued)
Mapping of Actions in Place in Africa for Capacity Building for RES

Designation	Duration	Description	Link
Several, ranging from electricity generation, distribution and system operation to leadership and safety courses		The Centre offers training in five (5) core focus areas as listed in the tabs below, The scheduled training programmes range in duration from two days to 13 weeks and are conducted on a residential basis. All the courses are certified at skills award level for competency under the Technical Education, Vocational, and Entrepreneurship Training Authority (TEVETA) in Zambia. In addition, the Centre offers tailor-made training programmes and conducts onsite training at clients' premises in order to cost effectively meet the specific needs of individual organizations.	www.kgrtc.org.zm
G-Res Tools Assessor of Hydropower Sustainability Tools		The Hydropower Sustainability Training Academy currently offers three professional training courses. Two of these courses are designed to help participants develop the skills and knowledge to become either a certified user or accredited assessor of the Hydropower Sustainability Tools. The third course caters to practitioners who want to know how to accurately estimate greenhouse gas (GHG) emissions from hydro-electric projects using the G-res Tool. The courses will soon be available in a variety of languages with plans to expand the courses on offer in 2021	www.esi-africa.com/industry-sectors/generation/ihas-new-training-academy-to-advance-sustainable-hydropower/
East African Regional Training Course on Operation and Maintenance of Small-scale Hydro Power Plants (SHPP)	5 days	The course is aimed to provide competence of the specialized aspect of operation and maintenance of hydropower station including basics of the electro-mechanical equipment, upkeep, operation and maintenance during the operation of the plant. This course will explore the operations and maintenance (O&M) of Small Hydropower Plants (SHPP) and provides participants with an understanding of the key issues, challenges and coping strategies specific to small hydropower installations in the East African Community (EAC). The course material is draw heavily from Intergovenmental Hydrological Program's (IHP) experience with the design and installation as well as training on SHP globally. After completing the course, participants will have a strong understanding of small hydropower O&M issues, enabling them to undertake important roles associated with plant oversight, management, and upkeep.	www.eacreee.org/event/east-african-regional-training-course-operation-and-maintenance-small-scale-hydro-power-plants
Enel Foundation Advanced Renewable Energy Training for Africa		Building skills and competences for the deployment of RE technologies	www.enelfoundation.org/topics/articles/2021/07/renewable-energy-for-africa--advanced-training-course

The Developing Developers programme		The South African Photovoltaic Industry Association (SAPVIA) and the South African Wind Energy Association (SAWEA) has launched its Developing Developers programme, aiming to enhance local skills development across the renewable energy value chain. Boosted by a proactive clean energy policy, the country is looking to reap the full benefits of a lower carbon economy	www.engineeringnews.co.za/article/sapvia-sawea-partner-to-assist-local-renewables-project-developers-2020-09-21/rep_id:4136
Several	5 days generally	Structured around three- to five-day learning modules closely aligned to real-world industry applications, to cater for the varied requirements of the intended capacity building audience. The modules are designed in such a way that candidates can obtain a variety of qualifications while attending the same contact session, with the differentiation between qualifications determined through the additional assessments, assignments, and projects	www.sagen.org.za/publications/capacity-building-technology-innovation/123-power-systems-planning-and-operations-training-overview-brochure/file
		Skills development/technical and vocational education and training sub-programme of the larger Beyond the Grid Fund for Africa (BGFA) programme	https://beyondthegrid.africa/news/promoting-the-job-creation-and-skills-development-agenda-in-the-off-grid-sector-in-uganda/
Power for All Powering Jobs Campaign		PoweringJobs is a global campaign to ensure that the needed skills and jobs in clean, distributed energy are created to achieve universal electricity access for 1 billion people, and to employ the energy workforce of the future, especially women and youth. "Awareness of the vast opportunity to create new jobs and positive economic impact by meeting the energy access employment needs of private and public sector stakeholders." "Behavior Change among funders and governments, as well as academic, training and private sector institutions that recognizes workforce training as a benefit, not a cost." Market Activation through increased financial, policy and programmatic support to develop new, equitable, diverse, and inclusive training and employment opportunities	www.powerforall.org/resources/calls-to-action/action-agenda-build-workforce-skills-needed-universal-energy-access
Transforming Energy Access Learning Partnership		The TEA Learning Partnership (TEA-LP) aims to support the achievement of SDG 7 through skills and capacity building to 8 partner universities in Africa, recognizing the importance of human capital for universal access to energy. TEA-LP provides training through workshops, webinars, etc.	https://tea-lp.org

(continued)

TABLE 16.1 (Continued)
Mapping of Actions in Place in Africa for Capacity Building for RES

Designation	Duration	Description	Link
Skills for Energy in Southern Africa (SESA) project		Kafue Gorge Regional Training Centre (KGRTC) will establish a new expanded training portfolio for renewable energy and energy efficiency, with support from the International Labour Organization (ILO) and private sector partners. The trainings will be based on demand from the industry and build on KGRTCs strategic position. More power technicians and managers in the region skilled in Renewable Energy, Energy Efficiency, and Regional Energy Integration. This will be reached through training RE/EE/REI programmes for in total around 1600 trainees. This will be done through face-to-face trainings, around 140 trainees per year, blended face-to- face and e-learning, around 75 trainees per year, and e-learning trainings for around 240 trainees per year. In addition, dedicated training for 240 young female engineers will be conducted	https://cdn.sida.se/app/uploads/ 2020/12/16073033/skills-for-energy-in-southern-africa.pdf
Skills for the Renewable Energy Sector (SkiDRES)		The Skills Development for the Renewable Energy Sector (SkiDRES) was a 19 months pilot project aimed at developing and building partnerships with the private sector, assess market needs, develop and test demand-driven training, and prepare for a three-year Public Private Development Partnership, with sub-regional coverage in Africa	www.ilo.org/wcmsp5/groups/ public/---africa/---ro-abidjan/ ---ilo-lusaka/documents/ publication/wcms_761097.pdf
Modernizing vocational training for renewable energies		To increased specialized local technical expertise and management skills available on the market for renewable energies and energy efficiency in Côte d'Ivoire. The project aims to strengthen the skills of teachers at vocational schools and universities, enabling them to act as multipliers for the dissemination of practical skills	www.giz.de/en/worldwide/79018. html
Schneider Electric Vocational training		1400+ courses from our 91 training centres worldwide with practical face-to-face session, digital programmes and electrical installation simulators.	www.se.com/ww/en/product-range/ 62301-technical-training-course-finder/18247010579-trainings/ ?N=96675186+3045984154+ 1734072006+3124198887+ 2492419691+1915941756+ 1854449066+1663370252+ 1380133721
Higher Education Programme for Renewable Energy and Energy Efficiency		Senegalese universities offer practice and labour-market-oriented degree courses as well as further training courses for professionals in renewable energies and energy efficiency and strengthen skills for business start-up in this field. The Initiative for Sustainable Energy Policy (ISEPs) are advised on developing practical degree courses on the productive use of renewable energies and energy efficiency in locally relevant sectors. To complement this, the programme supports the piloting of short-term ISEP training courses relating to energy, also for people without university entrance qualifications (including returnees)	www.giz.de/en/worldwide/39287. html

Energy Management	3 days	This programme is designed for mid to senior level management who are interested in energy matters within their organizations. Participants will also be able to learn about what opportunities are available in their facilities and practical ways of tapping into those opportunities	https://serc.strathmore.edu/programmes/
Capacity Building consultation services		KEREA has developed training for solar technicians and equipped training institutions with necessary materials to provide proper training. They focus on solar, hydro, geothermal, biogas, biomass, and wind	https://kerea.org/capacity-building-consultancy/
Energy Statistics, Residential Sector, Energy Efficiency Data Collection and Initial Step towards the Creation of Energy Efficiency Database for Industrial Sector	3 days	Participants received hands-on training on the methodologies and tools on energy statistics and energy efficiency data collection and how to organize energy data at the national level through different resources and sectors of the economy. The workshop also stressed the lack of statistical data from AU Member States countries and focal point's difficulties in collecting and reporting data to AFREC. The focus was therefore on how to identify the best energy efficiency indicators for residential and Industry sectors and how to use the basic energy data to establish comprehensive and accurate energy balances through a coherent definitions and units, thus, enable consistent regional and international reporting	https://au-afrec.org/en/news-and-media-events/latest-news/regional-training-workshop-southern-african-countries-energy
Several	5 days on average	The IESR is a Regional Centre of Excellence in Energy Training and Capacity Building. The institute offers Professional courses to corporate organizations, private companies and contractors in the fields of Energy, Electrical, Mechanical, Fiber Optics and Management. The Institute also offers tailor made courses in any topic within the fields	www.iesr.ac.ke/index.php/capacity-building
Ecowas Energy Efficiency Technical Assistance Facility		The initiative was designed to create and operationalize a technical assistance facility for Micro, Small and Medium Enterprises (MSMEs) whose business model includes provision of energy audit services. It is a follow up of the first energy audits training jointly organized by ECREEE and the National Renewable Energy Laboratory (NREL) in December 2015 in Praia, Cabo Verde, where local companies expressed their concern on the serious challenges they are experiencing in selling their energy audit services to Clients	www.ecreee.org/page/ecowas-energy-efficiency-technical-assistance-facility

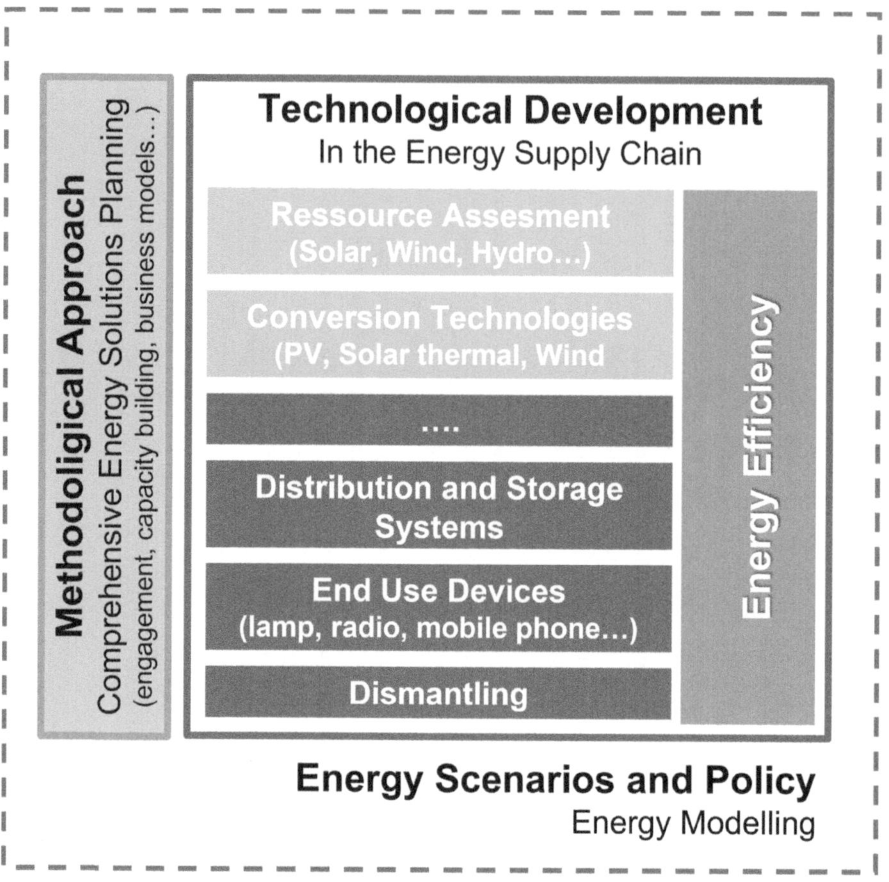

FIGURE 16.4 Comprehensive energy solution planning (CESP) as a methodological approach of analysis.

- Adopting a multidisciplinary approach encourages the development of locally appropriate scenarios that can support policymakers. Additionally, this approach necessitates the implementation of capacity building activities to empower and engage local stakeholders. Thus, Human and Institutional Capacity Building activities are included across all Multi-Annual Roadmaps (MARs) as part of LEAP-RE.

Based on the ecosystem study and assistance from the European Commission (EC), an interim list of 13 MARs covering the major RE development themes was produced. Following stakeholder input, these 13 roadmaps were trimmed down to 6, MARs (Figure 16.5). In terms of societal issues, research scope, anticipated outputs, results, and implications (Figure 16.4), these roadmaps are explained.

Overall, two have been the main outputs of the ecosystem analysis:

- The *6 MARs* as a programmatic framework of research.
- The *comprehensive planning approach* is a methodological approach to analysis.

The study conducted within the LEAP-RE project revealed the importance of research, innovation, and capacity development initiatives in Africa's renewable energy industry. The findings highlighted the need for a multifaceted strategy to address the possibilities and problems within the renewable energy market. Collaboration and synergy can be achieved by combining various disciplines and engaging stakeholders, leading to significant advancement. Key areas for improvement

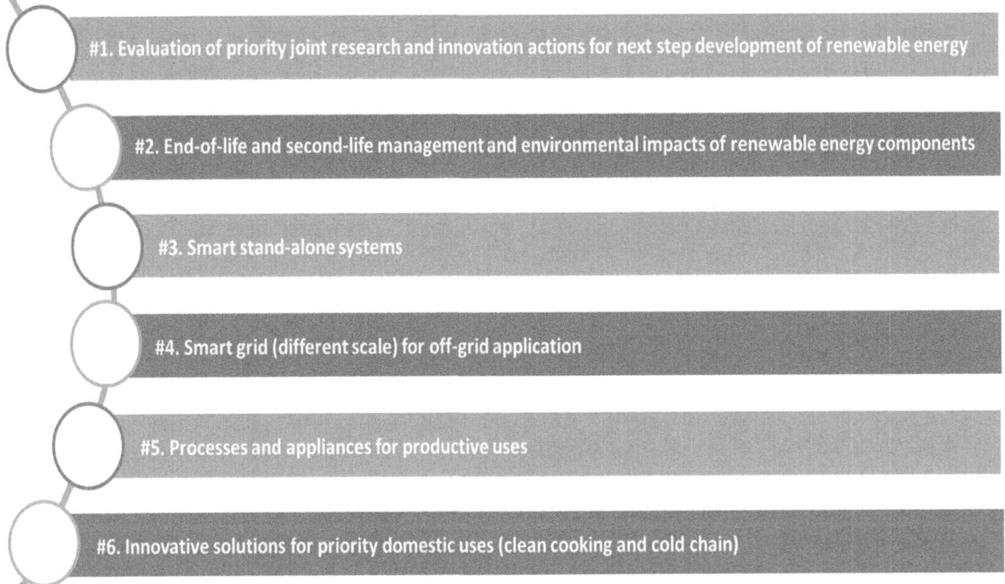

FIGURE 16.5   LEAP-RE multiannual roadmaps (MARs).

include advancing technical developments, resource assessment, and smart grid integration. A comprehensive methodological approach considering social demands, market analysis, and long-term sustainability is crucial.

Energy scenarios, regulatory frameworks, and capacity-building initiatives are also essential for directing renewable energy solutions. Empowering local stakeholders and encouraging ownership ensures sustainable energy solutions are based on African communities' needs. The findings serve as a basis for stakeholders, decision-makers, and researchers to evaluate the current environment, identify areas for cooperation, and design focused actions. By utilizing these findings, renewable energy solutions can be developed and used more quickly in Africa, promoting access to sustainable energy, job creation, and environmental stewardship.

16.3.1　Identifying Key Elements from Interview Abstracts

Based on each conducted interview between June and September 2021, the topics have been divided into two sections:

- *Mapping* the projects in terms of 4 main areas:
  - Technology,
  - Modelling tools,
  - Methodologies used,
  - Capacity building.

Comprehensive Planning: an overview of the current stage of the project throughout different subsequent steps following the methodological approach of comprehensive energy solution planning (CESP) as a methodology to increase the success of any energy-related project. CESP is reinterpreted as a useful tool to analyse the cross-sectoral actions of each project (including capacity-building activities), thus contributing to the definition of horizontal skill needs in this case study.

16.3.1.1   Geothermal Atlas 4 Africa Project

The "Geothermal Atlas 4 Africa" project aims to identify and develop geothermal systems across Africa by addressing factors such as high and low enthalpy applications, infrastructural constraints, and social dimensions. It will establish a framework for integrating geothermal projects with other RES, evaluate standalone installations' feasibility, and examine sustainability aspects in diverse African regions. The project will transfer knowledge and expertise from European counterparts to Africa, promote knowledge exchange, and establish regional networks to foster collaborations between African and European researchers and institutions. The project aims to augment investment opportunities for foreign enterprises in Africa's geothermal sector, contributing to the continent's sustainable energy future [5].

16.3.1.1.1   Mapping

The main *technology* involved in this project is geothermal technology, which is in accordance with the main goal, which is to lay out a complete and detailed atlas of geothermal resources on the African continent. It will also describe current possibilities of exploitation, thanks to the state-of-the-art available technology. Particular attention will be given to the opportunity to exploit geothermal resources for heating and cooling sectors, exploring the possibilities for a safe and efficient cold chain.

Some *modelling tools* will be used to map the whole continent, while some other software will be employed: The Global Positioning System (GPS) information to evaluate distances and locations, while Engineering Equation Solver to design plants and technologies for the exploitation of the resource. Various aspects will be linked using python-based tools, to favour open-source access to the model and to the final atlas.

The working *methodology* of the project begins with the assessment of the current situation, then by addressing the presence and quality of geothermal resource and finally outlining which current technology is best to exploit it. The output will be the Geothermal African Atlas, with different layers to highlight resource availability, its utilization, and technological possibilities.

Finally, some *capacity building* events will be organized to teach to researchers how to use the modelling tools properly and to exploit the atlas. This is crucial for the success of the project, which aims at providing the necessary tools to the scientists for efficient creation and design of optimized geothermal plants and mini grids integrated with geothermal sources.

16.3.1.1.2   CESP

The whole idea of this project started from the lack and necessity of a holistic mapping of the geothermal resources in the African continent, which could be crucial in the challenge to bring sustainable energy to unelectrified communities. This resource could be exploited to run local businesses, for irrigation water pumps, and for safe drinkable water for the population.

The assessment of the resource is the first step for the correct exploitation and spread of geothermal technologies. In Africa the major candidates are Kenya and the Rift Valley, thanks to the elevate geological activity of the area.

The project will provide to policy makers and scientists a complete atlas and the guidelines for the optimization of different components of geothermal power plants. It will provide also lessons and capacity building for the modelling tools and for the different possibilities of exploitation of this resource, suggesting also innovative ways of utilization, such use cooling and heating.

In the end, the major outcome will be the diffusion and dissemination of the Geothermal African Atlas, that can be a milestone to favour the use of this under exploited renewable source for a more capillary electrification of the continent.

16.3.1.2   Project PURAMS

"PURAMAS" (Productive Use in Rural African Markets using Standalone Solar) aims at engaging policymakers and exploring business models for electric cooking promotion, where they will be educated on the benefits of electric cooking, particularly in relation to healthcare, and provided with recommendations for enabling its growth potential using standalone solar devices. Additionally, diverse business models will be explored and adapted for the selected geographies to ensure the most suitable approaches are developed. By effectively engaging policymakers and implementing appropriate business models, the promotion of electric cooking can be advanced, leading to improved healthcare outcomes, and sustainable energy practices [6].

16.3.1.2.1 Mapping

The main *technology* analysed within this project is solar cooking. The current data collection will be developed in three countries: Rwanda, Kenya, and Mozambique. Based on the information from surveys and smart metres on-side, the appropriate technology to develop will be chosen. Electric pressure cookers and induction cookers emerged as viable options during the interview.

The selection of the *modelling* tools is pending and contingent upon the chosen technology. The decision on which modelling tools to utilize will primarily be based on the specific technology that will be employed in the project.

In terms of the *methodology*, an initial requirement is to establish the involvement of the partner organizations, particularly in Mozambique. This collective decision-making process aims to determine the most appropriate technology for each specific context. Insights gained from previous projects will be leveraged during this process to enhance the participation of the various partners in terms of coordination and data collection. By drawing on past knowledge and experiences, the project aims to optimize collaboration and ensure effective implementation across all partner organizations.

Capacity building plays a crucial role in educating and training the beneficiaries about the potential benefits of innovative cooking technologies within the community. It is also important to actively engage women in these awareness activities to gain insights into their specific needs and requirements concerning households energy consumption and cooking practices. A comprehensive understanding of the community's energy and cooking requirements can be obtained by involving women, ensuring that the capacity-building efforts effectively address their unique needs and promote inclusive and sustainable solutions.

16.3.1.2.2 CESP

The data collection activity will understand the energy needs of three communities in Rwanda, Kenya, and Mozambique. While Kenya's community is on-grid, Rwanda and Mozambique are powered with mini-grid technologies. Over a period of five months, this measurement process will involve 100 households. By utilizing kettles, induction hobs, and smart metres, valuable insights can be gained regarding the cooking and eating habits within these households, enabling a comprehensive understanding of their specific energy needs for cooking purposes.

The solution identification phase plays a crucial role in this project as it relies on the findings from the data collection phase to determine the most suitable pilot project for meeting the energy needs of the community. From the interview, solar cooking technology emerged as the most viable option. Additionally, the potential of a technology such as a fire cookstove was emphasized, considering the specific context in which the project will operate. However, no final decision has been made regarding the technology to be adopted at this stage.

Technology optimization is a crucial aspect in determining the ideal configuration for the selected technology. The interview highlighted that the project is currently in search of the most appropriate software to effectively carry out this vital task. This software will play a pivotal role in achieving the desired outcomes of the optimization process.

The primary ancillary tasks revolve around determining the most suitable business model for the selected technology and implementing capacity-building initiatives to enhance local awareness regarding the utilization of alternative cooking sources at home.

The anticipated impact of this project primarily centres around the adoption and utilization of the developed technology. For instance, if the prototype successfully extends beyond the project's intervention area, it has the potential to penetrate the pan-African market. Additionally, the project's impact on the academic and research landscape is noteworthy, as it will generate interest and discussions within academic circles beyond the scope of LEAP-RE.

16.3.1.3   Project Geothermal Village

The "Geothermal Village" project is expected to deliver various outcomes, such as underscoring the value of geothermal energy as a RES and economic asset in remote and underdeveloped environments. Additionally, it aims to demonstrate to potential investors, both public and private, the feasibility of utilizing stand-alone geothermal energy systems in isolated regions of Africa. The project also seeks to enhance knowledge and expertise in implementing community-based renewable energy initiatives in off-grid areas while simultaneously empowering local beneficiary communities by equipping them with a diverse set of skills through their involvement in the Geothermal Village concept's implementation [7].

16.3.1.3.1   Mapping

The objective of the project is to utilize geothermal *technology* to establish a pilot mini-grid that can provide the required energy for a small village. The project aims to develop a replicable model that can be implemented in similar settings, ensuring reliable and sustainable energy access for rural communities.

The primary *modelling* tools used for developing and optimizing will be geographic information system (GIS) databases that connect geological and technological factors. These databases will serve as essential resources for integrating and analysing data related to geological characteristics and technological considerations, enabling effective model development and optimization.

The implemented *methodology* focuses on developing effective approaches to accurately identify and extract hot geothermal fluids. These fluids are utilized in optimized plants to meet the social and economic requirements of the communities. The aim is to employ efficient methods that enable the sustainable harnessing of geothermal resources, resulting in positive social and economic outcomes for the communities involved.

Capacity building plays a critical role in this project as it encompasses the African partners and beneficiaries of the pilot project. The interview highlighted that an exchange programme for African students in European institutions has been proposed, set to start in September. This initiative is vital for enhancing knowledge and skills among the project stakeholders, empowering the African partners, and the local beneficiaries to actively contribute to the project's success.

16.3.1.3.2   CESP

The initial stage involves the identification of suitable locations within the Eastern Rift Valley, where high enthalpy geothermal resources exist, to establish two geothermal villages. Surveys and questionnaires will be conducted on-site to comprehensively understand the energy requirements and social aspects of the local population.

The subsequent step focuses on designing the geothermal power plant based on specific uses and needs, considering the feasibility of the project, and optimizing various components. Social considerations will be given priority, as different community types influence energy utilization and thermal water applications.

Regarding the selection of the energy village solution, no definitive answers have been reached yet. The interview revealed that the decision on the specific solution to be adopted would be planned after conducting fieldwork and gathering on-site data.

As mentioned during the interview, a partnering institution in France will commence capacity-building programmes in September. The training needs for the entire project will be assessed and documented, ensuring comprehensive skill development within the team.

16.3.1.4 Project RE4AFAGRI

"RE4AFAGRI" (Renewable Energy for African Agriculture – Modelling Excellence and Robust Business Models) seeks to empower African research institutions and decision-makers by providing them with the tools and expertise needed to operate a multiscale modelling platform. This platform will enable the design and implementation of integrated solutions for the energy and water nexus in rural areas. Additionally, RE4AFAGRI aims to foster a multi-stakeholder discussion platform to explore business models and create an enabling environment through policies and regulations. By promoting public-private partnerships, the project aims to contribute to the transformation of smallholder agriculture, ensuring sufficient food production to address persistent threats of hunger. The involvement of diverse stakeholders ensures the project's broad relevance to the energy access challenges faced in Africa [8].

16.3.1.4.1 Mapping

The project does not specifically concentrate on a singular technology. However, based on the insights revealed in the interview, the project will prioritize solar and wind energy for power generation. Geospatial data will be utilized to assess the energy potential of a given region, enabling a better understanding of the available renewable resources within a specific territory.

The *modelling* tools utilized in this project consist of three energy models developed by the involved partners. These models will be shared with other partner organizations across the African continent. The primary objective is to integrate these models into a comprehensive framework that considers the interconnectedness of energy and water resources, as this nexus plays a pivotal role in driving rural development. By incorporating the energy-water nexus into the integrated model, the project aims to address the multifaceted challenges and opportunities that arise in the context of sustainable rural development.

The *methodology* involves continuously comparing the three entities responsible for integrating the energy models. The interview highlighted the crucial role of African partners in comprehending the practicality of the different models in real-world scenarios and validating their effectiveness. Their contribution is integral to ensuring the applicability of the models and conducting rigorous field testing to assess their validity.

Substantial *capacity-building* initiatives are essential, specifically targeting entrepreneurs and stakeholders using the developed tool. Simultaneously, training activities will be required for researchers at African institutes, enabling them to effectively leverage the unified model, and devise optimal solutions for local contexts. The aim is to equip both entrepreneurs and researchers with the necessary skills and knowledge to maximize the potential of the developed tool and ensure its successful application in addressing local challenges.

16.3.1.4.2 CESP

During the interview, a prominent concern that emerged was the lack of appropriate modelling for correctly sizing agricultural equipment used for irrigation. Data collection for the project may involve field interviews, while geospatial software will be utilized to assess the energy potential of the targeted area. The identification of the most suitable solution for the local context will heavily rely on the feedback received during upcoming meetings, emphasizing the need for a flexible solution that can be adapted according to evolving needs. Although the optimization of technology is yet to be fully defined, solar and wind technologies are expected to be prioritized in the model's development. Capacity-building activities will play a vital role in ensuring the long-term success of the project, enabling African researchers to effectively use the models and facilitating discussions with stakeholders to identify the most suitable business model. The primary anticipated impact of the project is the extensive dissemination of the models among African universities and the widespread adoption of the developed model by future users.

16.3.1.5 Project SETADiSMA

"SETADISMA" (Sustainable Energy Transition and Digitalization of Smart Mini-Grids for Africa) aims to achieve multiple objectives, including increasing energy access in rural areas and promoting the use of Renewable Energy Services. By doing so, it strives to improve living conditions, foster social inclusivity, and stimulate local economic growth. The project also seeks to generate innovation and job opportunities within the mini-grid sector while enhancing economic development and income-generating activities. Additionally, it aims to strengthen local governance structures by involving the community in decision-making processes related to the energy system. The project aims to establish long-term collaboration between African and European funders in research, innovation, and human capacity building. Furthermore, it seeks to foster collaboration between academia, small and medium-sized enterprises (SMEs), energy utilities, and governmental entities, promoting synergy and knowledge exchange among these stakeholders [9].

161.3.1.5.1 Mapping

The *technologies* under analysis encompass different power sources for rural communities, including grid-connected systems, diesel generators, as well as RETs like photovoltaics and wind power.

The *modelling* tools to be utilized consist of optimization tools based on Excel or Python. These tools will be instrumental in modelling both the supply side of the mini-grid and the associated business model. As the programming advances into more intricate stages, the possibility of employing MATLAB as an optimization tool may also be considered.

The establishment of a unified *methodology* for data collection, including the development of questionnaires, is crucial among the diverse project partners. The interview revealed that the project aims to create and disseminate a methodological tool for sizing mini-grids, starting from resource assessment. Additionally, a business delivery model will be developed to enhance the economic sustainability of future projects. The implementation of these tools and models is intended to facilitate the effective planning and execution of mini-grid projects, contributing to their long-term viability and success.

The *capacity-building* initiatives will commence with a series of training sessions conducted in collaboration with partner universities. These sessions aim to equip researchers and students participating in the project with new skills and knowledge. A GitHub repository will be established to facilitate the dissemination of the developed model, serving as a platform for sharing and accessing project-related resources. This repository will enable wider accessibility and knowledge transfer, promoting collaborative learning and facilitating the adoption of the produced model.

16.3.1.5.2 CESP

The process of identifying priorities commences with an examination of previously developed mini grids, both in greenfield and brownfield settings. Simultaneously, geospatial models are employed to assess the energy potential and demand of specific areas. The selection of the optimal solution will be context-driven, considering the unique circumstances of each project. Technological optimization will be conducted to determine the most suitable approach. As indicated in the interview, sharing the projects on GitHub will facilitate the dissemination of the models within the academic community, fostering knowledge exchange and collaboration. Finally, the evaluation of impact will extend beyond the project itself, considering the extent to which the model has contributed to enriching the scientific community at large.

16.3.1.6 Project Energy Village Concept

The "Energy Village" project aims to achieve 100+% renewable energy self-sufficiency for villages through regional programmes, smart energy islands, innovative market models, and capacity building. It emphasizes testing tools, optimizing energy storage, and establishing a network of smart energy islands across Africa. The project focuses on knowledge transfer and researcher exchange to ensure system compatibility and usability [10].

16.3.1.6.1　Mapping

The primary objective of the project is to create a comprehensive and accurate framework for the design and establishment of energy villages in rural areas. These energy villages will operate as off-grid systems, supplying clean electricity to the local community. The project aims to develop a precise and thorough blueprint that ensures the effective implementation of these sustainable energy solutions, enabling access to clean and reliable power sources for rural communities.

The *technology* recommended for the development of this toolkit and methodology primarily relies on the most accessible resources within the region, namely local biomass and photovoltaic (PV) panels. These technologies have been identified as the most readily available and suitable options for harnessing renewable energy within the territory. By leveraging local biomass and PV panels, the project aims to optimize the utilization of indigenous resources and maximize the effectiveness of the implemented energy solutions.

Excel will play a pivotal role as a *modelling* tool for optimizing loads and managing data within the project. It will serve as a fundamental tool in effectively analysing and optimizing energy loads while efficiently handling and organizing relevant data.

The demonstration projects of the Energy Village Concept will be disseminated to foster the establishment of a network. The creation of this network may require *capacity-building* activities to enhance the skills and knowledge of the participating researchers. The aim is to facilitate knowledge sharing, collaboration, and the replication of successful energy village projects across different regions.

16.3.1.6.2　CESP

By conducting interviews with local communities and African partners, the project will identify the most pressing needs of villagers. Resource assessment will be carried out using an adapted version of the classical modelling tool, Excel, tailored to the African context. Simulations will be performed to accurately evaluate the energy demand of the Energy Village. The selection of the most promising technology and strategy for providing clean electricity will be data-driven, considering resource availability and quality. The entire village will be optimized using modelling tools to appropriately size the components of the mini-grid. Training sessions will be crucial to familiarize stakeholders with the technical tools while exploring various business models that align with the local and specific circumstances. The successful implementation of a viable business model is essential to enhance replication potential. The project aims to establish an African-wide network of smart energy villages and define guidelines for the development of mini-grids in rural communities, laying the foundation for sustainable energy access and rural electrification.

16.3.1.7　Project EURICA

"EURICA" (Europe & Africa cooperation, from grid digitization to sustainable energy for all) focuses on enhancing energy reliability for Africans who already have access to electricity by improving the System Average Interruption Duration Index (SAIDI) by 20%. It builds upon existing individual electrification efforts in rural areas, fostering a bottom-up approach that enables the productive use of energy, with up to 30% available for appliances like fridges and electric motors. The availability of productive electricity contributes to increased women's employment opportunities. The project emphasizes the involvement of local operators and offers capacity-building training to enhance their skills and expertise. Furthermore, the project promotes research exchange between the Institute of Science and Technology (IST) in Madagascar and G2ELAB in France, fostering collaboration and knowledge sharing between the two institutions [11].

16.3.1.7.1　Mapping

The project revolves around two contrasting *technologies*. First, it aims to expand the national grid to underprivileged families residing on the outskirts of major cities in Burkina Faso by implementing secure electrical connections. Simultaneously, the project focuses on interconnecting multiple nanogrids in the northern region of Madagascar, testing the integration of diverse grid systems. These nanogrids, which can cater to five or six families, primarily rely on rooftop solar panels, and lithium-ion batteries to generate and store energy.

Previous projects in Madagascar have already implemented *capacity-building* programmes aimed at providing training to on-field technicians. These initiatives have focused on enhancing the knowledge and skills of technicians involved in practical tasks and operations.

The methodology proposed relies on a solid base: thanks to smart metres in Burkina Faso and historical data in Madagascar, numerous data will be retrieved: this will allow us to correctly evaluate the load and implement an efficient technological solution. The project will use its own modelling tools to optimize the energy system, sizing the components according to the availability of resources and demand.

16.3.1.7.2　CESP

In Madagascar, the necessity for a more widespread network of nanogrids became evident through previous projects conducted in the same area. In Burkina Faso, community priorities were explored through dialogue with villagers and African partners. Resource assessment will be conducted utilizing Smart Metres and employing Machine Learning algorithms for the top-down approach, while previous data will be utilized for the bottom-up approach. The project has already determined the strategy for providing electricity to the communities, employing the national grid for Burkina Faso, and implementing photovoltaic panels and batteries for Madagascar.

Capacity-building activities will be organized for nano grids in Madagascar to train local technicians. The primary impact of the project lies in the acquired know-how from the two pilot projects, which can be replicated to facilitate wider access to electricity across the respective regions.

16.3.1.8　Project LEOPARD

"LEOPARD" (Micro-grid technology for widespread use of RES in Africa.) is focused on raising awareness of stand-alone solar installations in West Africa through a multifaceted approach. The initiative aims to stimulate and enhance collaboration between public and private partners from Europe and Africa, fostering a collective effort towards sustainable energy solutions. A key aspect of the project involves implementing a user-centric approach to decentralized solar production. This entails utilizing modular, affordable, and replicable container-based solutions that can be easily scaled up. To validate the feasibility and operability of the proposed solution, in situ pilot projects will be conducted, allowing for on-the-ground testing and local maintenance by trained technicians. By pursuing these strategies, LEOPARD seeks to drive the widespread adoption and implementation of stand-alone solar installations, facilitating increased access to clean and reliable energy in West Africa [12].

16.3.1.8.1　Mapping

The pilot project is planned for implementation in Benin and Senegal, offering a containerized solution to meet essential electricity needs. The *methodology* employed focuses on identifying the optimal solution by utilizing modular containers and leveraging the expertise and knowledge of multiple stakeholders involved. The aim is to develop a comprehensive and effective approach that addresses the specific requirements and challenges of the project, ensuring the successful deployment of containerized solutions for reliable electricity supply.

The chosen solutions will be built upon solar *technology*, capitalizing on the abundant solar resources available on the continent. By leveraging this RES, the project aims to harness the vast potential of solar technology to meet energy needs in a sustainable and environmentally friendly manner.

The project includes plans for *capacity-building* programmes targeting installers and users to ensure the sustainable and long-term utilization of the technological solution. Recognizing the off-grid nature of the system, the *modelling* tool HOMER will be employed to optimize and appropriately size the mini-grids. This tool enables accurate and efficient modelling of the off-grid system, facilitating optimal configuration and performance for a reliable and efficient energy supply.

16.3.1.8.2   CESP

The project aims to address the needs of the local population through an inclusive approach that involves engaging in dialogue with the communities. This dialogue will help determine their preferences, habits, and willingness to adopt the containerized solution. Data collection for solar panel sizing will commence using local measurement tools such as electrical and smart metres. Through field measurements and the utilization of HOMER, the optimal number of modules and PV panels will be determined, tailoring the solution to the specific requirements of the community. Developing an efficient business model will be crucial, one that can be adaptable to different scenarios while ensuring affordable electricity for end-users. Leveraging electricity for agricultural and business activities will play a vital role in sustaining the project economically. Additionally, capacity-building initiatives will be planned to empower villagers to utilize the containerized solution effectively. The project aims to develop a clean and portable solution that provides accessible electricity and can be replicated in various locations.

16.3.2   Main Horizontal Skill Needs

In the process of building a more precise taxonomy that will be developed later in this book chapter, the horizontal skills are here generally identified as all the skills needed beyond the technical aspects associated with the energy sector. For instance, managerial skills, which may be argued to be very technical skills, are included in the horizontal skills. This analysis will be of crucial relevance for the identification and formulation of the taxonomy: it identifies the needs expressed in the different technical reports as dedicated training or provided as transversal skills within academic programmes.

Through the analysis of the technical reports of the projects, common horizontal skills were highlighted, such as *project management, social sciences, communication skills, entrepreneurship, etc.*

Many initiatives in Africa dedicated for investing in renewable energies, such as "Solar 1000 MW" in Algeria requires management skills and other related skills (Table 16.2). As declared in the African Startup conference (Link: http://africanstartupconference.org/), the entrepreneurship is a highly prioritized axis in all initiatives at a continental level (Figure 16.6).

16.3.3   Actions in Place

A comprehensive figure has been formulated to outline a list of initiatives (Consisting of 19 ongoing actions). These initiatives specifically target the horizontal skills discussed earlier in this chapter.: *management, sales/business skills, quality assurance, and entrepreneurship.*

Referring to the information presented in Figure 16.7, among the total of 19 identified actions, 10 initiatives were primarily centred around fostering entrepreneurship, while 5 actions aimed to enhance sales and business skills. Additionally, 3 actions were specifically designed to improve management capabilities, and only 1 action was dedicated to quality assurance. Notably, three entrepreneurship programmes—Women Entrepreneurship for Africa, Digital Africa, and the

TABLE 16.2
Mapping of the Programmes in Place by Horizontal Skills

Designation	Duration	Description	Link
Household Solar Workforce Challenge		Solar training grants for third party training provides 'for initiative to manage training staff, technicians, or sales agents for the off-grid energy sector'	www.usaid.gov/householdsolar/call-for-proposals
Programme of: Enlight Institute	8 weeks	Technical and sales training through certified skills courses (solar design, installation, maintenance & servicing, safety, project management, solar water pumps, and solar water heaters. Also provides on the job coaching for 'frontline staff and agents and employers' for continuous development, sales, and professional skills	www.enlightinstitute.org
T1 and T2	20 days	This course is the entry point to become a licensed Solar technician or Solar products vendor. Participants receive comprehensive know-how for photovoltaics, hands-on training with technical components, designing optimized PV stand-alone systems to apply for a T2 license from the Energy Regulatory Commission	https://serc.strathmore.edu/programmes/
Certification of Solar PV Installers		For successful RE and EE installations and projects, there is a need for quality assurance at various levels. This includes quality assurance of equipment, e.g., through the establishment and enforcement of product standards. However, high-quality products will only provide the desired services if RE and EE systems are designed, installed, and maintained by highly qualified individuals. ECREEE and its partners, therefore, decided to support the development of the regional market for RE and EE services by establishing a scheme for certifying the skills of solar PV installers and other sustainable energy professionals. The objective is to introduce a quality mark for sustainable energy skills that is recognized by professionals and end users across borders in all 15 ECOWAS member states	www.ecreee.org/certification
ENEL Green Power Photovoltaic Skills Training		Solar PV Rooftop installer course and PV Enterprise & Sales Development Course (free course funded by ENEL green power)	www.green-cape.co.za/assets/Uploads/Course-Brochure-2018.pdf
Public Private Development Partnership for Renewable Energy Skills Training and Women's Economic Empowerment in Somalia		The project has two objectives: to economically empower women, especially women entrepreneurs in the MSME sector, and to develop skills to support an expanded electrical system with a greater proportion of renewable energy sourcing. The project will also establish a common curriculum for the design and installation of solar PV systems. At least 10% of the participants in each training are women	www.ilo.org/africa/countries-covered/somalia/ppdp-wee/lang--en/index.htm

Programme of: The Micro-grid Academy IESR, formerly Kenya Power Training School		The Micro-grid Academy is the main programme of RES4Africa focused on providing the necessary skills and knowledge for mini-grid development. It is a vocational capacity building programme aiming at creating a skilled and conscious workforce, to deploy decentralized renewable energy solutions and business skills. Local partners including the IESR (or formerly the Kenya Power Training School), the University of Strathmore, St Kizito Vocational Training Institute, and AVSI, a non-governmental organization.	www.res4africa.org/ micro-grid-academy
Several, ranging from electricity generation, distribution and system operation to leadership and safety courses		The Centre offers training in five (5) core focus areas as listed in the tabs below, the scheduled training programmes range in duration from two days to 13 weeks and are conducted on a residential basis. All the courses are certified at skills award level for competency under the TEVETA in Zambia. In addition, the Centre offers tailor-made training programmes and conducts onsite training at clients' premises to cost effectively meet the specific needs of individual organizations	www.kgrtc.org.zm
Modernizing vocational training for renewable energies		To increased specialized local technical expertise and management skills available on the market for renewable energies and energy efficiency in Côte d'Ivoire. The project aims to strengthen the skills of teachers at vocational schools and universities, enabling them to act as multipliers for the dissemination of practical skills	www.giz.de/en/worldwide/79018. html
Higher Education Programme for Renewable Energy and Energy Efficiency		Senegalese universities offer practice and labour-market-oriented degree courses as well as further training courses for professionals in renewable energies and energy efficiency and strengthen skills for business start-up in this field. The ISEPs are advised on developing practical degree courses on the productive use of renewable energies and energy efficiency in locally relevant sectors. To complement this, the programme supports the piloting of short-term ISEP training courses relating to energy, also for people without university entrance qualifications (including returnees)	www.giz.de/en/worldwide/39287. html
Energy Management	3 days	This programme is designed for mid to senior level management who are interested in energy matters within their organizations. Participants will also be able to learn about what opportunities are available in their facilities and practical ways of tapping into those opportunities	https://serc.strathmore.edu/ programmes/
Programme of: The Akilah Institute for Women		The Davis College model is a unique hybrid of a liberal arts education with applied and technical curricula, combining competency-based academic programmes with a customized, data-driven learning model to actualize practical experience and equip students with the tools needed to thrive in the fastest growing sectors of the economy. As East Africa's pre-eminent higher education institution, our graduates have launched careers in finance, clean energy, eco-tourism, agribusiness, conservation, technology, and more. The Akilah Women's Center experience combines the Davis academic model with an intensive focus on women's leadership and career development	https://akilah.org/
Programme of: Ashesi University		Ashesi University, "aims to educate a new generation of ethical, entrepreneurial leaders in Africa," with a strong emphasis on cultivating critical thinking and leadership skills among its students	www.ashesi.edu.gh/

(continued)

TABLE 16.2 (Continued)
Mapping of the Programmes in Place by Horizontal Skills

Designation	Duration	Description	Link
Programme of: The African Leadership Academy (ALA)		With the goal of building the next generation of African leaders, the ALA, has been educating students since 2008. Its training curriculum couples standard secondary school classes with several entrepreneurship courses and activities	www.africanleadershipacademy.org
Programme of: The Meltwater Entrepreneurial School of Technology		This Ghanaian based institute provides aspiring African entrepreneurs with a fully sponsored 12 month- long intensive programme aimed at equipping them with the skills to take their ventures to the next level. Subjects include computer programming, software development, product management, finance, marketing, sales, and leadership training	https://meltwater.org/
Programme of: Women Entrepreneurship for Africa		The virtual Acceleration Programme, implemented by SAFEEM, seeks to provide 100 female entrepreneurs from the TEF alumni network, with access to €10.000 in grant funding that will be paired with 3 months of technical support	https://safeem.org/ women-entrepreneurship-4-africa/
Programme of: Digital Africa		As a super-aggregator of data, capacities, and opportunities, Digital Africa have a unique ability to intensify entrepreneur financing, training, support, and promotional activities. Digital Africa develop expertise, create knowledge-based communities, provide technical assistance, finance projects and businesses, ease market access and the creation of a regulatory environment that supports African innovation. Digital Africa use these drivers to nurture innovative projects and help them grow into successful initiatives that can transform the everyday lives of African people and boost the international competitiveness of Africa's digital industries	https://digital-africa.co/
Programme of: African German Entrepreneurship Academy (AGEA)		AGEA is a joint initiative consisting of a dynamic network coordinated by the International Small Enterprises Promotion and Training (SEPT) Competence Center at Leipzig University (Germany). The network collaborates with its dedicated academic and business partners in Africa and Germany. The aim is to promote cutting-edge practice-oriented entrepreneurship education, entrepreneurship promotion and start-ups and businesses development in Africa. AGEA contributes to improving graduate employability through the inclusion of a high level of hands-on training in African partner universities. AGEA empowers Higher Educational Institutions in entrepreneurship promotion activities, especially entrepreneurship promotion initiative, encourages the establishment of vibrant university-business linkages and knowledge sharing between universities and business associations in Germany and African partner countries	www.ageacademy.de/
Renewable Energy Management and Finance Course	2 days	This course has been revised to take account of the changes in government support policy for all types of Renewables. The training will fully equip delegates with the latest information on financing all types of renewable energy projects to allow them to continue to participate successfully in the Renewables Industry, both in the UK and internationally	www.renewableinstitute.org/ training/renewable-energy-management-and-finance-course/

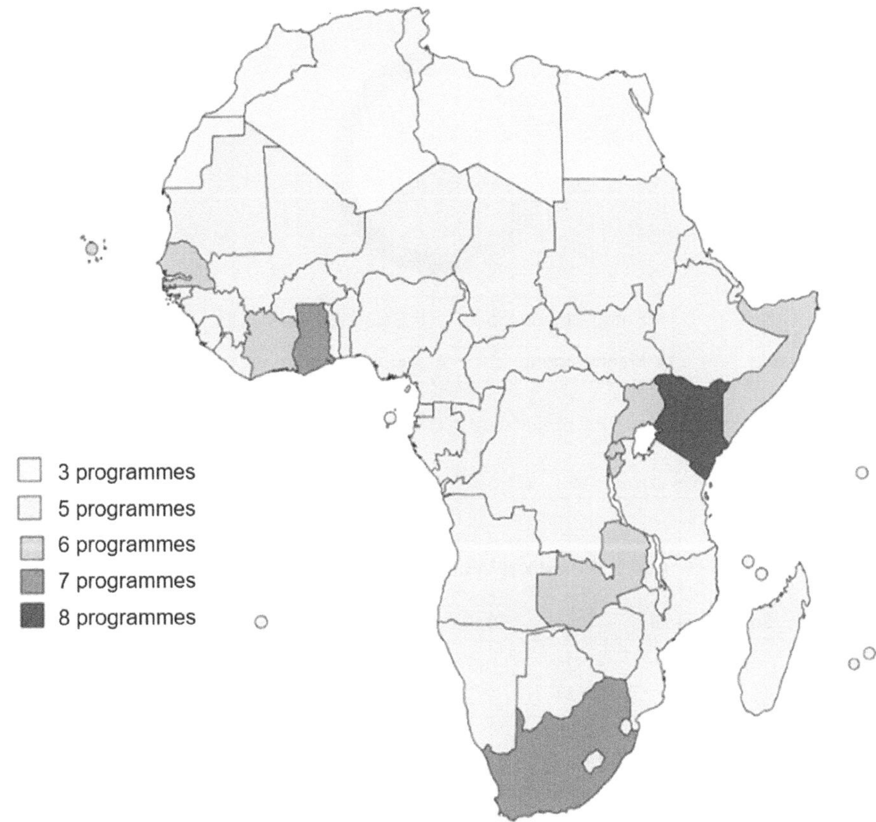

FIGURE 16.6 Mapping of the programmes in place by horizontal skills (general definition).

African German Entrepreneurship Academy (AGEA)—have been established to support entrepreneurship in Africa. These programmes are implemented by Switzerland, France, and Germany, respectively.

16.4 TOWARDS A TAXONOMY FOR CAPACITY BUILDING ACTIVITIES IN RES IN AFRICA

As stated in the Social and Environmental Assessment Report (SEAR) Report 2017 by the World Bank [13], within the energy challenge the cross-cutting role of human capital (individually and collectively, as communities and institutions) is increasingly crucial both as a catalyst and a booster. Indeed, without the proper human resources, it will be impossible to achieve a transformative change in energy access – one that is efficient, effective, equitable, empowering, and long-lasting. This means that capacity building is much more than delivering training hours and embracing various levels. The following paragraphs advance a common taxonomy for capacity-building activities.

16.4.1 THE MULTI-LEVEL CONCEPT OF CAPACITY BUILDING

As reported by the WB in 2017, the idea of capacity building has undergone significant revision over the past 20 years [14]. First, the idea of "capacity" has changed from one that emphasizes self-reliance to one that focuses on an organization or person's ability to be resilient and effective.

FIGURE 16.7   Mapping of the actions in place by horizontal skills.

TABLE 16.3
EU/AU Research Papers on RE Based on WoS Database

	Wind	Solar Thermal	PV	Biofuel	Geothermal	Green Hydrogen	Hydropower
EU	138945	24168	34624	16077	9760	7053	11051
AU	15058	4298	8554	1738	962	818	865

Here, the emphasis is on people's capacity for action, self-sustainment, and self-renewal as well as their capacity for setting and achieving their own development goals. In this context, the process of releasing, enhancing, and maintaining these capacities is known as capacity building. Capacity building also serves as a tactical tool for long-term and independent growth. Additionally, there is a need to expand on the direct analogy that characterized capacity building as training, view education as a fundamental human right, and to broaden the functional dependency of capacity building.

Such a comprehensive vision calls for several different sets of actions, such as:

- Developing people's abilities, relationships, and values.
- Enhancing the systems and laws that influence both collective and individual behaviour.
- Improving people's technical competences, soft skills, and attitudes to enable them to be pro-active players in development.

At the UN level, the UN Sustainable Development Agenda 21 from 1992 suggests that the capacity of a country's people and institutions, which complements its ecological and geographic factors, influences its capacity to pursue sustainable development paths [15]. In turn, UN Agenda 2030 for Sustainable Development – Transforming Our World includes Sustainable Development Goal 17 – Revitalizing the Global Partnership for Sustainable Development [16]. This accounts for several targets concerning capacity building, such as increasing technology and innovation in least-developed countries and improving data collection and monitoring for the achievement of the Sustainable Development Goals (SDGs) themselves. Universities are referred to as capacity-building institutions through research, innovation, and data collection and analysis. Particularly, Target 17.9 aims to enhance international support for implementing effective and targeted capacity-building in developing countries to support national plans to implement all the Sustainable Development Goals, including through North-South, South-North, and triangular cooperation.

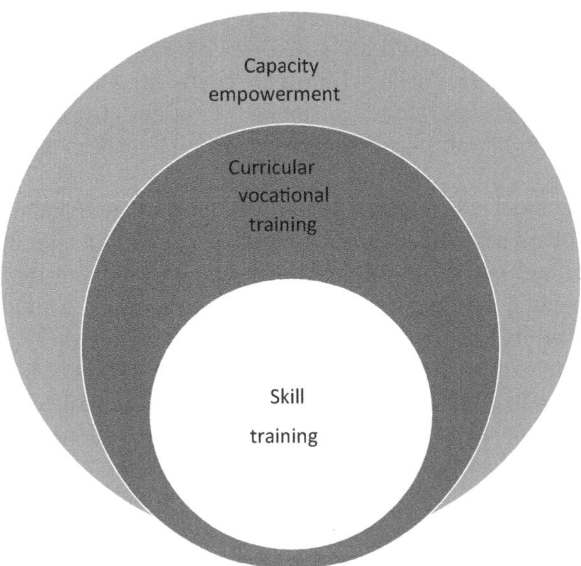

FIGURE 16.8   Venn vertical diagram of capacity building levels.

Given the variety of approaches to tackle the topic of capacity building and to bring clarity to the discussion, the following definitions are preliminarily provided and will be used as a taxonomy to aggregate the mapped initiatives (Figure 16.8):

- *Capacity empowerment* is defined, in accordance with the UN, as "the process of developing and strengthening the skills, instincts, abilities, processes and resources that organizations and communities need to survive, adapt, and thrive in a fast-changing world" [17], able to support and sustain transformation over time. Concerning RES and RETs, the concept can be rephrased as the implementation of all those activities aimed to foster the process of integration of sustainable energy sources and technologies in the social, technical, and economic contexts of interest.
- *Curricular or Vocational training* is defined as the programmatic training supplied to future professionals to provide them with sector-specific skills. This level of capacity building provides training programmes, including high school programmes, career, and technical education (CTE), but also training and apprenticeship programmes as well as bachelor's, master's, and Ph.D.
- *Skill training* is defined as the training needed to develop precise skills for a specific workplace. Work skills are divided into *hard* and *soft skills*. In the RES and RETs sectors, *hard skills* include technical knowledge (e.g., energy modelling skills, operation and maintenance skills, renewable resource assessment skills, etc.) and management skills (e.g., project management, project cycle management, financial reporting activities, etc.). *Soft skills*, on the other hand, space from communication skills to leadership, flexibility, and multitasking abilities. Skill training activities can be considered a sub-group of capacity empowerment actions.

16.4.2   STATED URGENCY AND THREE PILLARS FOR CAPACITY BUILDING IN AFRICA IN THE RES SECTOR

Suitable energy options and technological decisions should consider the requirements, abilities, and aspirations of people and be assimilated into the local culture or improved by the recipient community [18,19]. To grant self-empowerment and community engagement, it is hence necessary to briefly track the existing and already-stated needs for capacity building in the RE sector in Africa. Many African governments acknowledged the contribution that renewable energy makes to sustainable development over time. However, among the various obstacles that still prevent renewable energy from reaching its full potential in the planning agendas, capacity-building needs are mentioned as a priority.

Africa's efforts to achieve a sustainable transition involve cooperation from all players in the energy ecosystem, and capacity building may play a key role in facilitating cross-fertilization and partnership. National power utilities are expected to facilitate the transition to a more carbon-neutral future and contribute to preserving energy security, which is the main source of electricity. Capacity-building options in this sector must be made readily available to define and commit to the energy transition paths. This will help to boost the mix of RES, which Africa needs to adapt at a reasonable rate. Additionally, it must be acted in accordance with the socioeconomic priorities and development mandate of the continent, as reported in the agenda 2063 – The Africa We Want [20]. Based on the taxonomy presented in Figure 16.8, the following paragraph proposes a thematic distinction of the pathways that can be undertaken by each of the levels of capacity building.

According to the World Economic Forum [21], capacity-building actions to undertake energy transition pathways span three thematic pillars ("Triple D") (Figure 16.9):

- **Decarbonization**: Switching from fossil fuels to renewable energy will not be a one-size-fits-all answer; Africa's strategy and objectives must consider its own energy security circumstances and *hard skills* are required to advise the policymaking community scientifically.
- **Decentralization**: the energy transition in Africa is dependent on the state of its electricity grid. Historically, power utilities have held centre stage in managing the grid, which enables the generation, transmission, and distribution of electricity. Reaching non-grid consumers requires scaling up a new concept of decentralization, which can promote further connection and also offer service to the national grid.
- **Digitalization**: the continent has immense opportunities to leverage its digital skills and capabilities to enable the energy transition; the uptake of smart mini-grid and digital off-grid solutions within Africa's rural communities continues to support improvements in energy access. In a report of 2017, Tambo et al. evidenced how cutting-edge educational technologies may enhance instruction and speed the development of human capital abilities in the RE sector [22].

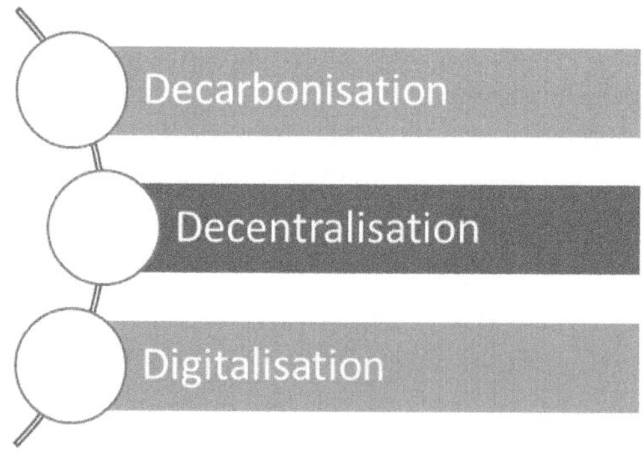

FIGURE 16.9   The Triple-D of capacity building actions in the RES sector.

16.4.3 Skills Definition and Nomenclature

In any professional domain of skill development, it is essential to obtain the best results and increase productivity as well as to improve them, to engage in discussion about *skills*, one needs to first know the diverse types of skills that can complement each other, and their definitions as stated by different authors:

- Knowing how to identify and implement one or two skills within a conceptual or disciplinary determined domain. To be more precise, a skill is associating a range of problems precisely identified with a programmed and determined assessment [23].
- A skill is part of knowing how to mobilize [24].
- A skill permits to confront a complex situation, to construct an adapted answer without drawing it from a programmed response [25].
- A skill is a complex "know how to act" based on the effective mobilization and combination of several situations [26].

16.4.3.1 Technical Skills

The Web of Science (WoS) Core Collection database is a selected citation index of scientific and academic publications comprising journals, conferences, books, and data compilations. It is the earliest citation index for the sciences, having been released commercially by the Institute of Scientific Information (ISI) in 1964 [27], first as an information retrieval tool named the Science Citation Index (SCI).

Nowadays, it is considered the world's leading scientific citation search and analytical information platform. It is utilized as both a research tool enabling a wide variety of scientific activities across multiple knowledge fields and a dataset for large-scale, data-intensive investigations spanning multiple academic subjects. WoS has tens of millions of bibliographic entries with billions of citation links and extra information fields, and many thousands of more items are consumed on a regular basis. The WoS platform also contains software productivity functions, namely EndNote and InCites [28]. Between 1998 and 2018, there has been a huge and constant surge in the number of articles discussing WoS, as illustrated in Figure 16.10 [29].

Considering the characteristics, the use of the WoS database in this deliverable is deemed necessary and will constitute the basis of the presented results in this deliverable.

The number of publications provided from 2010 till November 2022 is recapitulated through the following tables and figures (Table 16.3).

A preliminary aggregated data can be analysed in the comparison shown in Figure 16.11. The below figure shows how most of the scientific research in AU belongs to solar thermal energy and biofuels, while Wind covers more than half of the sources in EU.

In the context of the AU-EU cooperation, apart from hydropower, there is a recognition of the need to investigate six specific types of renewable energy (RE) technologies further. These technologies are deemed significant and warrant in-depth analysis as part of the cooperation framework:

- Biofuel
- Geothermal
- Solar Photovoltaic
- Wind
- Green Hydrogen
- Solar Thermal

For each of them, the more relevant skills necessary to a curricular or vocational training are depicted according to the achievement of the WoS analysis:

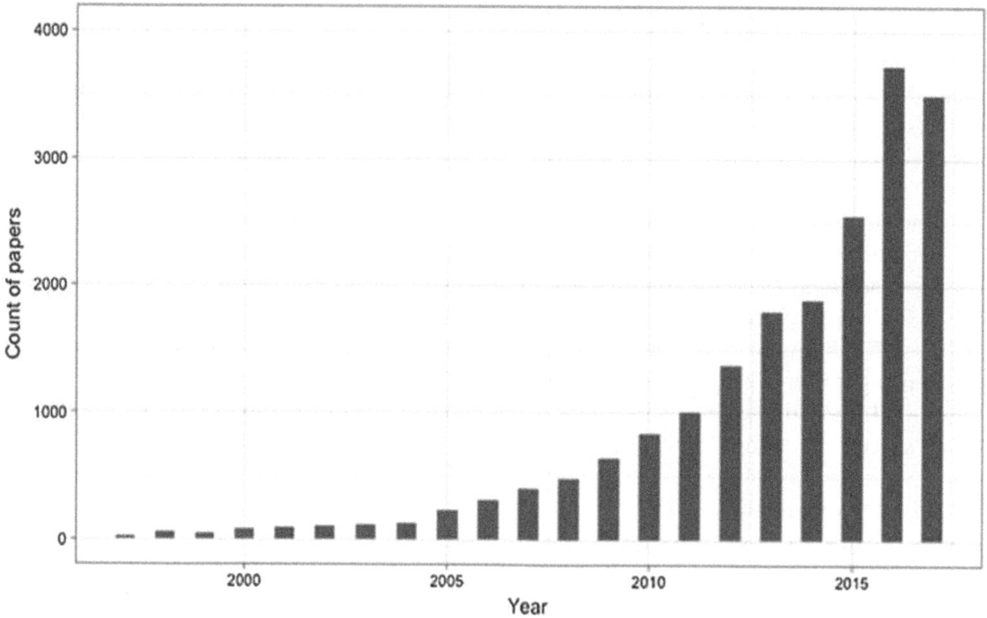

FIGURE 16.10   Yearly distribution of all papers mentioning web of science.

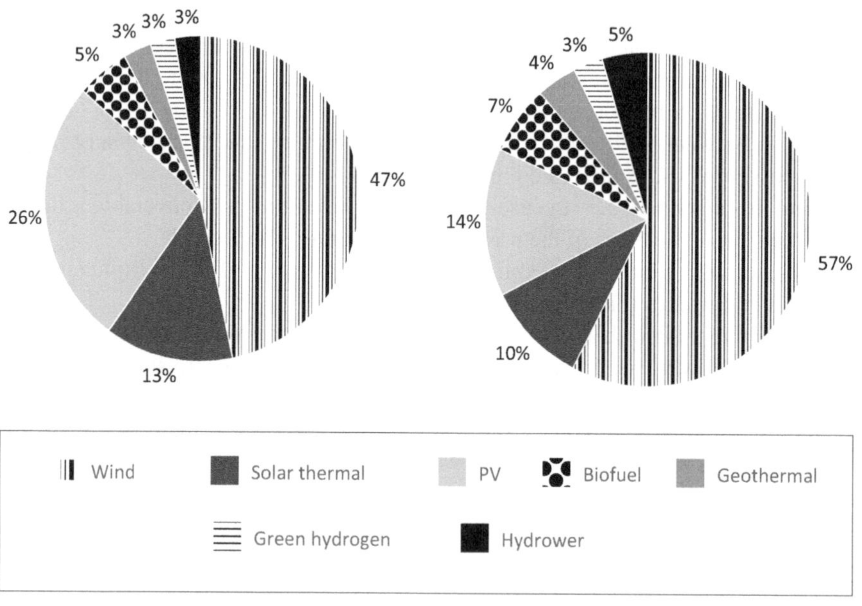

FIGURE 16.11   AU (left) and EU (right) research papers, divided by RE of pertinence.

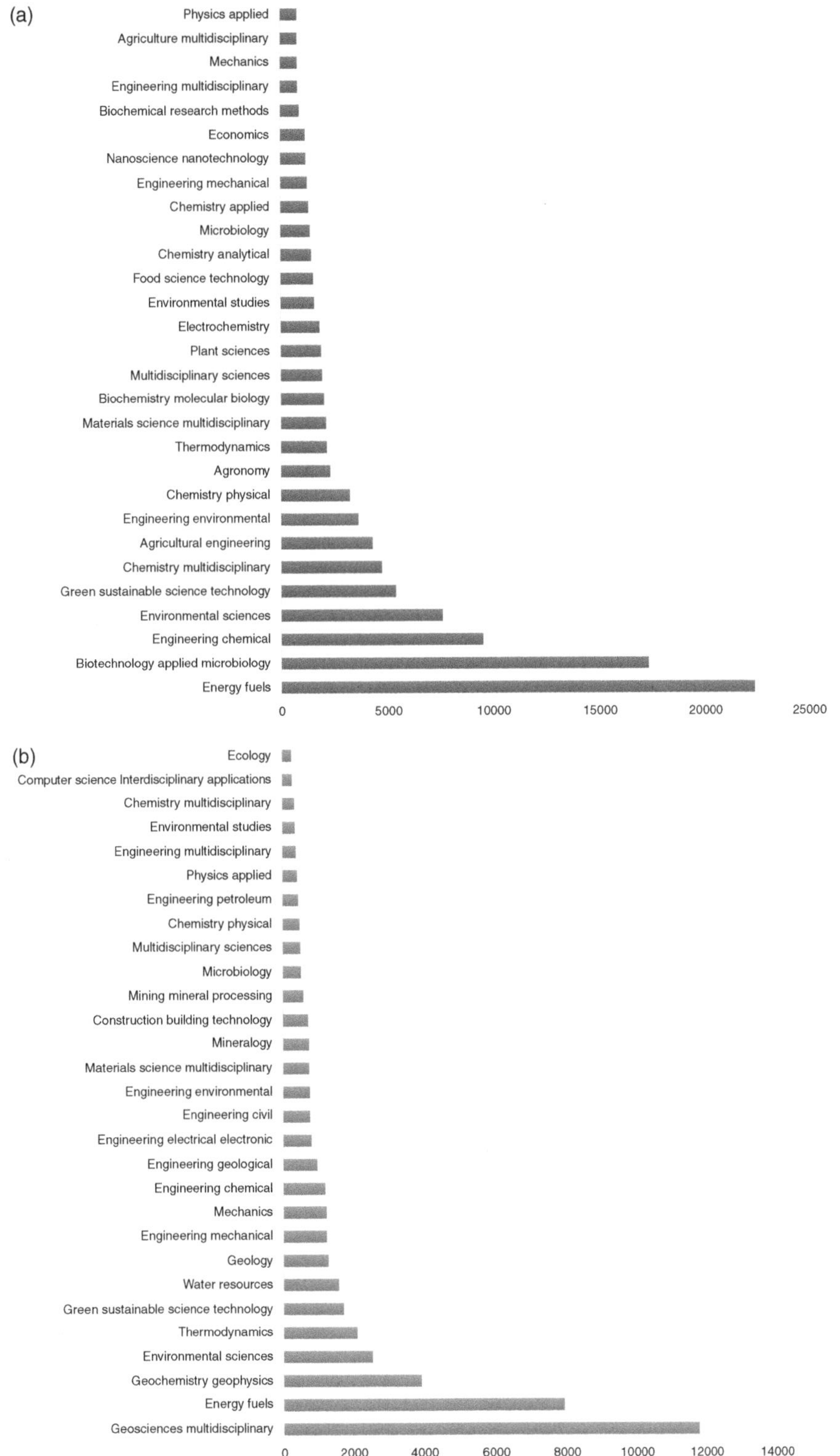

FIGURE 16.12 Skills needed in the biofuel (a), geothermal (b), green H2 (c), solar PV (d), wind (e), thermal solar (f).

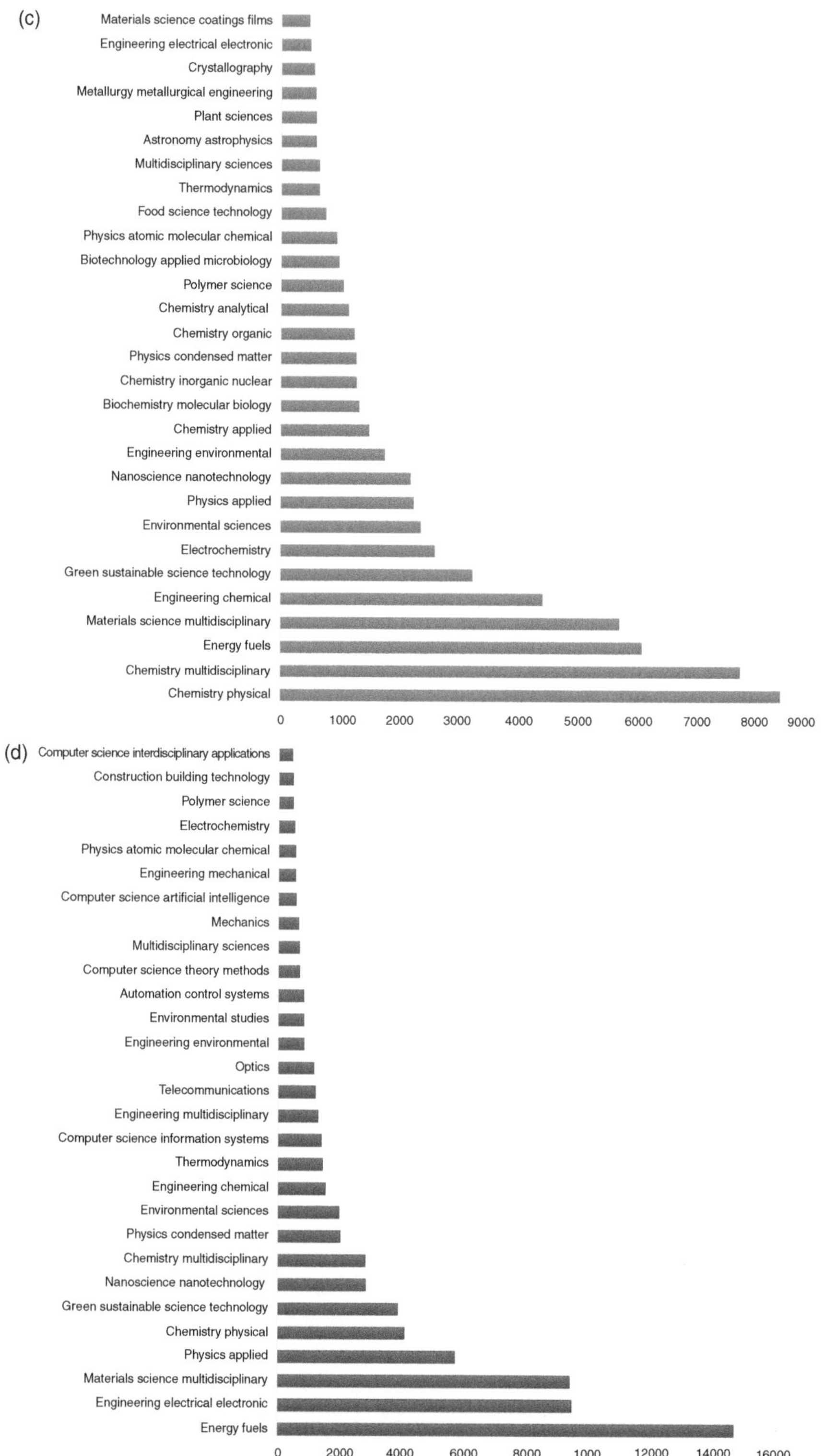

FIGURE 16.12 (Continued)

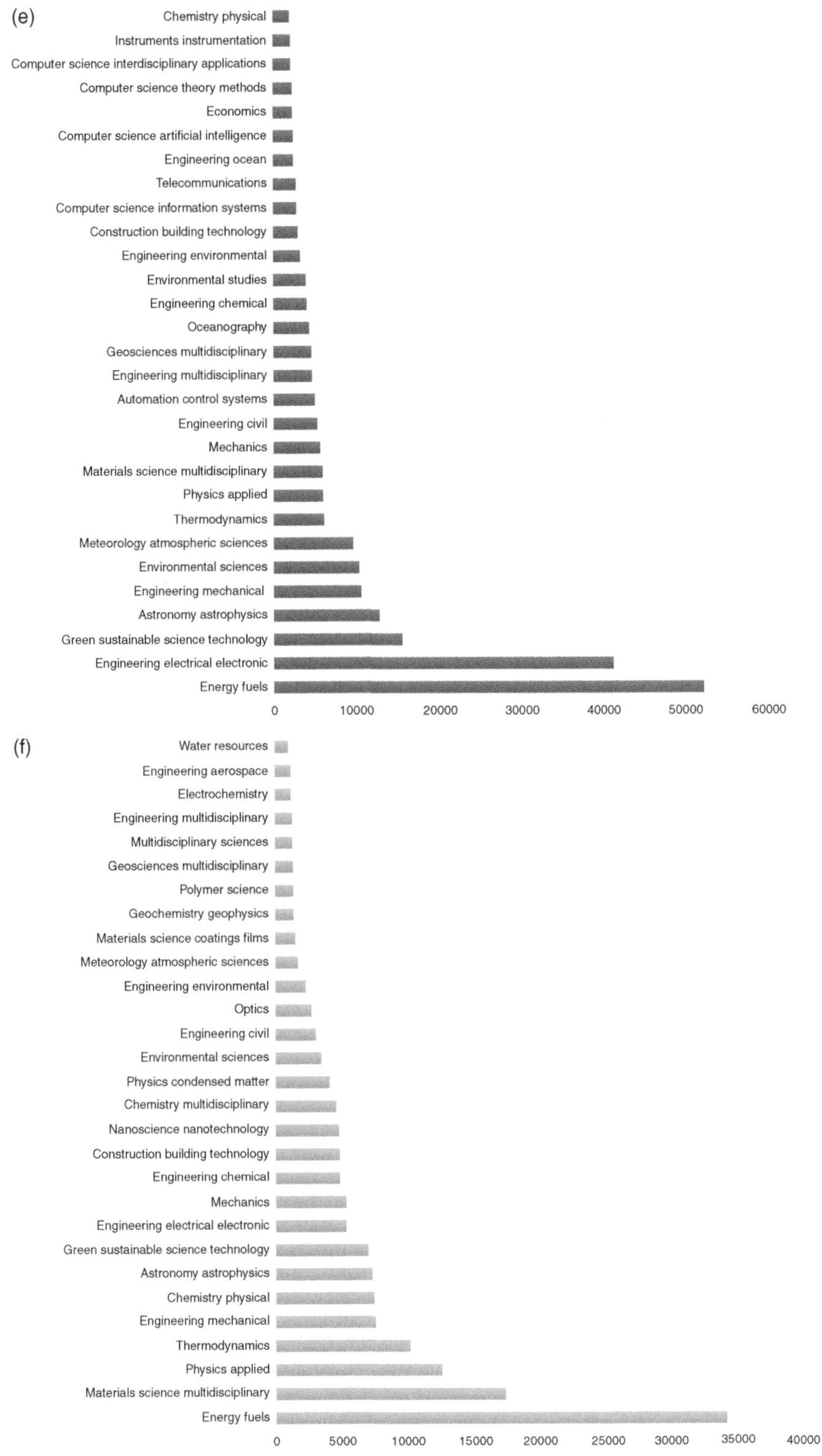

FIGURE 16.12 (Continued)

Competence refers to a comprehensive and operational combination of knowledge, expertise, and skills that enables individuals to adapt, address challenges, and successfully undertake projects within a specific domain of circumstances.

Extracting from Figure 16.12, a compilation of technical skills can be derived from the shared disciplines associated with the six renewable energy (RE) technologies. The collected technical skills are as follows (Table 16.4):

The table can be used to drive *syllabus and curricula and syllabus definition for skill development* in the RES sector for both curricula and Vocational Activities.

16.4.3.2　Horizontal Skills

The definition of horizontal skills can be interpreted in different ways. To avoid any confusion, all non-technical skills will be considered as horizontal/transversal ones such as entrepreneurship, communication skills, project management, ethics, etc.

TABLE 16.4
Top 5 Technical Skills by RE Sources

RE Sources	Top 5 Disciplines
Biofuel	Energy Fuels
	Biotechnology Applied Microbiology
	Engineering Chemical
	Environmental Sciences
	Green Sustainable Science Technology
Geothermal	Geosciences Multidisciplinary
	Energy Fuels
	Geochemistry Geophysics
	Environmental Sciences
	Thermodynamics
Green Hydrogen	Chemistry Physical
	Chemistry Multidisciplinary
	Energy Fuels
	Materials Science Multidisciplinary
	Engineering Chemical
Wind	Energy Fuels
	Engineering Electrical Electronic
	Green Sustainable Science Technology
	Astronomy Astrophysics
	Mechanical Engineering
Photovoltaic	Energy Fuels
	Engineering Electrical Electronic
	Materials Science Multidisciplinary
	Applied Physics
	Physical Chemistry
Solar thermal	Energy Fuels
	Materials Science Multidisciplinary
	Physics Applied
	Thermodynamics
	Engineering Mechanical

This consideration is in line with the learning outcomes approach supported by the EU Commission and many National Education systems, which are commonly referred to as the Dublin Descriptors.

They are general statements about the ordinary outcomes that are achieved by students after completing a curriculum of studies and obtaining a qualification. They are neither meant to be prescriptive rules nor represent benchmarks or minimal requirements since they are not comprehensive. The descriptors are conceived to describe the overall nature of the qualification. Furthermore, they are not to be considered disciplines, and they are not to be limited to specific academic or professional areas. The Dublin Descriptors consist of the following elements:

- Knowledge and understanding;
- Applying knowledge and understanding;
- Making judgements;
- Communication skills;
- Learning skills.

Indeed, any learning outcomes need to include in addition to *know how to* also do two other aspects: *know how to be and know how to become*.

Therefore, the following association can be created and used within the frame of this case study to generate a positioning of soft skills (Figure 16.13).

16.5   RESULTS ANALYSIS

Through this analysis, different aspects related to RE development in the African continent, including research and innovation, dedicated infrastructure for research, topics and disciplines, and published patents were elaborated using different databases (Web of Science, the technical reports of each project of the case study as well as the research results conducted by the Capacity Building cluster group, through an online investigation as well as qualitative interviews).

In this context the main recommendation within this case study and beyond is the use of a new taxonomy for capacity building that can match with the complexity analysed. The taxonomy may

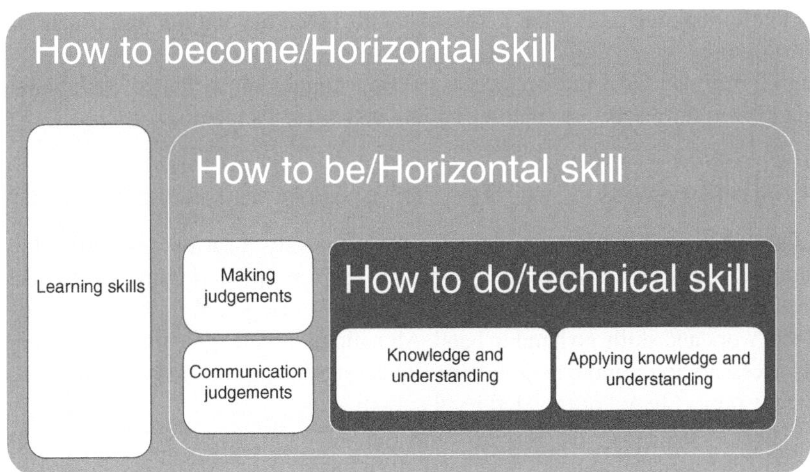

FIGURE 16.13   Overall taxonomy of horizontal and technical skills.

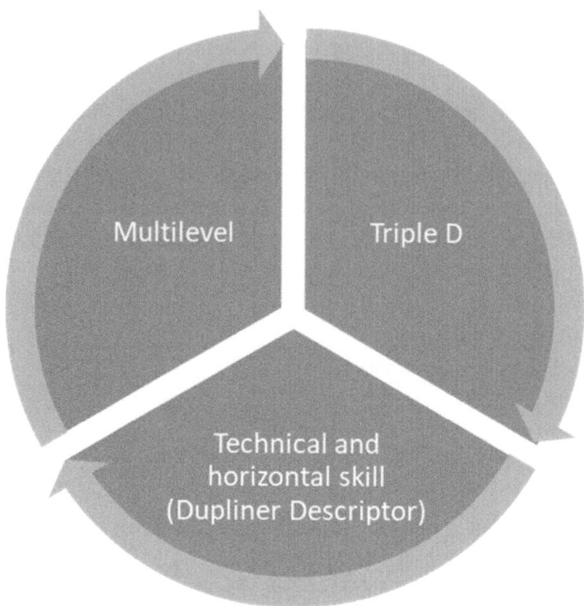

FIGURE 16.14 The three pillars of the capacity building taxonomy within LEAP-RE.

be graphically represented with three main pillars each of which is then briefly re-summarized from the previous session of the document (Figure 16.14).

16.5.1 Multilevel Capacity Building

A comprehensive strategy based on human, scientific, organizational, and institutional capacities should be used to enhance capacity in the Energy sector for Africa (Figure 16.15). The most active research institutions in the field of RE should relate to their counterparts from other continents for a long-term EU-AU partnership. A variety of local, national, and international stakeholders should be involved due to the diversity of the necessary skills (even beyond the traditional players of the educational systems).

Recipient groups may have varying access to opportunities for technical, vocational, or institutional training. Therefore, capacity building should take these factors into account.

16.5.2 Conceptualization of the "Triple D" in CB Actions in the Energy Sector

Change is sparked by and is driven by people. Their capacity must be developed throughout the design solution's supply chain where the "Triple D" finds its space (Figure 16.16). Their capacity must be developed throughout the design solution's supply chain, and this strategy must be guided by the idea that work and skills go hand in hand. All nations should agree that strengthening national capacities is necessary to promote national priority definition and regional coordination, and to provide funding for project based or targeted local activities.

16.5.3 Technical and Horizontal Skill Needs to be Aligned to the Dubliner Descriptor

Interventions for developing capacity should be varied to suit the various skill requirements present at various levels of the energy supply chain and within various local contexts—and be in line with

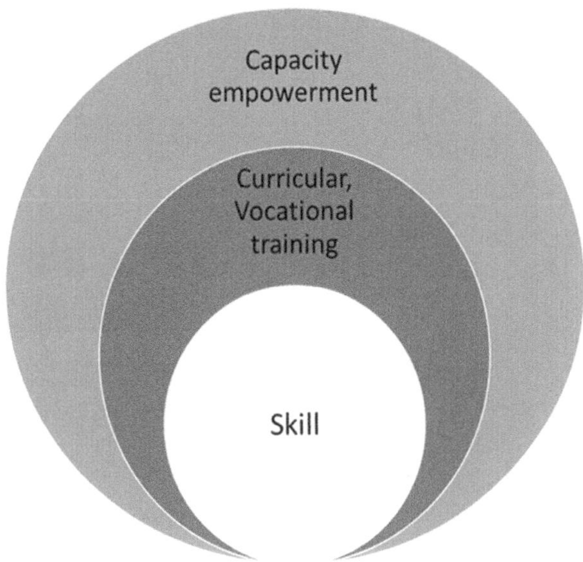

FIGURE 16.15 Proposed taxonomy for multilevel capacity building.

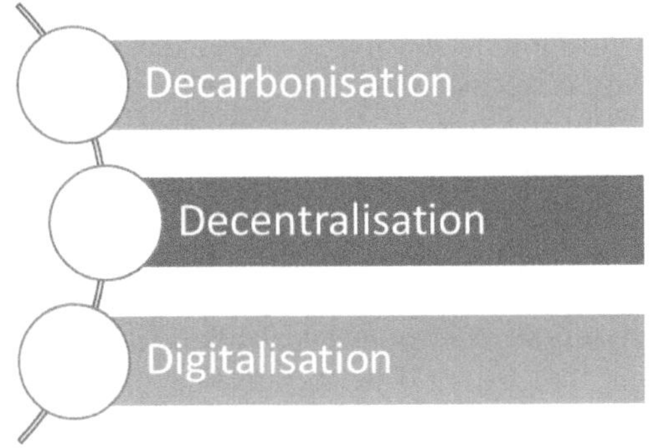

FIGURE 16.16 Proposed thematic taxonomy ("Triple D").

the capabilities of the various target groups (Figure 16.17). Various tools may be utilized depending on the objectives and anticipated learning results (including training, seminars, workshops, on-the-job tutoring, and site visits).

16.6 CONCLUSION

The growth of renewable energy (RE) in Africa has been thoroughly examined in this book chapter, which draws information from a variety of databases, research sources, technical papers, and interviews. With a heavy emphasis on building up capacity in the energy industry, major findings and suggestions have been drawn from the assessment of several areas of RE growth.

The main conclusion resulting from this research is to put in place a strong plan to boost the institutional, organizational, scientific, and human resources within Africa's energy industry. Establishing long-term collaborations among active RE research institutes in Europe and Africa is

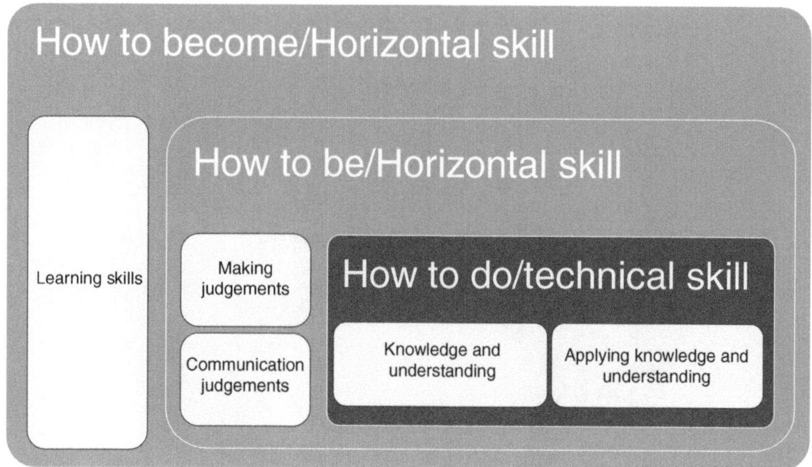

FIGURE 16.17   Proposed alignment of technical and horizontal skills, accordingly to the Dubliner descriptor.

essential to achieving this. Such partnerships will promote information sharing, technology transfer, and expertise sharing, all of which are essential for developing a workforce with the necessary skills to meet the region's energy concerns.

Beyond conventional educational techniques, the energy sector's unique skill needs must be addressed. Therefore, it is crucial to include several stakeholders at the local, national, and international levels. A more inclusive and thorough approach to capacity building may be undertaken by cooperating with a wide variety of players, such as governments, commercial businesses, academic institutions, and local communities.

The "Triple D" strategy, which stands for Design, Develop, and Deliver, is at the heart of the chapter's focus on capacity-building programmes. This method emphasizes the need to develop capabilities across the whole supply chain to create solutions while highlighting the interconnectivity of work and skill development. Africa can promote innovation, creativity, and adaptation in the renewable energy industry by using the "Triple D" strategy, making it possible to successfully implement sustainable solutions.

Additionally, enhancing national capabilities is essential for advancing Africa's energy transformation. African countries may establish their national goals, promote regional cooperation, and win finance for specific local projects through strengthening the capabilities of local institutions and organizations. Africa can take the lead in directing the development of its renewable energy future by giving national capacity development top priority.

In conclusion, this chapter emphasizes the crucial part that capacity building plays in accelerating the expansion of renewable energy in Africa. It outlines a thorough strategy for capacity building, promotes global partnerships, and emphasizes the importance of the "Triple D" approach in pushing sustainable energy solutions on the continent. Africa can pioneer the way to a greener and more sustainable energy future by investing in human capital, knowledge transfer, and international collaborations, promoting equitable growth and environmental stewardship across the continent.

ACKNOWLEDGEMENTS

We extend our heartfelt appreciation to the European Union for their essential support in the implementation of the LEAP-RE project, which has been instrumental in the successful completion of the present study. This study is funded by the European Union (EU) as part of the LEAP-RE project, a significant initiative that underscores the EU's commitment to fostering sustainable energy development and collaborative research efforts. The EU's involvement in the LEAP-RE project has enabled us to conduct comprehensive research and develop insights that are vital for the growth of renewable energy in our focus areas. We are immensely grateful for the opportunity to contribute to this important field, made possible through the EU's funding and the collaborative framework established by the LEAP-RE project.

GLOSSARY

AU	African Union
CESP	comprehensive energy solution planning
CB	capacity building
DoA	description of action
EURICA	Europe & Africa cooperation, from grid digitization to sustainable energy for all
RE	renewable energy
MS	milestone
M&E	monitoring & evaluation
MARs	multiannual roadmaps
MOOCs	massive open online courses
O&F	organizational & funding
tbd	to be determined
UN	United Nations
EU	European Union
PURAMS	productive use in rural African markets using standalone solar
SETADISMA	sustainable energy transition and digitalization of smart mini-grids for Africa
RE	renewable energy
RE4AFAGRI	renewable energy for African agriculture: modelling excellence and robust business models
RETs	renewable energy technologies
RES	renewable energy sources
R&I	research and innovation
SSA	Sub-Saharan Africa
WoS	Web of Science

REFERENCES

[1] S. Tabrizian, "Technological innovation to achieve sustainable development—Renewable energy technologies diffusion in developing countries," *Sustainable Development*, vol. 27, no. 3, pp. 537–544, 2019, DOI: 10.1002/sd.1918.

[2] M. Khezri, A. Heshmati, and M. Khodaei, "The role of R&D in the effectiveness of renewable energy determinants: A spatial econometric analysis," *Energy Economics*, vol. 99, p. 105287, 2021. DOI: 10.1016/J.ENECO.2021.105287.

[3] B. Vallejo and U. I, "Capacity development evaluation: The challenge of the results agenda and measuring return on investment in the global south," *World Development*, vol. 79, pp. 1–13, 2016, March. DOI: 10.1016/J.WORLDDEV.2015.10.044.

[4] LEAP-RE, "About LEAP-RE," [Online]. Available: www.leap-re.eu/. [Accessed 01 July 2023].

[5] LEAP-RE, "Geothermal Atlas 4 Africa," [Online]. Available: www.leap-re.eu/geothermal-atlas-4-africa/. [Accessed 01 July 2023].

[6] LEAP-RE, "PURAMS," [Online]. Available: www.leap-re.eu/purams/. [Accessed 01 July 2023].

[7] LEAP-RE, "Geothermal Village" [Online]. Available: www.leap-re.eu/geothermal-village/. [Accessed 01 July 2023].

[8] LEAP-RE, "RE4AFAGRI" [Online]. Available: www.leap-re.eu/re4afagri/. [Accessed 01 July 2023].

[9] LEAP-RE, "SETADISMA" [Online]. Available: www.leap-re.eu/setadisma/. [Accessed 01 July 2023].

[10] LEAP-RE, "Energy Village Concept" [Online]. Available: www.leap-re.eu/energy-village/. [Accessed 01 July 2023].

[11] LEAP-RE, "EURICA" [Online]. Available: www.leap-re.eu/eurica/. [Accessed 01 July 2023].

[12] LEAP-RE, "LEOPARD" [Online]. Available: www.leap-re.eu/leopard/. [Accessed 01 July 2023].

[13] WB, *State of Electricity Access Report 2017*," World Bank, Washington, DC, 2. [Online]. Available on: https://documents.worldbank.org/en/publication/documentsreports/documentdetail/36457149451 7675149/full-report

[14] E. Colombo, L. Mattarolo, S. Bologna, and D. Masera, "The Power of Human Capital Multi-level Capacity building for Energy Access," 2017.

[15] UN, *UN Conference on Environment and Development – Rio de Janeiro, Brazil, 3 to 14 June 1992*," 1992. DOI: 10.4135/9781412971867.n128.

[16] UN, *Transforming our world: the 2030 Agenda for Sustainable Development*," 2010. DOI: 10.1163/157180910X12665776638740.

[17] United Nations, *Capacity-Building | United Nations*," 2022. www.un.org/en/academicimpact/capacity-building (accessed 30 June 2022).

[18] E. Colombo, S. Bologna, and D. Masera, *Renewable Energy for Unleashing Sustainable Development.* Springer International Publishing, 2013. DOI: 10.1007/978-3-319-00284-2.

[19] X. Wang et al., "A capabilities-led approach to assessing technological solutions for a rural community," *Energies*, vol. 14, no. 5, 2021. DOI: 10.3390/en14051398.

[20] African Union, "Agenda 2063 – The Africa We Want," 2015.

[21] World Economic Forum, "Finding financing solutions for the future of energy for Africa | World Economic Forum," 2022. www.weforum.org/agenda/2021/09/financing-solutions-forthe-future-of-energy-for-africa/ (accessed 30 June 2022).

[22] E. G. Tambo, L. Larbi, D. Paulus, and J. Szarzynski, "eLearning for renewable energy higher education in Africa: Role," *Potential and Outlook*, 2017.

[23] P. Meirieu, *Apprendre … oui, mais comment?*, ESF, Paris, France, 1989.

[24] G. Le Boterf, *De la compétence: essai sur un attracteur étrange, Les Éditions d'Organisation.* Paris, France, 1994.

[25] P. Perrenoud, *Construire des compétences dès l'école*, ESF, Paris, France, 1999.

[26] J. Tardif, *L'évaluation des compétences: de la nécessité de documenter un parcours de formation*," Sherbrooke, Canada, 2006.

[27] E. Garfield, '*"Science Citation Index"—A New Dimension in Indexing,*' *Science*, vol. 144, no. 3619, pp. 649–654, DOI: 10.1126/science.144.3619.64.

[28] 'Web of Science product webpage,' Clarivate Analytics, 2017. https://clarivate.com/products/web-of-science/

[29] L. Kai, J. Rollins, and E. Yan, 'Web of Science use in published research and review papers 19972017: A selective, dynamic, cross-domain, content-based analysis', *Scientometrics*, vol. 115, no. 1, pp. 1–20, 2018.

Index